Manual for
Theoretical
Chemistry

Manual for
Theoretical
Chemistry

Dmitry Matyushov
Arizona State University, USA

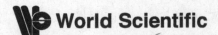

World Scientific

NEW JERSEY · LONDON · SINGAPORE · BEIJING · SHANGHAI · HONG KONG · TAIPEI · CHENNAI · TOKYO

Published by

World Scientific Publishing Co. Pte. Ltd.
5 Toh Tuck Link, Singapore 596224
USA office: 27 Warren Street, Suite 401-402, Hackensack, NJ 07601
UK office: 57 Shelton Street, Covent Garden, London WC2H 9HE

Library of Congress Control Number: 2020046389

British Library Cataloguing-in-Publication Data
A catalogue record for this book is available from the British Library.

ISBN 978-981-122-889-6 (hardcover)
ISBN 978-981-123-011-0 (paperback)
ISBN 978-981-122-890-2 (ebook for institutions)
ISBN 978-981-122-891-9 (ebook for individuals)

For any available supplementary material, please visit
https://www.worldscientific.com/worldscibooks/10.1142/12061#t=suppl

Printed in Singapore

A theorist must know everything.

Conversation in a coffee shop.

Introduction

This study guide aims at explaining theoretical concepts encountered by practitioners applying theory to molecular science. By no means this is another thick textbook covering all topics of theoretical chemistry. Rather, this is a collection of short essays, a Manual, attempting to walk the reader through two types of topics: (i) those that are usually covered in standard courses but are difficult to grasp and (ii) topics not usually covered, but are essential for successful theoretical research. The main focus is on the latter. The philosophy of this short Manual is not to cover a complete theory of each phenomenon, but instead to provide a simple study case helping to illustrate the main idea. Significant effort was put into finding the shortest derivation for each result. The focus is on simplicity. Each section is made deliberately short, something one can digest in a coffee shop or in the airport terminal. Sections are combined to chapters carrying common themes, but the expectation is that one should be able to study each section separately, with minimal consulting with other parts of the book. Perhaps an optimal pace might be to work out one section per day to have the whole chapter covered in one-two weeks. Deriving all equations is the best way to learn theory.

A few comments on the structure of the chapters. They are kept short by avoiding long explanations of parameters entering the equations. In the modern world of open searchable databases, nearly everything can be quickly looked up. Some sections start with framed boxes as reminders of the basic material that might be required for reading without cross-referencing with other chapters. Avoiding repetition is not the goal here and some of the boxes are used to repeat the concepts covered in other chapters. The same applies to equations which are often repeated in the text. The purpose is to avoid long searches through the text to find the right equation or concept. The most significant equations are framed.

A graduate student in a chemistry or chemical engineering PhD program is potentially the main beneficiary of these notes. The topics covered by traditional graduate programs are increasingly limited to very basic concepts of Quantum and Statistical Mechanics. Problems related to statistics and dynamics of condensed materials, where the large body of chemical research is now active, are hardly even touched. This note also applies to molecular biophysics where elasticity, electrostatics, and molecular dynamics form the foundations for many presently active research areas. Covering these subjects was the main focus of these notes. Some of more challenging sections can be skipped at first reading, and they are marked with asterisk. The approach to more traditional topics was to explain difficult concepts rather than to lay out detailed derivations usually covered by standard textbooks. The responsibility for these decisions, as well as for the selection of the topics, lies fully with the author.

Topics covered in this volume are usually taught as separate courses. One of the goals of this collection was to provide an integrated picture and the sense of their mutual interlinkage when it comes to the observable properties discussed in molecular science. The book is split roughly in two equal parts, with the foundations explained in the first five chapters followed by applications covered in the second half.

A number of significant subjects have been omitted to keep this volume sufficiently brief. Elasticity of solids allowing shear deformation is included, but the entire subjects of electronic structure of crystalline materials and solid-state chemistry are omitted. The same applies to electronic structure calculations and methods of classical numerical simulations. Much of chemistry and biology happens in solutions and properties of polar liquids and electrolytes take a prominent stage here. Electrostatic interactions are essential for understanding molecular biophysics and interfacial chemistry. Significant effort was invested in covering polarization of dielectric materials. Dielectric relaxation is considered through a number of sections as a realistic application of theoretical tools of statistical mechanics and dynamics discussed in the corresponding chapters. The goal of this volume is to teach how to arrive at analytical results within simplified models to gain physical insight. Getting the numbers right by realistic computational algorithms is a whole different area not touched upon here.

The citation strategy adopted for this Manual is to ease the learning curve for a student. Only citations to textbooks helping the reader to expand on a particular topic and to data collections are included. Many excellent sources that have helped the author to derive the essential results over many years of work in the field have been omitted. With the gratitude to many unnamed colleagues who taught me to do theory come my sincere apologies for omitting much of the original work.

Dmitry Matyushov
December 2020

Contents

Chapter 1

Vectors and Tensors

This short Chapter introduces the *vector and tensor* notations used throughout the text [1–4]. The tensorial form of physical quantities reflects the fact that they change under change of coordinate systems, while leaving the form of physical laws invariant to such changes. Two forms of tensorial quantities are particularly important: first-rank tensors (vectors) and second-rank tensors (3×3 matrices). A brief summary of their properties and notation convention are presented in this Chapter.

1.1 Vectors

A vector is represented by bold type, $\mathbf{A}$, and its Cartesian components are given by Greek subscripts A_α, where α takes one of three values: x, y, and z. Cartesian components define the geometric vector in a given coordinate frame when, additionally, the unit vectors $\hat{\mathbf{i}}, \hat{\mathbf{j}}, \hat{\mathbf{k}}$ along the x, y, z coordinate axes are provided (Fig. 1.1)

$$\mathbf{A} = A_x\hat{\mathbf{i}} + A_y\hat{\mathbf{j}} + A_z\hat{\mathbf{k}}. \tag{1.1}$$

The scalar product of two vectors is given as contraction over common Cartesian indices for which summation is assumed

$$\mathbf{A} \cdot \mathbf{B} = A_\alpha B_\alpha = A_x B_x + A_y B_y + A_z B_z. \tag{1.2}$$

The middle equation here represents the summation convention assuming that the sum sign can be dropped and summation performed over every pair of common indices appearing in an expression. An index on which the summation is performed is called the dummy index. Any repeated letter can be used for it and $A_\alpha B_\alpha = A_\beta B_\beta$. This summation convention is applied throughout the book.

1

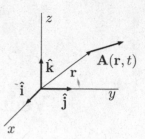

Fig. 1.1   Vector field.

The scalar product of a vector with itself defines its magnitude

$$A = |\mathbf{A}| = \sqrt{\mathbf{A} \cdot \mathbf{A}} = \sqrt{A_\alpha A_\alpha}. \tag{1.3}$$

The unit vector specifying the direction of $\mathbf{A}$ is given with a hat: $\hat{\mathbf{A}} = \mathbf{A}/A$. For instance, the unit radial vector often used below is $\hat{\mathbf{r}} = \mathbf{r}/r$. We will define its Cartesian components as $\hat{\mathbf{r}}_\alpha$, $\alpha = x, y, z$ such that the unit vector can be written as

$$\hat{\mathbf{r}} = \hat{\mathbf{r}}_x + \hat{\mathbf{r}}_y + \hat{\mathbf{r}}_z. \tag{1.4}$$

The component vectors can be identified with the axial base vectors of the coordinate system (Fig. 1.1): $\hat{\mathbf{r}}_x = \hat{\mathbf{i}}$, $\hat{\mathbf{r}}_y = \hat{\mathbf{j}}$, $\hat{\mathbf{r}}_z = \hat{\mathbf{k}}$. We prefer to use $\hat{\mathbf{r}}_\alpha$ since it allows us to apply indexing of tensor components. For instance, the Cartesian projections of the vector $\mathbf{A}$ are given as scalar products

$$A_\alpha = \hat{\mathbf{r}}_\alpha \cdot \mathbf{A} \tag{1.5}$$

and one can write

$$\mathbf{A} = A_\alpha \hat{\mathbf{r}}_\alpha. \tag{1.6}$$

1.2   Coordinate transformations

Physical quantities can be classified by the manner how they change when the system of coordinates changes. Consider a Cartesian system xyz which changes into a primed system $x'y'z'$. Two types of transformations are considered: rotations and inversion. Inversion means inverting the directions of all Cartesian axes. Inversion changes the right-handed coordinate frame to a left-handed system of coordinates.

A scalar is defined as a quantity that does not change by the coordinate transformations. Therefore, a scalar is a tensor of rank zero. There are also pseudoscalars which do not change under rotations, but change the sign under inversion. A vector is a tensor of rank one. It changes the sign of all its components under inversion: $\mathbf{A} \to -\mathbf{A}$. It is transformed by the rotational matrix $\mathbf{R}$ when the system of coordinates is rotated

$$A'_\alpha = R_{\alpha\beta} A_\beta = \sum_{\beta=x,y,z} R_{\alpha\beta} A_\beta. \tag{1.7}$$

Since the length of the vector is preserved by rotation, the rotation matrix represents a unitary transformation. This implies that

$$\mathbf{R}^T \cdot \mathbf{R} = \mathbf{I}, \tag{1.8}$$

where $\mathbf{R}^T$ is the transpose of the rotation matrix and $\mathbf{I}$ is the unitary matrix. The determinant of the transformation matrix is $+1$ for a rotation and -1 for an inversion. A transformation satisfying Eq. (1.8) is also called an orthogonal transformation. Thus, the rotation of a coordinate frame is an orthogonal transformation.

The unitary matrix $\mathbf{I}$ in Eq. (1.8) is composed of zeros off the diagonal and ones for the diagonal elements

$$\mathbf{I} = \begin{pmatrix} 1 & 0 & 0 \\ 0 & 1 & 0 \\ 0 & 0 & 1 \end{pmatrix}. \tag{1.9}$$

The matrix elements of the unitary matrix $I_{\alpha\beta} = \delta_{\alpha\beta}$ are also known as the Kronecker delta.

A typical coordinate transformation used in applications is the rotation of the coordinate system from the laboratory frame x, y, z to the body frame x', y', z'. The transformation is given as a matrix product

$$\hat{\mathbf{r}}'_\alpha = R_{\alpha\beta} \hat{\mathbf{r}}_\beta \tag{1.10}$$

with the rotation matrix $\mathbf{R}$ and all vectors expressed in the same coordinate system. This notion implies that Eq. (1.10) expresses the coordinates of the body-frame axes in the laboratory system.

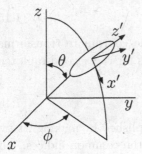

Fig. 1.2 The laboratory system of coordinates x, y, z and the body frame x', y', z'. The body frame is symmetric about the z'-axis.

An example of such a rotational transformation is shown in Fig. 1.2. The unit vectors $\hat{\mathbf{r}}'_\alpha$ associated with the body can be transformed to the unit vectors associated with the laboratory frame $\hat{\mathbf{r}}_\alpha$ by using two rotation angles θ, ϕ. This transformation is simplified by assuming the axial symmetry of the body in respect to rotations about the z'-axis, which eliminates the third angle of rotation. The transformation matrix is calculated by directly expressing the coordinates of the unit vectors $\hat{\mathbf{r}}'_\alpha$ in the laboratory frame

$$\mathbf{R} = \begin{pmatrix} \cos\theta\cos\phi & \cos\theta\sin\phi & -\sin\theta \\ -\sin\phi & \cos\phi & 0 \\ \sin\theta\cos\phi & \sin\theta\sin\phi & \cos\theta \end{pmatrix}. \tag{1.11}$$

Transposing the matrix will produce the inverse transformation $\hat{\mathbf{r}} = \mathbf{R}^T \cdot \hat{\mathbf{r}}'$ with

$$\mathbf{R}^T = \begin{pmatrix} \cos\theta\cos\phi & -\sin\phi & \sin\theta\cos\phi \\ \cos\theta\sin\phi & \cos\phi & \sin\theta\sin\phi \\ -\sin\theta & 0 & \cos\theta \end{pmatrix}. \tag{1.12}$$

As an example, one finds from Eqs. (1.10) and (1.11)

$$\hat{\mathbf{r}}'_z = \sin\theta\cos\phi\,\hat{\mathbf{r}}_x + \sin\theta\sin\phi\,\hat{\mathbf{r}}_y + \cos\theta\,\hat{\mathbf{r}}_z. \tag{1.13}$$

Taking the dot product of both sides of this equation with $\mathbf{A}$, one obtains according to Eq. (1.5)

$$A'_z = \sin\theta\cos\phi\,A_x + \sin\theta\sin\phi\,A_y + \cos\theta\,A_z. \tag{1.14}$$

This result is a special case of Eq. (1.7), which specifies the transformation of Cartesian components of a vector under rotation of the coordinate frame.

1.3 Second-rank tensors

The nine components of the second-rank tensor are obtained be all possible products of the components of two vectors

$$C_{\alpha\beta} = A_\alpha B_\beta. \tag{1.15}$$

Transformations of vector components under the coordinate transformations establish the corresponding transformations for the second-rank tensors

$$C'_{\alpha\beta} = R_{\alpha\gamma} R_{\beta\delta} C_{\gamma\delta}. \tag{1.16}$$

Contraction of a second-rank tensor is accomplished by making its two components equal and performing the sum over the common indices. The scalar (dot) product in Eq. (1.2) can be viewed as the contraction of the second-rank tensor given by Eq. (1.15)

$$\mathbf{A} \cdot \mathbf{B} = C_{\alpha\alpha}. \tag{1.17}$$

Special contractions of tensors have the same value in all coordinate systems. They are called the invariants of the tensor. For instance, the dot product is a scalar (zero-rank tensor) not affected by the coordinate transformations. It is an invariant of the second-rank tensor $C_{\alpha\beta}$ also known as its trace

$$\text{Tr}[\mathbf{C}] = C_{\alpha\alpha}. \tag{1.18}$$

The tensor with zero trace is called the traceless (deviatoric) tensor. Any second-order symmetric tensor $C_{\alpha\beta} = C_{\beta\alpha}$ can be decomposed in its isotropic part $\frac{1}{3}\mathbf{I}\text{Tr}[\mathbf{C}]$ and the deviatoric part $\bar{\mathbf{C}}$

$$C_{\alpha\beta} = \tfrac{1}{3}\delta_{\alpha\beta}\text{Tr}[\mathbf{C}] + \bar{C}_{\alpha\beta}, \quad \bar{C}_{\alpha\beta} = C_{\alpha\beta} - \tfrac{1}{3}\delta_{\alpha\beta}\text{Tr}[\mathbf{C}]. \tag{1.19}$$

The last equation introduces one of two special tensors broadly used in tensor analysis. The second-rank Kronecker delta $\delta_{\alpha\beta}$ is zero when $\alpha \neq \beta$ and is equal to unity otherwise. This is the tensorial representation of the identity matrix corresponding to the identity transformation.

The permutation constant (Levi-Civita tensor) $\epsilon_{\alpha\beta\gamma}$ is zero if any two indices coincide. It is equal to $+1$ for $\alpha\beta\gamma = xyz$ and for any even permutation of this sequence. An odd permutation produces the value of -1. The permutation constant is used to define the vector product as

$$(\mathbf{A} \times \mathbf{B})_\alpha = \epsilon_{\alpha\beta\gamma} A_\beta B_\gamma = \epsilon_{\alpha\beta\gamma} C_{\beta\gamma}. \tag{1.20}$$

It is seen that the vector product is a contraction of the permutation index with the second-rank tensor. It does not change its sign under inversion and is, therefore, considered a pseudovector.

The rotational transformation between the laboratory and body frames considered in the previous section can be used to bring a second-rank tensor to the diagonal form. The body frame that diagonalizes the second-rank tensor $\mathbf{C}$ is called its principal frame. The only nonzero values at the diagonal of the 3×3 matrix are the principal values λ_i

$$\mathbf{C}' = \begin{pmatrix} \lambda_1 & 0 & 0 \\ 0 & \lambda_2 & 0 \\ 0 & 0 & \lambda_3 \end{pmatrix}. \tag{1.21}$$

The axes of the principal frame are principal axes of $\mathbf{C}$ characterized by the principal directions $\mathbf{r}'_\alpha$. If the tensor $\mathbf{C}$ is symmetric, $C_{\alpha\beta} = C_{\beta\alpha}$, the principal values λ_i are real numbers. If all principal numbers are positive, $\mathbf{C}$ is a positive definite tensor.

The tensor $\mathbf{C}$ is changed by rotations of the coordinate frame. There are, however, four invariants of the coordinate transformations: (1) the trace, $\lambda_1 + \lambda_2 + \lambda_3$, (2) the minor $\lambda_1\lambda_2 + \lambda_2\lambda_3 + \lambda_3\lambda_1$, (3) the determinant $\lambda_1\lambda_2\lambda_3$, and (4) the tensor magnitude $\lambda_1^2 + \lambda_2^2 + \lambda_3^2$.

The example considered in Fig. 1.2 assumes axial symmetry for the particle. This assumption implies $C'_{xx} = C'_{yy}$ and the following diagonal form for the tensor in its principal axes

$$\mathbf{C}' = \begin{pmatrix} c_\perp & 0 & 0 \\ 0 & c_\perp & 0 \\ 0 & 0 & c_\parallel \end{pmatrix}, \tag{1.22}$$

where $c_\perp = C'_{xx} = C'_{yy}$ and $c_\parallel = C'_{zz}$.

From the transformation in Eq. (1.16), $\mathbf{C}' = \mathbf{R} \cdot \mathbf{C} \cdot \mathbf{R}^T$, one obtains the inverse transform

$$\mathbf{C} = \mathbf{R}^T \cdot \mathbf{C}' \cdot \mathbf{R}. \tag{1.23}$$

Equations (1.11) and (1.12) can be used to establish the elements of $\mathbf{C}$ in the laboratory frame. For instance, one obtains for the zz component

$$C_{zz} = \begin{pmatrix} -\sin\theta & 0 & \cos\theta \end{pmatrix} \begin{pmatrix} c_\perp & 0 & 0 \\ 0 & c_\perp & 0 \\ 0 & 0 & c_\parallel \end{pmatrix} \begin{pmatrix} -\sin\theta \\ 0 \\ \cos\theta \end{pmatrix} = \cos^2\theta c_\parallel + \sin^2\theta c_\perp.$$

$$\tag{1.24}$$

1.4   Tensor fields

In many applications of tensor analysis to physics and chemistry, the components of tensors depend on position and time. One therefore can determine the scalar field $\phi(\mathbf{r}, t)$, the vector field $A_\alpha(\mathbf{r}, t)$, and the second-rank tensor field $T_{\alpha\beta}(\mathbf{r}, t)$.

A field that specifies only one number at each point in space $\mathbf{r}$ and at a given time t is a scalar field. The electrostatic potential $\phi(\mathbf{r}, t)$ is a scalar field. If three numbers, which are three Cartesian components of a vector at a given point $\mathbf{r}$, are specified, one deals with a vector field $A_\alpha(\mathbf{r}, t)$. Cartesian components specify the vector field in a given system of coordinates, called the laboratory frame, with the unit axes vectors $\hat{\mathbf{i}}, \hat{\mathbf{j}}, \hat{\mathbf{k}}$ (Fig. 1.1). The vector field is written as

$$\mathbf{A}(\mathbf{r}, t) = A_x(\mathbf{r}, t)\hat{\mathbf{i}} + A_y(\mathbf{r}, t)\hat{\mathbf{j}} + A_z(\mathbf{r}, t)\hat{\mathbf{k}} \tag{1.25}$$

or, alternatively, as

$$\mathbf{A}(\mathbf{r},t) = A_\alpha(\mathbf{r},t)\hat{\mathbf{r}}_\alpha. \tag{1.26}$$

The differential vector calculus is based on acting on the tensorial fields with the del operator

$$\nabla = \hat{\mathbf{i}}\frac{\partial}{\partial x} + \hat{\mathbf{j}}\frac{\partial}{\partial y} + \hat{\mathbf{k}}\frac{\partial}{\partial z}. \tag{1.27}$$

The del operator is defined by acting on functions of coordinates. Since del is defined as a vector operator, acing on a scalar field $\phi(\mathbf{r})$ produces a vector with Cartesian components given as partial derivatives of ϕ

$$\nabla\phi = \hat{\mathbf{i}}\frac{\partial\phi}{\partial x} + \hat{\mathbf{j}}\frac{\partial\phi}{\partial y} + \hat{\mathbf{k}}\frac{\partial\phi}{\partial z} = \hat{\mathbf{r}}_\alpha\partial_\alpha\phi, \tag{1.28}$$

where

$$\boxed{\partial_\alpha = \partial/\partial r_\alpha} \tag{1.29}$$

represents derivative over the Cartesian coordinates $\alpha = x, y, z$. If the scalar potential depends on the spherical coordinates r, θ, φ, the corresponding del operator reads

$$\nabla\phi = \hat{\mathbf{r}}\frac{\partial\phi}{\partial\theta} + \hat{\boldsymbol{\theta}}\frac{1}{r}\frac{\partial\phi}{\partial r} + \hat{\boldsymbol{\phi}}\frac{1}{r\sin\theta}\frac{\partial\phi}{\partial\varphi}, \tag{1.30}$$

where $\hat{\mathbf{r}}, \hat{\boldsymbol{\theta}}, \hat{\boldsymbol{\phi}}$ are the unit vectors of the spherical coordinates.

The divergence of a vector field $\mathbf{A}$ is $\nabla\cdot\mathbf{A}$

$$\nabla\cdot\mathbf{A} = \partial A_x/\partial x + \partial A_y/\partial y + \partial A_z/\partial z = \partial_\alpha A_\alpha. \tag{1.31}$$

Finally, $\nabla\times\mathbf{A}$ is the curl of $\mathbf{A}$ which denotes the cross product

$$\nabla\times\mathbf{A} = \begin{vmatrix} \hat{\mathbf{i}} & \hat{\mathbf{j}} & \hat{\mathbf{k}} \\ \partial_x & \partial_y & \partial_z \\ A_x & A_y & A_z \end{vmatrix} = \hat{\mathbf{r}}_\alpha\epsilon_{\alpha\beta\gamma}\partial_\beta A_\gamma. \tag{1.32}$$

More specifically, the x-projection of the curl is

$$(\nabla\times\mathbf{A})_x = \partial_y A_z - \partial_z A_y. \tag{1.33}$$

It is easy to see that if $\nabla\times\mathbf{A} = 0$, one gets

$$\partial_\beta A_\alpha = \partial_\alpha A_\beta. \tag{1.34}$$

This property of irrotational fields with zero curl is used in electrostatic derivations below.

The last operator required for the vector calculus is the Laplacian operator (or Laplacian)

$$\nabla^2 = \Delta = \frac{\partial^2}{\partial x^2} + \frac{\partial^2}{\partial y^2} + \frac{\partial^2}{\partial z^2} = \partial_\alpha \partial_\alpha. \tag{1.35}$$

In addition to Cartesian coordinates, the Laplacian operator is often used in spherical coordinates in which the radial operator is separated from the operator acting on azimuthal angle ϕ and polar angle θ

$$\boxed{\nabla^2 = \nabla_r^2 + \frac{1}{r^2}\nabla_{\theta\phi}^2.} \tag{1.36}$$

The radial component is

$$\nabla_r^2 = \frac{\partial^2}{\partial r^2} + \frac{2}{r}\frac{\partial}{\partial r} \tag{1.37}$$

and the angular part is given in terms of the derivatives in θ and ϕ

$$\nabla_{\theta\phi}^2 = \frac{1}{\sin\theta}\frac{\partial}{\partial\theta}\left(\sin\theta\frac{\partial}{\partial\theta}\right) + \frac{1}{\sin^2\theta}\frac{\partial^2}{\partial\phi^2}. \tag{1.38}$$

1.5   Gauss's theorem

Consider a volume V enclosed by the surface S. The divergence theorem replaces the volume integral of the divergence of the vector field $\mathbf{A}$ with the surface integral of this field projected on the outward unit vector $\hat{\mathbf{n}}$ at each point where the surface integral is evaluated

$$\boxed{\int_V d\mathbf{r}\nabla \cdot \mathbf{A} = \oint_S dS\hat{\mathbf{n}} \cdot \mathbf{A}.} \tag{1.39}$$

A more general statement is in terms of the contraction between the del operator and the second-rank tensor field $C_{\alpha\beta}(\mathbf{r})$

$$\int_V d\mathbf{r}\partial_\beta C_{\alpha\beta} = \oint_S dS\hat{n}_\beta C_{\alpha\beta}. \tag{1.40}$$

A similar statement can be made for the gradient of the scalar field $\phi(\mathbf{r})$

$$\int_V d\mathbf{r}\partial_\alpha\phi = \oint_S dS\hat{n}_\alpha\phi. \tag{1.41}$$

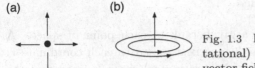

Fig. 1.3 Examples of the longitudinal (irrotational) (a) and solenoidal (transverse) (b) vector fields.

1.6   Longitudinal and transverse fields

Any vector field can be decomposed into divergence-free and curl-free components. This is a general mathematical result known as the Helmholtz theorem [3,5]. For instance, from Maxwell equations (Sec. 2.6), the magnetic field is a non-divergent field, $\nabla \cdot \mathbf{B} = 0$. In contrast, the electric field is a non-curl field in the absence of the magnetic field, $\nabla \times \mathbf{E}$. The latter case applies to electrostatics, which will be the main focus of Sec. 2.

According to the Helmholtz theorem, any vector field $\mathbf{A}$ can be separated into the longitudinal (irrotational) $\mathbf{A}_L$ and transverse (solenoidal) $\mathbf{A}_T$ components (Fig. 1.3)

$$\mathbf{A} = \mathbf{A}_L + \mathbf{A}_T, \tag{1.42}$$

where

$$\nabla \times \mathbf{A}_L = 0, \quad \nabla \cdot \mathbf{A}_T = 0. \tag{1.43}$$

The reason for this terminology is that the field produced by a point source and given as $\mathbf{A}_L = \nabla\varphi$ (φ is an unspecified scalar potential) is directed along the radius-vector, i.e., is longitudinal in respect to the radial direction. Correspondingly the curl, $\nabla \times \mathbf{B}$, of a radial field $\mathbf{B}(\mathbf{r}) = B(r)\hat{\mathbf{r}}$ has zero projection along the radial unit vector $\hat{\mathbf{r}}$ and is transverse (perpendicular) to the radial direction.

The longitudinal and transverse components of the vector fields can be obtained from the field itself by applying the following mathematical identities [5,6]

$$\begin{aligned}
\mathbf{A}_L &= -\frac{1}{4\pi} \nabla \int d\mathbf{r}' \frac{\nabla' \mathbf{A}'}{|\mathbf{r} - \mathbf{r}'|}, \\
\mathbf{A}_T &= \frac{1}{4\pi} \nabla \times \int d\mathbf{r}' \frac{\nabla' \times \mathbf{A}'}{|\mathbf{r} - \mathbf{r}'|},
\end{aligned} \tag{1.44}$$

where $\mathbf{A}' = \mathbf{A}(\mathbf{r}')$. A more elegant approach to producing longitudinal and transverse projections through the corresponding longitudinal and transverse dyads is discussed in Sec. 1.8 below. This is achieved by transforming the tensor fields to reciprocal space.

1.7   Fields in reciprocal space

Tensor fields assign a tensor of a given rank to each point of space. A second-rank field $C_{\alpha\beta}(\mathbf{r})$ is a 3×3 matrix assigned to each coordinate $\mathbf{r}$. Many calculations can be performed in reciprocal space $\mathbf{k}$ instead of the direct space $\mathbf{r}$. The direct and reciprocal-space fields are connected by spatial Fourier transformation

$$\tilde{C}_{\alpha\beta}(\mathbf{k}) = \int d\mathbf{r} C_{\alpha\beta}(\mathbf{r}) e^{i\mathbf{k}\cdot\mathbf{r}}. \qquad (1.45)$$

Here, and in the following chapters the space integral without specifying the integration volume is understood as the integral over the entire space

$$\int d\mathbf{r} \cdots = \iiint_{-\infty}^{\infty} dx dy dz \cdots = \int_0^{\infty} dr r^2 \int_0^{2\pi} d\phi \int_0^{\pi} d\theta \sin\theta \ldots, \quad (1.46)$$

where $d\mathbf{r} = dV = dx dy dz$ is the volume element in the Cartesian coordinates. This volume element can be alternatively taken in spherical coordinates when it becomes $d\mathbf{r} = r^2 dr d\Omega$, where $d\Omega = \sin\theta d\theta d\phi$ is the element of the solid angle expressed in coordinates of the polar angle θ and the azimuthal angle ϕ.

In applications, the transformation to reciprocal space corresponds to representing the corresponding fields as functions of the wave vectors $\mathbf{k}$. The inverse Fourier transform to direct space is given by the integral in reciprocal space

$$C_{\alpha\beta}(\mathbf{r}) = \int \frac{d\mathbf{k}}{(2\pi)^3} \tilde{C}_{\alpha\beta}(\mathbf{k}) e^{-i\mathbf{k}\cdot\mathbf{r}}. \qquad (1.47)$$

This result is proved by substituting Eq. (1.47) to Eq. (1.45) and using the identity defining the delta-function in terms of the Fourier integral

$$(2\pi)^3 \delta(\mathbf{k}) = \int d\mathbf{r} e^{i\mathbf{k}\cdot\mathbf{r}}. \qquad (1.48)$$

Two properties of vector fields in reciprocal space are important for applications: vector calculus and the convolution theorem. In application to vector calculus, the transition to reciprocal space transforms differential operations with vector and scalar field to algebraic operations. For example the Fourier transform of the vector $\mathbf{A} = \nabla\phi$ results in the algebraic relation between $\tilde{A}(\mathbf{k})$ and $\tilde{\phi}(\mathbf{k})$

$$\int \frac{d\mathbf{k}}{(2\pi)^3} \tilde{\mathbf{A}}(\mathbf{k}) e^{-i\mathbf{k}\cdot\mathbf{r}} = \nabla \int \frac{d\mathbf{k}}{(2\pi)^3} \tilde{\phi}(k) e^{-i\mathbf{k}\cdot\mathbf{r}}. \qquad (1.49)$$

The relation for the reciprocal-space fields is

$$\tilde{\mathbf{A}}(\mathbf{k}) = -i\mathbf{k}\tilde{\phi}(\mathbf{k}). \tag{1.50}$$

Similarly, one obtains for $\psi = \nabla \cdot \mathbf{A} = \partial_\alpha A_\alpha$

$$\tilde{\psi}(\mathbf{k}) = -i\mathbf{k} \cdot \tilde{\mathbf{A}}(\mathbf{k}) = -ik_\alpha \tilde{A}_\alpha(\mathbf{k}). \tag{1.51}$$

For brevity, we will also use the subscript to indicate reciprocal-space fields, $\tilde{\phi}(\mathbf{k}) = \phi_\mathbf{k}$. Therefore, for $\psi = \nabla^2\phi$ one obtains

$$\psi_\mathbf{k} = -k^2\phi_\mathbf{k}. \tag{1.52}$$

An important result for reciprocal-space fields in the convolution theorem. Consider a scalar formed by tensor contraction and integration of two vector fields, $\mathbf{A}(\mathbf{r})$ and $\mathbf{B}(\mathbf{r})$, with the second-rank tensor field $\mathbf{C}(\mathbf{r} - \mathbf{r}')$

$$K = \int d\mathbf{r}d\mathbf{r}' \mathbf{A}(\mathbf{r}) \cdot \mathbf{C}(\mathbf{r} - \mathbf{r}') \cdot \mathbf{B}(\mathbf{r}') = \int d\mathbf{r}d\mathbf{r}' A_\alpha(\mathbf{r})C_{\alpha\beta}(\mathbf{r} - \mathbf{r}')B_\beta(\mathbf{r}'). \tag{1.53}$$

The fact that the second-rank tensor $\mathbf{C}(\mathbf{r} - \mathbf{r}')$ depends only on the vector $\mathbf{r} - \mathbf{r}'$ connecting two points in space allows one to apply the convolution theorem. The integral involving $\mathbf{C}(\mathbf{r} - \mathbf{r}')$ is called the convolution integral and the theorem states that transforming each of the fields to reciprocal space leads to a single integral in reciprocal space in place of two integrals in direct space

$$\boxed{K = \int \frac{d\mathbf{k}}{(2\pi)^3} \tilde{A}_\alpha(\mathbf{k})\tilde{C}_{\alpha\beta}(\mathbf{k})\tilde{B}_\beta(-\mathbf{k}).} \tag{1.54}$$

In some applications considered below, we will be replacing the integral over the continuous values of the three-dimensional vector $\mathbf{k}$ with a sum over the lattice vectors $\mathbf{k}$ defined on a cube with the side length L. The lattice vectors are defined by three integers n, m, k such that $\mathbf{k} = (2\pi/L)\{n, m, k\}$. The continuous integral over $\mathbf{k}$ will be replaced with the lattice sum according to the convention

$$\int \frac{d\mathbf{k}}{(2\pi)^3} \longrightarrow V^{-1} \sum_\mathbf{k}, \tag{1.55}$$

where $V = L^3$ is the lattice volume and summation is performed over the integer numbers $\{n, m, k\}$.

1.8   *Longitudinal and transverse dyads

The transformation of vector fields to reciprocal space also renders their splitting into longitudinal and transverse projections particularly simple. Performing the Fourier transform of $\nabla \times \mathbf{A}_L = 0$ yields

$$\mathbf{k} \times \tilde{\mathbf{A}}_L(\mathbf{k}) = 0. \tag{1.56}$$

Similarly, from the relation $\nabla \cdot \mathbf{A}_T = 0$, one gets

$$\mathbf{k} \cdot \tilde{\mathbf{A}}_T(\mathbf{k}) = 0. \tag{1.57}$$

The longitudinal and transverse components of the reciprocal-space vector field can be written by using the longitudinal and transverse dyads

$$\mathbf{J}_L = \hat{\mathbf{k}}\hat{\mathbf{k}}, \quad \mathbf{J}_T = \mathbf{I} - \hat{\mathbf{k}}\hat{\mathbf{k}}, \tag{1.58}$$

where $\mathbf{I}$ is the unitary matrix (Eq. (1.9)). In terms of these dyads, the longitudinal and transverse components of a reciprocal-space vector field $\tilde{\mathbf{A}} = \tilde{\mathbf{A}}(\mathbf{k})$ are

$$\tilde{\mathbf{A}}_L = \mathbf{J}_L \cdot \tilde{\mathbf{A}}, \quad \tilde{\mathbf{A}}_T = \mathbf{J}_T \cdot \tilde{\mathbf{A}}. \tag{1.59}$$

The dyads satisfy the multiplication relations

$$\mathbf{J}_L \cdot \mathbf{J}_L = \mathbf{J}_L, \quad \mathbf{J}_T \cdot \mathbf{J}_T = \mathbf{J}_T, \quad \mathbf{J}_L \cdot \mathbf{J}_T = 0. \tag{1.60}$$

Based on these equations, the requirement of orthogonality is satisfied

$$\tilde{\mathbf{A}}_L \cdot \tilde{\mathbf{A}}_T = 0. \tag{1.61}$$

One can inverse Fourier transform $\tilde{\mathbf{A}}_\gamma$, $\gamma = L, T$ to find the corresponding fields in direct space

$$\mathbf{A}_\gamma = \int \frac{d\mathbf{k}}{(2\pi)^3} \mathbf{J}_\gamma \cdot \tilde{\mathbf{A}} e^{-i\mathbf{k}\cdot\mathbf{r}} = \int \frac{d\mathbf{k}}{(2\pi)^3} \int d\mathbf{r}' \mathbf{J}_\gamma \cdot \mathbf{A}(\mathbf{r}') e^{-i\mathbf{k}\cdot(\mathbf{r}-\mathbf{r}')}. \tag{1.62}$$

Corresponding to $\mathbf{J}_\gamma$, one defines the longitudinal and transverse dyads in direct space

$$\boldsymbol{\delta}_\gamma(\mathbf{r}) = \int \frac{d\mathbf{k}}{(2\pi)^3} \mathbf{J}_\gamma e^{-i\mathbf{k}\cdot\mathbf{r}}. \tag{1.63}$$

The direct-space projections of the vector field become

$$\mathbf{A}_\gamma(\mathbf{r}) = \int d\mathbf{r}' \boldsymbol{\delta}_\gamma(\mathbf{r} - \mathbf{r}') \cdot \mathbf{A}(\mathbf{r}'). \tag{1.64}$$

To calculate the direct-space dyads, one observes that for $\phi = (4\pi r)^{-1}$, $\phi_\mathbf{k} = k^{-2}$. This result follows from the Poisson equation for the Coulomb potential (Eq. (2.14)). One therefore has

$$\nabla_\alpha \nabla_\beta \frac{1}{4\pi r} = -\int \frac{d\mathbf{k}}{(2\pi)^3} \hat{k}_\alpha \hat{k}_\beta e^{-i\mathbf{k}\cdot\mathbf{r}} = -\delta_L^{\alpha\beta}(\mathbf{r}). \tag{1.65}$$

The longitudinal dyad can therefore be identified with the second-rank dipolar tensor

$$\delta_L = -\frac{1}{4\pi}\mathbf{T}, \quad \mathbf{T} = \nabla\nabla\frac{1}{r}. \tag{1.66}$$

In terms of Cartesian projections, the dipolar tensor becomes

$$T_{\alpha\beta} = \frac{1}{r^3}\left(3\hat{r}_\alpha\hat{r}_\beta - \delta_{\alpha\beta}\right). \tag{1.67}$$

For the transverse dyad, one obtains from Eq. (1.63)

$$\delta_T(\mathbf{r}) = \mathbf{I}\delta(\mathbf{r}) - \delta_L(\mathbf{r}). \tag{1.68}$$

One finally obtains two real-space dyads:

$$\boxed{\begin{aligned}\delta_L(\mathbf{r}) &= -\frac{1}{4\pi}\mathbf{T}(\mathbf{r}), \\ \delta_T(\mathbf{r}) &= \mathbf{I}\delta(\mathbf{r}) + \frac{1}{4\pi}\mathbf{T}(\mathbf{r}).\end{aligned}} \tag{1.69}$$

From this definition, the longitudinal projection of the field in direct space becomes

$$\mathbf{A}_L(\mathbf{r}) = -\frac{1}{4\pi}\int d\mathbf{r}'\mathbf{T}(\mathbf{r} - \mathbf{r}') \cdot \mathbf{A}(\mathbf{r}'). \tag{1.70}$$

The most significant application of Eq. (1.70) is for the calculation of the electric field of bound charges in dielectric materials. When the polarization density $\mathbf{P}$ is used in place of $\mathbf{A}$ in Eq. (1.70), the convolution integral with the dipolar tensor describes the electric field produced by the medium polarization. This is the field of bound charges $\mathbf{E}_b$, which is identically equal to $-4\pi\mathbf{P}_L$ according to Eq. (1.70).

It is easy to prove that the convolution of two longitudinal dyads leads to the longitudinal dyad again

$$\int d\mathbf{r}\delta_L(\mathbf{r}' - \mathbf{r}) \cdot \delta_L(\mathbf{r} - \mathbf{r}'') = \delta_L(\mathbf{r}' - \mathbf{r}''). \tag{1.71}$$

This property allows one to prove the orthogonality of the longitudinal and transverse projections of a vector field in real space. In contrast to the local dot product in reciprocal space (Eq. (1.61)), orthogonality in real space requires integration of the dot product of orthogonal projections over the entire space

$$\int d\mathbf{r}\mathbf{A}_L(\mathbf{r}) \cdot \mathbf{A}_T(\mathbf{r}) = 0. \tag{1.72}$$

1.9   Dipolar tensor and spherical Bessel functions

Here we present the calculation of the spatial Fourier transform of the Cartesian dipolar tensor $T_{\alpha\beta}$ given by Eq. (1.67). To have a broadly applicable result, we will perform the Fourier transform over the volume Ω excluding a small sphere of the radius a around the origin

$$\tilde{T}_{\alpha\beta}(\mathbf{k}) = \int_\Omega d\mathbf{r} T_{\alpha\beta} e^{i\mathbf{k}\cdot\mathbf{r}}. \tag{1.73}$$

It is convenient to start with the angular integration over the orientations of the radial unit vector $\hat{\mathbf{r}}$

$$\int d\Omega_r \left[3\hat{r}_\alpha\hat{r}_\beta - \delta_{\alpha\beta}\right] e^{i\mathbf{k}\cdot\mathbf{r}} = -4\pi \left[3r^{-2}\partial_{k\alpha}\partial_{k\beta} + \delta_{\alpha\beta}\right] j_0(kr), \tag{1.74}$$

where $\partial_{k\alpha}$ is the α-projection of the del operator in reciprocal space and

$$j_0(x) = \frac{\sin x}{x} \tag{1.75}$$

is the zeroth-order spherical Bessel function. Calculation of the derivative leads to

$$\int d\Omega_r \left[3\hat{r}_\alpha\hat{r}_\beta - \delta_{\alpha\beta}\right] e^{i\mathbf{k}\cdot\mathbf{r}} = -4\pi j_2(kr) \left[3\hat{k}_\alpha\hat{k}_\beta - \delta_{\alpha\beta}\right], \tag{1.76}$$

where $j_2(x)$ is the second-order spherical Bessel function.

The Fourier transform of the dipolar tensor becomes

$$\tilde{T}_{\alpha\beta}(\mathbf{k}) = -4\pi \left[3\hat{k}_\alpha\hat{k}_\beta - \delta_{\alpha\beta}\right] \int_{ka}^\infty \frac{dx}{x} j_2(x). \tag{1.77}$$

One can apply the following identity connecting the first- and second-order spherical Bessel functions

$$\frac{1}{x} j_2(x) = -\frac{d}{dx} \left(\frac{j_1(x)}{x}\right) \tag{1.78}$$

to obtain

$$\boxed{\tilde{T}_{\alpha\beta}(\mathbf{k}) = -4\pi \frac{j_1(ka)}{ka} \left[3\hat{k}_\alpha\hat{k}_\beta - \delta_{\alpha\beta}\right].} \tag{1.79}$$

In the limit $ka \to 0$, one applies $j_1(x)/x \to 1/3$ to obtain

$$\tilde{T}_{\alpha\beta}(ka \to 0) = -\frac{4\pi}{3} \left[3\hat{k}_\alpha\hat{k}_\beta - \delta_{\alpha\beta}\right]. \tag{1.80}$$

This Chapter ends with a brief overview of spherical Bessel functions [7]. These functions appear in many applications involving Fourier transforms of tensor fields depending on spherical coordinates. The general definition of the spherical Bessel function of nth order is given by Rayleigh's formula

$$j_n(x) = x^n \left(-\frac{1}{x}\frac{d}{dx} \right)^n j_0(x),$$ (1.81)

where the zeroth-order function $j_0(x)$ is from Eq. (1.75). The next two orders follow by direct differentiation

$$j_1(x) = \frac{\sin x}{x^2} - \frac{\cos x}{x},$$

$$j_2(x) = \left(\frac{3}{x^3} - \frac{1}{x} \right) \sin x - \frac{3}{x^2} \cos x.$$ (1.82)

The functions $j_n(x)$ obey the following orthogonality equations

$$\int_0^\infty dt\, t^2 j_m(at) j_n(bt) = \frac{\pi}{2ab}\delta(a-b),$$

$$\int_0^\infty dt\, j_m(t) j_n(t) = \frac{\pi}{2(2n+1)}\delta_{mn}.$$ (1.83)

They also satisfy the sum rules

$$\sum_{n=0}^{\infty}(2n+1)j_n^2(x) = 1,$$

$$\sum_{n=0}^{\infty}(-1)^n(2n+1)j_n^2(x) = j_0(2x).$$ (1.84)

Chapter 2

Electrostatics

This Chapter is mostly focused on *electrostatics*, that is the interaction of the stationary electric field with electric charges in vacuum and in condensed materials forming dielectrics (nonconducting media) [5, 6, 8–11]. In addition, equations for electromagnetic waves are derived from Maxwell's equations to provide the background required for considering spectra in Chap. 11. The main focus of the Chapter is on physics of dielectric polarization and thermodynamics of polarized dielectrics. Throughout this book, Gaussian units are used for electromagnetism. This convention provides more transparent physical picture of electromagnetic phenomena since electric and magnetic fields come on equal footing. It is also convenient to have the electric field and electric displacement to carry the same units, which makes it easier to appreciate their connection. The rules of conversion between SI and Gaussian units are discussed in many standard textbooks [6]. A short primer on converting between Gaussian and SI units is given at the end of the Chapter. The Maxwell equations of electromagnetism are only briefly discussed assuming previous exposure of the reader to the subject. The electrostatic part of the discussion is, however, complete and can be studied from this Chapter without referring to other sources.

The standard coverage of electrostatics adopted in many textbooks is through the solution of the boundary-value problem of Maxwell's equations. Electrostatics from this perspectives becomes a mathematical exercise of solving the Poisson equation with the appropriate boundary conditions. A somewhat different, and more intuitive, approach is adopted here. The problem of polarization of dielectric boundaries is considered from the viewpoint of the surface charge created at the dividing dielectric surface and the postulate of locality of the Maxwell field combining the fields of external and internal (bound) charges. This approach allows one to introduce self-consistency of dielectric polarization directly through physically motivated arguments. This approach is also applied in Sec. 12.2 to derive the Born formula for ion solvation.

2.1   Coulomb Law

The Coulomb law describes the force between electrical charges. If one separates one single charge q, observations suggest that the force $\mathbf{f}_{el}$ is proportional to q

$$\mathbf{f}_{el} = q\mathbf{E}. \tag{2.1}$$

The vector field $\mathbf{E}$ (not pseudovector) is the electric field. Except for a few special cases, $\mathbf{E}(\mathbf{r})$ is an inhomogeneous vector field depending on the position $\mathbf{r}$ in space. The electric field is produced by all charges external to charge q. The probe charge q is thus excluded from the calculation of the field. This condition mathematically defines the field as the limit of zero probe charge: $\mathbf{f}_{el}/q \to \mathbf{E}$ at $q \to 0$ [12]. However, this definition applies only to macroscopic bodies.

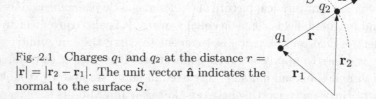

Fig. 2.1   Charges q_1 and q_2 at the distance $r = |\mathbf{r}| = |\mathbf{r}_2 - \mathbf{r}_1|$. The unit vector $\hat{\mathbf{n}}$ indicates the normal to the surface S.

The Gauss law (Sec. 1.16) is equivalent to the Coulomb law, but provides a convenient algorithm for evaluating electric fields. This is an integral formulation of the Coulomb law, which states that the integral of the normal projection of the field over a closed surface S is equal to the total electric charge inside that surface

$$\oint_S dS\, \mathbf{E} \cdot \hat{\mathbf{n}} = 4\pi \sum_i q_i. \tag{2.2}$$

Here, the electric field $\mathbf{E}$ at each point of the surface is projected on the unit vector $\hat{\mathbf{n}}$ directed normally and outward from the surface at the point where the electric field is evaluated (Fig. 2.1). On the right-hand side of this equation is the sum over all charges q_i inside the surface. For a continuous charge density $\rho_q(\mathbf{r})$, one replaces the sum with the integral over the volume Ω enclosed by S

$$\sum_i q_i = \int_\Omega d\mathbf{r} \, \rho_q(\mathbf{r}). \tag{2.3}$$

In the case of a single charge q_1 separated from charge q_2 by the distance r, one can choose the spherical surface with the radius r around the first charge, with the result

$$4\pi r^2 E_n = 4\pi q_1, \tag{2.4}$$

where the field $\mathbf{E}$ is produced by charge q_1 and

$$E_n = \hat{\mathbf{n}} \cdot \mathbf{E} = \hat{\mathbf{r}} \cdot \mathbf{E}. \tag{2.5}$$

The force at the position of the second charge is $f_{el} = E_n q_2$. This yields the standard formulation of the Coulomb law for the force magnitude

$$\boxed{f_{el} = \frac{q_1 q_2}{r^2}.} \tag{2.6}$$

In addition to specifying the magnitude, the Gauss theorem asserts that the force acting from charge 1 on charge 2 is a vector aligned along the radial direction connecting two charges

$$\mathbf{f}_{el} = \frac{q_1 q_2}{r^2} \hat{\mathbf{r}}. \tag{2.7}$$

The Gauss law allows one to arrive at the Coulomb expression for the force in terms of the electric field as the entity transmitting forces (Faraday's lines of force). The formulation of electrostatic forces in terms of fields is preferable given the central role of the electric and magnetic fields in electromagnetism. Equation (2.1) involves the product of the charge and the field of all charges external to the chosen charge to calculate the force. The need to include the electric charge in the force calculation can be bypassed in terms of the Maxwell stress tensor considered in Sec. 2.9. The force acting on a given charge is calculated in this formulation by integrating the electromagnetic stress tensor over an arbitrary surface enclosing the charge. The field becomes the sole medium both transmitting the force and determining its magnitude.

If the charge q_1 is located at $\mathbf{r}_1$ and q_2 is located at $\mathbf{r}_2$, the distance between the charges is $r = |\mathbf{r}_2 - \mathbf{r}_1|$ (Fig. 2.1). The normal direction at the point of charge q_2 is $\hat{\mathbf{n}} = (\mathbf{r}_2 - \mathbf{r}_1)/r$. The field of charge q_1 at the position of q_2 is now obtained from Eq. (2.4) by substituting these new equations for r and $\hat{\mathbf{n}}$

$$\mathbf{E}(\mathbf{r}_2) = q_1 \frac{\mathbf{r}_2 - \mathbf{r}_1}{|\mathbf{r}_2 - \mathbf{r}_1|^3}. \tag{2.8}$$

If the charge q_1 is distributed with the charge density $\rho'_q = \rho_q(\mathbf{r}')$, one replaces the previous equation with the integral $(\mathbf{r} = \mathbf{r}_2)$

$$\mathbf{E}(\mathbf{r}) = \int d\mathbf{r}' \rho'_q \frac{\mathbf{r} - \mathbf{r}'}{|\mathbf{r} - \mathbf{r}'|^3}. \tag{2.9}$$

One can further recognize that

$$\frac{\mathbf{r} - \mathbf{r}'}{|\mathbf{r} - \mathbf{r}'|^3} = -\nabla \frac{1}{|\mathbf{r} - \mathbf{r}'|} \tag{2.10}$$

and re-write the expression for the field as the gradient of the electrostatic potential ϕ_q

$$\mathbf{E} = -\nabla \phi_q, \tag{2.11}$$

where

$$\phi_q(\mathbf{r}) = \int d\mathbf{r}' \frac{\rho'_q}{|\mathbf{r} - \mathbf{r}'|}. \tag{2.12}$$

The divergence of the electric field is now found by applying the del operator to both sides of Eq. (2.11)

$$\nabla \cdot \mathbf{E} = -\nabla^2 \phi_q = -\nabla^2 \int d\mathbf{r}' \frac{\rho'_q}{|\mathbf{r} - \mathbf{r}'|}, \tag{2.13}$$

where ∇^2 is the Laplacian operator (Eq. (1.35)). One can next use the mathematical result

$$\nabla^2 \frac{1}{|\mathbf{r} - \mathbf{r}'|} = -4\pi \delta(\mathbf{r} - \mathbf{r}'), \tag{2.14}$$

where $\delta(\mathbf{r})$ is the delta-function in three-dimensional Cartesian coordinates (Box in Sec. 2.2). Substituting this relation to Eq. (2.13), one obtains

$$-\nabla^2 \int d\mathbf{r}' \frac{\rho'_q}{|\mathbf{r} - \mathbf{r}'|} = 4\pi \int d\mathbf{r}' \rho_q(\mathbf{r}')\delta(\mathbf{r} - \mathbf{r}') = 4\pi \rho_q(\mathbf{r}). \tag{2.15}$$

With this identity, Eq. (2.13) turns into the differential form of the Coulomb law, also known as the Poisson equation

$$\boxed{\nabla^2 \phi_q = -4\pi \rho_q.} \tag{2.16}$$

Electric field expressed as the gradient of the electrostatic potential implies that $\mathbf{E}$ is a conservative field. No work is done on a probe particle moved around a closed path and the line integral of the field on any such closed path is zero

$$\oint \mathbf{E} \cdot d\boldsymbol{\ell} = 0. \tag{2.17}$$

2.2 Polarization of dielectrics

Delta-function (Cartesian coordinates):

$$\delta(\mathbf{r} - \mathbf{r}') = \delta(x - x')\delta(y - y')\delta(z - z')$$

$$\delta(\mathbf{r}) = \int \frac{d\mathbf{k}}{(2\pi)^3} e^{i\mathbf{k}\cdot\mathbf{r}}$$

$$\nabla^2 r^{-1} = -4\pi\delta(\mathbf{r})$$

Delta-function (spherical coordinates):

$$\delta(\mathbf{r} - \mathbf{r}') = r^{-2}\delta(r - r')\delta(\phi - \phi')\delta(\cos\theta - \cos\theta')$$

Delta-function (expansion in spherical harmonics $Y_{lm}(\theta, \phi)$):

$$\delta(\mathbf{r} - \mathbf{r}') = r^{-2}\delta(r - r') \sum_{l=0}^{\infty} \sum_{m=-l}^{l} Y_{lm}^*(\theta', \phi')Y_{lm}(\theta, \phi)$$

Properties of delta-function:

$$\int_{-\infty}^{\infty} dx f(x)\delta(x - a) = f(a), \qquad \int_{-\infty}^{\infty} dx\delta(x - a) = 1$$

$$\delta(x) = dh(x)/dx, \quad h(x) = 1 \text{ at } x \geq 0, \quad h(x) = 0 \text{ at } x < 0$$

$h(x)$ is the Heaviside (unit step) function.

We now consider a dielectric material placed in the field of external charges ρ_q. In contrast to the previous section, we will designate the field produced by the external charges as the vacuum field $\mathbf{E}_0(\mathbf{r})$. The field inside of the dielectric is a sum of $\mathbf{E}_0(\mathbf{r})$ and the internal field in the material. To be more specific about the origin of the internal field, one can adopt a simple and intuitive view of partial charges q_{jk} assigned to atoms with coordinates $\mathbf{r}_{jk}$ within molecule j. This picture of distributed atomic charge allows one to define the distribution of bound charge and the bound charge density ρ_b within the material

$$\rho_b(\mathbf{r}) = \sum_j \sum_k q_{jk}\delta(\mathbf{r} - \mathbf{r}_{jk}), \tag{2.18}$$

where the sum runs over all molecules j and over all atoms k within each molecule (Fig. 2.2). This charge distribution reflects an instantaneous con-figuration of the material and fluctuates as the molecules in the material (such as in a polar liquid) translate and rotate by thermal agitation.

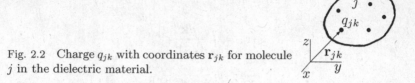

Fig. 2.2   Charge q_{jk} with coordinates $\mathbf{r}_{jk}$ for molecule j in the dielectric material.

The total microscopic field $\mathbf{E}_m$ is a sum of the external field and the field created by the material

$$\mathbf{E}_m = \mathbf{E}_0 - \nabla\phi_b, \tag{2.19}$$

where

$$\phi_b = \int d\mathbf{r}' \frac{\rho_b'}{|\mathbf{r} - \mathbf{r}'|} \tag{2.20}$$

is the microscopic electrostatic potential of the bound charges.

Divergence of $\mathbf{E}_m$ is obtained by using Eq. (2.14)

$$\boxed{\nabla \cdot \mathbf{E}_m = 4\pi\left[\rho_q + \rho_b\right].} \tag{2.21}$$

This is the differential form of the microscopic Coulomb law connecting the fluctuating microscopic field $\mathbf{E}_m$ to the fluctuating charge density ρ_b. It is important to stress the locality of this equation: the divergence of the field is fully determined by the charge density at the point where the divergence is calculated. For instance, $\rho_b = 0$ outside of the dielectric and the field $\mathbf{E}_m$ is fully determined by the density of the external charge and the boundary conditions required to solve the differential equation. For a practical example, defining the microscopic field in a cavity within the dielectric requires the distribution of free charges ρ_q inside the cavity and the boundary conditions applied to the field at the cavity boundary.

Statistical configurations of the material and the corresponding fluctuat-ing scalar field ρ_b are usually not available, except in numerical simulations. Different levels of averaging and coarse graining are therefore required to arrive at practical equations solving for electrostatics in terms of ρ_q only. One notes that the total charge of a neutral dielectric is zero. Integrating the charge density over the volume V of the dielectric gives zero charge

$$\int_V d\mathbf{r}\rho_b = 0. \tag{2.22}$$

This constraint implies that no charge flux j_b exists through the surface enclosing the dielectric sample. Locally, this requirement translates to an alternative representation of the scalar field ρ_b as the divergence of a vector field $\mathbf{P}$

$$\boxed{\rho_b = -\nabla \cdot \mathbf{P}.} \tag{2.23}$$

The vector field $\mathbf{P}$ is called the polarization density. The minus sign is chosen to make a direct link to the polarization density produced by point dipoles. However, the microscopic definition of $\mathbf{P}$ involves all molecular multipoles and is not restricted by the point-dipole approximation (Sec. 2.3).

The connection between the density of bound charge and the polarization field deserves special comment. Molecules in a dielectric material are uncharged and one might wonder what is the origin of the bound charge. Equation (2.23) suggests that a uniform polarization $\mathbf{P}$ will not produce a nonzero ρ_b: the polarization field ought to be non-uniform, with a non-zero divergence, for the bound charge to exist. Mathematically, this implies that for any small volume chosen within the dielectric, more charge should enter the volume than leave it (or vice versa). The bound charge interacts with the field of external charges by exactly the same rules of electrostatics as those applied to free charges. In other words, bound charge is as real as free charge [13], we just specify its origin from charges within the molecules with the label "bound".

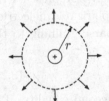

Fig. 2.3 Radial polarization produced by alignment of dipoles in the radial field of a positive point charge.

A specific example is instructive here. Figure 2.3 shows dipoles aligned by the radial field of a positive charge. If the alignment is perfect, one expects the radial polarization

$$\mathbf{P} = \hat{\mathbf{r}}P_r \propto \hat{\mathbf{r}}/r^2, \tag{2.24}$$

where $\hat{\mathbf{r}} = \mathbf{r}/r$. This type of polarization does not allow the bound charge because one gets for the divergence

$$\nabla \cdot \mathbf{P} = \frac{1}{r^2}\frac{\partial(r^2 P_r)}{\partial r} = 0. \tag{2.25}$$

However, fluctuations of the dipoles away from the perfect alignment will cause fluctuations of the bound charge.

The bound charge density defined by Eq. (2.23) allows one to rewrite Eq. (2.21)

$$\nabla \cdot [\mathbf{E}_m + 4\pi\mathbf{P}] = 4\pi\rho_q. \tag{2.26}$$

This equation shows that microscopic fluctuations of the sum of two vector fields in the brackets cancel out to produce no fluctuating charge: the distribution of the external charge is fixed by the charges placed on the conductors. Given this simplification, one defines the vector field of electric displacement

$$\mathbf{D} = \langle \mathbf{E}_m + 4\pi\mathbf{P} \rangle, \tag{2.27}$$

where the angular brackets $\langle \ldots \rangle$ specify an average over the statistical configurations of the material. Even when averaging over the statistical configurations is taken in Eq. (2.27), the vector field $\mathbf{D}$ will still carry strong oscillations of its amplitude caused by microscopic granularity of the material. Therefore, getting a smooth field for practical calculations requires an additional coarse graining eliminating microscopic oscillations of statistically averaged fields.

The combination of Eqs. (2.26) and (2.27) is the equation

$$\boxed{\nabla \cdot \mathbf{D} = 4\pi\rho_q.} \tag{2.28}$$

This result is the first out of four Maxwell equations. It represents the differential form of the Coulomb law when applied to polarizable dielectrics.

Applying the combined statistical average and coarse graining to the microscopic field $\mathbf{E}_m$ yields the Maxwell field

$$\mathbf{E} = \langle \mathbf{E}_m \rangle = -\nabla \left(\phi_q + \langle \phi_b \rangle \right) = -\nabla\phi, \tag{2.29}$$

where Eq. (2.19) and $\mathbf{E}_0 = -\nabla\phi_q$ were used in the second equality. The Maxwell field is given as the gradient of the electrostatic potential $\phi = \phi_q + \langle \phi_b \rangle$, which, according to Eq. (2.21), is specified by the Poisson equation

$$\nabla^2\phi = -4\pi \left[\rho_q + \langle \rho_b \rangle \right]. \tag{2.30}$$

The comparison between Eqs. (2.28) and (2.30) shows why the formulation in term of the electric displacement $\mathbf{D}$ is superior to the formulation in terms of the Maxwell field $\mathbf{E}$. The average distribution of the bound charge $\langle \rho_b \rangle$ is not available to us and only the density of external charge ρ_q can be controlled. In addition, the Maxwell field itself is not amenable to measurement. One instead can measure only the potential difference between two space points 1 and 2, which is a line integral of the Maxwell field [5] (Fig. 2.4)

$$\Delta\phi = \phi_2 - \phi_1 = -\int_1^2 \mathbf{E} \cdot d\boldsymbol{\ell}. \tag{2.31}$$

Only in the specific case of a uniform Maxwell field does the measurement of the potential give access to this property (plane capacitor, Sec. 2.12).

Fig. 2.4 Line integral of the Maxwell field between space points 1 and 2.

The fundamental inability to measure the inhomogeneous Maxwell field affects multiple applications. In the problem of ion solvation considered in Chap. 12, this fact severely limits the scope of dielectric theories of solvation thus requiring the development of microscopic electrostatic theories eliminating the locality of the Maxwell field (Sec. 12.6). As we briefly discuss in Sec. 2.16, small cavities carved in dielectrics in principle provide experimental access to the local inhomogeneous field in the dielectric. However, connecting the field in the dielectric to the field inside the cavity is a problem far from resolved, and that hampers the potential use of optical probes for accessing the field strength in terms of changes it makes to the optical bandshape (Sec. 11.5). Even though the concept of the Maxwell field is central to the theory of dielectric polarization, the quest to measure the inhomogeneous Maxwell field has never fully succeeded.

2.3 Multipolar interaction potential

Expansion of Coulomb potential ($r' < r$):
$|\mathbf{r} - \mathbf{r}'|^{-1} = r^{-1} \sum_{\ell=0}^{\infty} \left(r'/r\right)^{\ell} P_{\ell}(\hat{\mathbf{r}} \cdot \hat{\mathbf{r}}')$
Addition theorem:
$P_{\ell}(\hat{\mathbf{u}}_1 \cdot \hat{\mathbf{u}}_2) = \frac{4\pi}{2\ell+1} \sum_{m=-\ell}^{\ell} Y_{\ell m}^*(\hat{\mathbf{u}}_1) Y_{\ell m}(\hat{\mathbf{u}}_2)$

The polarization density $\mathbf{P}$ so far is an unspecified vector field the divergence of which produces the density of bound charge via Eq. (2.23). One can come up with a more specific meaning of this field by considering the multipolar expansion of the electrostatic potential. The polarization density is then specified by the density of dipoles and higher multipoles in the material.

Consider a molecule with the charge distribution $\rho_q(\mathbf{r})$ given by the distribution of partial atomic charges in the molecule (Eq. (2.18)). This molecule will produce the electrostatic potential outside of its molecular repulsive core according to Eq. (2.12)

$$\phi_q(\mathbf{r}) = \int_{\Omega_0} d\mathbf{r}' \frac{\rho_q(\mathbf{r}')}{|\mathbf{r} - \mathbf{r}'|}, \qquad (2.32)$$

where the integral is limited to the volume Ω_0 of the molecule. Since $r > r'$, one can expand the Coulomb potential in Legendre polynomials $P_\ell(\hat{\mathbf{r}} \cdot \hat{\mathbf{r}}')$ [7] of the dot product between the unit vectors $\hat{\mathbf{r}}$ and $\hat{\mathbf{r}}'$

$$\frac{1}{|\mathbf{r} - \mathbf{r}'|} = \frac{1}{(r^2 + r'^2 - 2rr'\hat{\mathbf{r}} \cdot \hat{\mathbf{r}}')^{1/2}} = \frac{1}{r} \sum_{\ell=0}^{\infty} \left(\frac{r'}{r}\right)^\ell P_\ell(\hat{\mathbf{r}} \cdot \hat{\mathbf{r}}'). \qquad (2.33)$$

By applying the addition theorem (see the Box), one obtains an expansion in spherical harmonics

$$\frac{1}{|\mathbf{r} - \mathbf{r}'|} = \frac{1}{r} \sum_{\ell=0}^{\infty} \sum_{m=-\ell}^{\ell} \frac{4\pi}{2\ell+1} \left(\frac{r'}{r}\right)^\ell Y_{\ell m}^*(\hat{\mathbf{r}}') Y_{\ell m}(\hat{\mathbf{r}}). \qquad (2.34)$$

Substituting Eq. (2.34) to Eq. (2.32) results in an expansion of the electrostatic potential of the molecule in spherical multipole moments $q_{\ell m}$

$$\phi_q(\mathbf{r}) = \sum_{\ell=0}^{\infty} \phi_\ell(\mathbf{r}) \qquad (2.35)$$

where $\phi_\ell(\mathbf{r})$ are successive multipole potentials

$$\phi_\ell(\mathbf{r}) = \frac{\sqrt{4\pi}}{2\ell+1} \sum_{m=-\ell}^{\ell} \frac{q_{\ell m}}{r^{\ell+1}} Y_{\ell m}(\hat{\mathbf{r}}). \qquad (2.36)$$

The spherical multipole moments are defined by integrating the charge distribution of the molecule with the complex conjugate spherical harmonic $Y_{lm}^*(\hat{\mathbf{r}})$

$$q_{\ell m} = \sqrt{4\pi} \int_{\Omega_0} d\mathbf{r}' r'^\ell \rho_q(\mathbf{r}') Y_{\ell m}^*(\hat{\mathbf{r}}'). \qquad (2.37)$$

For instance, in the case $\ell = m = 0$ one obtains

$$q_{00} = q = \int_{\Omega_0} d\mathbf{r}' \rho_q(\mathbf{r}'). \qquad (2.38)$$

The zero spherical multipole represents the total charge of the molecule. Correspondingly, one can show that three spherical multipoles q_{1m} yield three projection of the vector dipole moment $\mathbf{m}$ (see Sec. 2.4 for an example)

$$\mathbf{m} = \int_{\Omega_0} d\mathbf{r}' \mathbf{r}' \rho_q(\mathbf{r}'). \qquad (2.39)$$

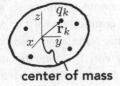

Fig. 2.5 Charge q_k with coordinates $\mathbf{r}_k$ for a molecule positioned with its center of mass at the origin of the Cartesian system of coordinates.

The corresponding dipolar potential becomes

$$\phi_1 = \frac{\mathbf{m} \cdot \mathbf{r}}{r^3}. \tag{2.40}$$

The multipole expansion can alternatively be derived in Cartesian coordinates. The molecule can be placed with its center of mass at the origin of the Cartesian coordinate system and the charges q_k assigned the coordinates $\mathbf{r}_k$ (Fig. 2.5). The distribution of the atomic charges of a single molecule can be written as

$$\rho_q(\mathbf{r}') = \sum_k q_k \delta \left(\mathbf{r}' - \mathbf{r}_k \right). \tag{2.41}$$

The electrostatic potential associated with this charge distribution follows from Eq. (2.32) as

$$\phi_q(\mathbf{r}) = \sum_k \frac{q_k}{|\mathbf{r} - \mathbf{r}_k|}. \tag{2.42}$$

If the distances to the atomic charges r_k are small compared to the distance r at which the potential is calculated, one can perform a series expansion in powers of Cartesian projections $r_{k\alpha}$

$$\frac{1}{|\mathbf{r} - \mathbf{r}_k|} = \frac{1}{r} - \sum_\alpha r_{k\alpha} \frac{d}{dr_\alpha} \frac{1}{r} + \frac{1}{2} \sum_{\alpha,\beta} r_{k\alpha} r_{k\beta} \frac{d^2}{dr_\alpha dr_\beta} \frac{1}{r} + \dots. \tag{2.43}$$

Given that the molecular charge density ρ is represented by atomic charges q_k, one can write Eq. (2.42) in the form

$$\phi_q = \frac{q}{r} - m_\alpha \nabla_\alpha \frac{1}{r} + \frac{1}{3} Q'_{\alpha\beta} \nabla_\alpha \nabla_\beta \frac{1}{r} + \dots. \tag{2.44}$$

Here, q is the charge of the molecule from Eq. (2.38) and m_α is the vector of molecular dipole given by Eq. (2.39). Further, the multipole quadratic in the distances from the molecular center to the atomic partial charges is the molecular quadrupole moment (a second-rank tensor)

$$Q'_{\alpha\beta} = \frac{3}{2} \sum_k q_k r_{k\alpha} r_{k\beta}. \tag{2.45}$$

It is clear that the quadrupole moment is a symmetric Cartesian tensor determined by six components. This is at odds with the spherical quadrupole tensor q_{2m} which has only five independent components arising from $m = -2, \ldots, 2$. There is, therefore, one component of the Cartesian tensor that can be defined in terms of the remaining five components. Achieving five independent components can be done by making the quadrupole tensor traceless (zero trace, see Sec. 1.3 for the definition and Sec. 2.4 for an example)

$$Q_{\alpha\beta} = \frac{1}{2} \sum_k q_k \left[3 r_{k\alpha} r_{k\beta} - r_k^2 \delta_{\alpha\beta} \right]. \tag{2.46}$$

It is achieved by subtracting the term $(1/2) \delta_{\alpha\beta} \sum_k q_k r_k^2$ from $Q'_{\alpha\beta}$. One obviously has $Q_{\alpha\alpha} = 0$. When this term is added to the potential of the molecule, the result for the quadrupolar potential (third summand in Eq. (2.44)) is

$$\phi_2 = \frac{1}{3} Q_{\alpha\beta} \nabla_\alpha \nabla_\beta \frac{1}{r} + \frac{1}{6} \sum_k r_k^2 \nabla^2 \frac{1}{r}. \tag{2.47}$$

However, the Poisson equation for the Coulomb potential (Eq. (2.14)) requires

$$\nabla^2 \frac{1}{r} = -4\pi \delta(\mathbf{r}), \tag{2.48}$$

which is equal to zero at $r > 0$. Note that positions $\mathbf{r}$ inside the volume Ω_0 of the molecule are excluded from the electrostatic potential and this condition always holds. The additional term making the quadrupole tensor traceless does not affect the overall electrostatic potential produced by the molecule.

One can next translate and rotate the molecule from the origin of the coordinate system to positions $\mathbf{r}_j$ in the condensed material to create the microscopic charge density given by Eq. (2.18). This charge density will interact with an external potential $\phi_0(\mathbf{r})$ with the energy

$$U = \int d\mathbf{r} \phi_0 \rho_q, \tag{2.49}$$

where ρ_q is now a sum of individual molecular charge densities each given by Eq. (2.41) upon translating the molecule by $\mathbf{r}_j$ and rotating it to the instantaneous nuclear configuration in the material. One can next take $\phi_0(\mathbf{r}_{jk})$ at the position of each molecular charge and expand it relative to the center of mass of each molecule in terms of small distance $|\mathbf{r}_{jk} - \mathbf{r}_j|$. This procedure will follow the steps leading to Eq. (2.44) with the result

$$U = \sum_j q_j \phi_0(j) - \int d\mathbf{r} \mathbf{E}_0 \cdot \mathbf{P}, \tag{2.50}$$

where $\mathbf{E}_0 = -\nabla\phi_0$ is the vacuum field and the polarization density is a sum of the dipolar and quadrupolar densities (and higher order terms)

$$\mathbf{P}(\mathbf{r}) = \sum_j \mathbf{m}_j \delta\left(\mathbf{r} - \mathbf{r}_j\right) - \tfrac{1}{3}\nabla \cdot \sum_j \mathbf{Q}_j \delta\left(\mathbf{r} - \mathbf{r}_j\right) + \ldots. \qquad (2.51)$$

For many problems of electrostatics, the combination of dipolar and quadrupolar polarization densities is sufficient for describing the interaction of external fields with the polarizable material. The polarization density field $\mathbf{P}$, introduced in Eq. (2.23) to represent bound molecular charge, carries the meaning of the density of molecular dipoles minus the divergence of the density of molecular quadrupoles.

2.4   Molecular multipoles

$k_B T \simeq 26$ meV, $\beta = (k_B T)^{-1} \simeq 39$ eV^{-1}
1 Å $= 10^{-8}$ cm, e $\times$ Å $= 4.8032$ D, $e = 4.803 \times 10^{-10}$ statcoulomb
$e^2/$Å $\simeq 14.4$ eV, $(1 \text{ D})^2/(1 \text{ Å})^3 \simeq 0.6242$ eV

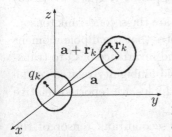

Fig. 2.6   Molecule with partial atomic charges q_k at coordinates $\mathbf{r}_k$.

The distribution of molecular charge $\rho_q(\mathbf{r})$ considered in the previous section is viewed here in terms of partial atomic charges q_k placed at atoms with coordinates $\mathbf{r}_k$. The molecule is still placed with its center of mass at the origin of the coordinate system as illustrated in Figs. 2.5 and 2.6. The total charge of the molecule becomes

$$q = \sum_k q_k \qquad (2.52)$$

and the total dipole moment is

$$\mathbf{m} = \sum_k q_k \mathbf{r}_k. \qquad (2.53)$$

For instance, for a diatomic with who charges, $q_1 = -e$, $\mathbf{r}_1 = -(d/2)\hat{\mathbf{z}}$ and $q_2 = +e$, $\mathbf{r}_2 = (d/2)\hat{\mathbf{z}}$, one obtains $\mathbf{m} = ed\hat{\mathbf{z}}$. The dipole moment is along the unit vector of the z-axis and is directed from the negative to the positive charge (Fig. 2.7).

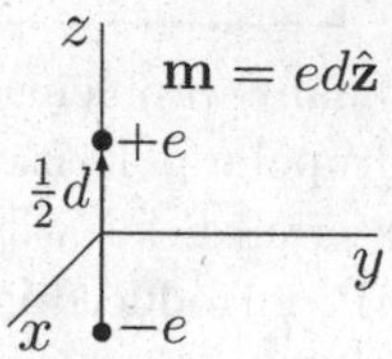

Fig. 2.7   Dipole moment of a diatomic with opposite charges.

If the molecule is translated by the vector $\mathbf{a}$ (Fig. 2.6), the charge q stays the same, but the dipole moment changes if $q \neq 0$. Denoting $\mathbf{m}'$ as the dipole moment of the shifted molecule, one obtains $\mathbf{r}'_k = \mathbf{r}_k + \mathbf{a}$ and

$$\mathbf{m}' = \mathbf{m} + q\mathbf{a}. \tag{2.54}$$

A similar transformation follows for the quadrupole moment in Eq. (2.46)

$$\mathbf{Q}' = \mathbf{Q} + \left[\tfrac{3}{2}(\mathbf{ma} + \mathbf{am}) - (\mathbf{m} \cdot \mathbf{a})\mathbf{I}\right] + \tfrac{1}{2}q\left[3\mathbf{aa} - a^2\mathbf{I}\right], \tag{2.55}$$

where the two last terms on the right-hand side are the second-rank tensors and $\mathbf{I}$ is the identity matrix (Eq. (1.9)). One notes that the dipole moment is invariant to molecular translations if $q = 0$ and invariance of $\mathbf{Q}$ to translations requires both $q = 0$ and $\mathbf{m} = 0$. To avoid ambiguity related to the origin of the coordinate system, molecular multipoles are reported relative to molecular center of mass.

The vector of the dipole moment and the second-rank tensor of the quadrupole moment also depend on the orientation of the coordinate frame (Sec. 3.5). In order to provide a measure of molecular polarity, one therefore needs properties invariant to rotations of the system of coordinates. The resolution for a vector is the scalar product, or truncation

$$m^2 = m_\alpha m_\alpha. \tag{2.56}$$

Magnitudes of molecular dipoles m are usually used for describing molecular polarity. The units typically used for dipole moments are debyes (D). To understand the origin of this unit, one can consider two elementary charges, $+e$ and $-e$, at the distance of 1 Å ($1\text{Å} = 10^{-8}$ cm). According to Eq. (2.53), the dipole moment is $\mathbf{m} = ed$, where the vector $\mathbf{d}$ is directed from the negative to the positive charge (Fig. 2.7). The elementary charge

Table 2.1 Molecular multipoles and diameters

Molecule	m, D[a]	Q, D×Å[b]	m^*	Q^*	σ_s, Å[a]
Water	1.83	2.96	1.85	1.10	2.87
Methanol	2.87	5.7	1.93	1.04	3.77
Acetonitrile	3.92	1.8	2.29	0.06	4.14
Benzene	0	8.69	0	0.45	5.27

[a]From Ref. [14]. [b]From Ref. [15].

is $e = 4.803 \times 10^{-10}$ statcoulomb (electrostatic unit of charge, esu). Multiplying this elementary charge with 1 Å measured in cm, one obtains 4.803 D. Dipoles of some molecules used as common solvents are listed in Table 2.1.

The problem with characterizing the magnitude of the quadrupole moment is more complex, but the idea is similar: one needs a rotationally invariant property. The clue how to proceed comes from spherical multipoles since one can produce the rotationally invariant quantity by tracing the second index m [15]

$$q_\ell^2 = \frac{1}{2\ell + 1} \sum_m q_{\ell m} q_{\ell m}^*, \qquad (2.57)$$

where $q_{\ell m}^*$ is the complex conjugate. In the case of the dipole moment, $\ell = 1$ and this equation leads to the dot product in Eq. (2.56). After transforming to the quadrupole moment in Cartesian coordinates, the resulting effective quadrupole is

$$Q^2 = \tfrac{2}{3} Q_{\alpha\beta} Q_{\beta\alpha}, \qquad (2.58)$$

where the usual summation over the common indices is involved. The value of the quadrupole multipole is usually given in D × Å. Some representative values of $Q = \sqrt{Q^2}$ are listed in Table 2.1.

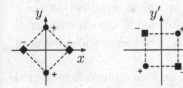

Fig. 2.8 Distribution of charge in x,y-coordinates and in the coordinate system x', y' rotated by 45°.

To illustrate the concept of the quadrupole moment and its dependence on the orientation of the coordinate frame, Fig. 2.8 shows four charges placed at the corners of a square in the x, y-plane. The diagonals of the square fall on the Cartesian axes of the x, y-system, which is rotated by $45°$ to produce the x', y' system of coordinates. The whole picture is then rotated back to make two systems of coordinate parallel. It is easy to convince oneself that both the total charge q (monopole) and the dipole moment $\mathbf{m}$ are equal to zero ($\mathbf{m} = 0$ when the charge distribution is invariant to inversion). The first nonzero multipole is the quadrupole moment specified by Eq. (2.46). If the length of the diagonal of the square is $2a$, the quadrupole moment in x, y, z-coordinates becomes (z-axis perpendicular to the plane is not shown)

$$\mathbf{Q} = \begin{pmatrix} -3ea^2 & 0 & 0 \\ 0 & 3ea^2 & 0 \\ 0 & 0 & 0 \end{pmatrix}. \tag{2.59}$$

It is clear that the trace of this matrix is zero and the quadrupole moment is a traceless tensor. Since the off-diagonal elements of the matrix are zero, the x, y system of coordinates is the principal frame of the second-rank tensor $\mathbf{Q}$ (Sec. 1.3). The matrix of the quadrupole moment is affected by rotation of the coordinate system and changes in x', y' coordinates to

$$\mathbf{Q}' = \begin{pmatrix} 0 & 3ea^2 & 0 \\ 3ea^2 & 0 & 0 \\ 0 & 0 & 0 \end{pmatrix}. \tag{2.60}$$

It is easy to observe that the quadrupole matrix is symmetric, $Q_{\alpha\beta} = Q_{\beta\alpha}$, and zero trace is maintained. While the individual quadrupole matrices are different in x, y- and x', y'-coordinates, the effective quadrupole Q from Eq. (2.58) is invariant to rotations and is equal to $Q = 2\sqrt{3}ea^2$ in both cases.

The values of the dipole and quadrupole moments do not provide a direct grasp of molecular polarity. These multipole moments eventually determine the interaction strength between molecules in the condensed phase. At the typical densities at normal conditions, the molecules in the material are closely packed with average distances close to the molecular diameter σ_s. The molecular diameter specifies an effective spherical repulsive core of a generally nonspherical molecule and is usually defined as the effective hard-sphere diameter [14, 16] (Sec. 6.8). Some of these values are also listed in Table 2.1.

The strength of multipolar interactions in condensed phase is measured relative to thermal energy $k_B T$, where k_B is the Boltzmann constant and T is the temperature. Many properties on the molecular scale are measured in electronvolt (eV) and one has for the Boltzmann constant $k_B \simeq 8.617 \times 10^{-5}$ eV/K. Correspondingly, the room-temperature ($T \simeq 300$ K) thermal energy is $k_B T \simeq 25.7$ meV. Given that dipole-dipole interactions decay as r^{-3} and quadrupole-quadrupole interactions decay as r^{-5}, one defines two dimensionless parameters to characterize the strength of the dipole and quadrupole moments in intermolecular interactions

$$(m^*)^2 = \beta m^2/\sigma_s^3, \quad (Q^*)^2 = \beta Q^2/\sigma_s^5, \tag{2.61}$$

where $\beta = (k_B T)^{-1}$ and the rotationally invariant values m^2 and Q^2 come from Eqs. (2.56) and (2.58), respectively. These parameters are also listed in Table 2.1. One observes that electrostatic interactions of acetonitrile molecules are nearly entirely determined by its dipole moment and molecular quadrupole can be mostly neglected. On the contrary, dipole and quadrupole moments contribute comparable strengths to electrostatic interactions in water.

For systems made of charged molecules, such as ionic liquids, the strength of Coulomb interactions between molecules carrying charge q is determined by the dimensionless parameter

$$(q^*)^2 = \beta q^2/\sigma_s. \tag{2.62}$$

In defining this parameter, the interaction energy between two unit charges $q = e$ at the distance of 1 Å is useful: $e^2/(1\text{ Å}) = 14.4$ eV. Since the energy of two charges at the distance of the Bohr radius $a_0 = 0.529$ Å is equal to 1 hatree (Ha), 27.211 eV, one also gets $e^2/(1\text{ Å}) = a_0 \times \text{Ha}$.

2.5 Surface charge density

If the molecules of the material carry no charge, the fist nonvanishing term in the interaction energy of the dielectric with the external field is the interaction of the vacuum field with the polarization density

$$U = -\int_V d\mathbf{r}\, \mathbf{E}_0 \cdot \mathbf{P}, \tag{2.63}$$

where, like in Eq. (2.22), integration is performed over the volume V of the dielectric.

One can write $\mathbf{E}_0 = -\nabla \phi_q$ and then apply the identity

$$\nabla \cdot (\phi_q \mathbf{P}) = -\mathbf{E}_0 \cdot \mathbf{P} + \phi_q \nabla \cdot \mathbf{P}. \tag{2.64}$$

The interaction energy becomes

$$U = \oint_S dS \hat{\mathbf{n}} \cdot \mathbf{P} \phi_q - \int_V d\mathbf{r} \phi_q \nabla \cdot \mathbf{P}, \tag{2.65}$$

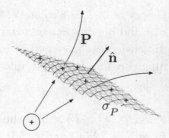

Fig. 2.9 Surface charge density σ_P.

where the surface integral is over the surface bounding the dielectric. By applying $\rho_b = -\nabla \cdot \mathbf{P}$ for the density of bound charge ρ_b, the interaction energy can be given as a sum of interactions with the bound charge in the volume and with the surface charge at the bounding surface.

The surface charge density is (Fig. 2.9)

$$\boxed{\sigma_P = \hat{\mathbf{n}} \cdot \mathbf{P}.} \tag{2.66}$$

The interaction energy becomes a sum of two contributions

$$U = \oint_S dS \phi_q \sigma_P + \int_V d\mathbf{r} \phi_q \rho_b. \tag{2.67}$$

The normal projection of the polarization density $P_n = \sigma_P$ defines the surface polarization charge, which, along with the volume bound charge, participates in the interaction of the dielectric with external charges.

The potential produced by the dielectric in the space outside of it is also a sum of these two components

$$\phi_b = \oint_S dS \frac{\sigma_P(\mathbf{r}_S)}{|\mathbf{r} - \mathbf{r}_S|} + \int_V d\mathbf{r}' \frac{\rho_b(\mathbf{r}')}{|\mathbf{r} - \mathbf{r}'|}. \tag{2.68}$$

The field of bound charges in the bulk

$$\mathbf{E}_b = -\nabla \int_V d\mathbf{r}' \frac{\rho_b(\mathbf{r}')}{|\mathbf{r} - \mathbf{r}'|} = \nabla \int_V d\mathbf{r}' \frac{\nabla' \cdot \mathbf{P}(\mathbf{r}')}{|\mathbf{r} - \mathbf{r}'|} = -4\pi \mathbf{P}_L, \tag{2.69}$$

where Eq. (1.44) for the longitudinal component of the vector field $\mathbf{P}$ was used in the last equality. Since $\mathbf{P}_L = 0$ outside of the dielectric, $\mathbf{E}_b = 0$. The bound charge in the volume creates a constant potential outside of the dielectric, but no electric field. All field comes from the surface polarization charge.

The physical origin of the surface charge is in most cases the electrostatic external field orienting the dipole moments (and higher multipoles) of the medium. The fields employed in most experiments are too weak to significantly affect the random statistics of molecular orientations. This is expressed by the requirement that the interaction energy of the external field E_0 with the molecular dipole m is substantially below the thermal energy k_BT (the low-field limit of the Langevin formula considered in Sec. 2.17)

$$mE_0 \ll k_BT. \tag{2.70}$$

However, a small fraction of molecules arriving, due to thermal motion, to the surface of a dielectric placed in the external field will have some orientational preferences imposed by the field (Fig. 2.10). This small induced orientational preference is ultimately responsible for the surface charge. This line of thought also allows the surface charge to spontaneously exist in an interface where its structure favors some specific orientations of the interfacial dipoles.

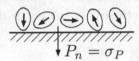

Fig. 2.10 Small preferential orientations of molecular dipoles in the interface affected by an external field yield, on average, the surface charge density σ_P.

2.6 Maxwell's equations

The set of four equations is known as Maxwell's equations describing classical macroscopic electromagnetic phenomena

$$\nabla \cdot \mathbf{D} = 4\pi\rho_q,$$
$$\nabla \cdot \mathbf{B} = 0,$$
$$\nabla \times \mathbf{E} = -\frac{1}{c}\frac{\partial \mathbf{B}}{\partial t}, \tag{2.71}$$
$$\nabla \times \mathbf{H} = \frac{4\pi}{c}\mathbf{j}_q + \frac{1}{c}\frac{\partial \mathbf{D}}{\partial t}.$$

The macroscopic electric and magnetic fields $\mathbf{E}$ and $\mathbf{B}$ are obtained by statistical averaging and coarse graining of fluctuating microscopic fields $\mathbf{E}_m$ and $\mathbf{B}_m$. As was specifically demonstrated in the case of electric displacement, adding the electric field and the polarization density $\mathbf{P}$ within the divergence mutually cancels their microscopic fluctuations (Eq. (2.26)). Similarly, one adds the magnetization $\mathbf{M}_m$ to produce the derived magnetic field $\mathbf{H}$. Taken together, these equations become

$$\mathbf{D} = \mathbf{E} + 4\pi\mathbf{P},$$
$$\mathbf{H} = \mathbf{B} - 4\pi\mathbf{M}_m. \tag{2.72}$$

These relations, however, do not provide the resolution of Maxwell's equations since closure relations, known as constitutive relations, are still required to connect $\mathbf{D}$ to $\mathbf{E}$ and $\mathbf{H}$ to $\mathbf{B}$.

The traditional formulation of Maxwell's equations in material media assumes linear local relations between these two pairs of vector fields

$$\mathbf{D} = \epsilon_s \mathbf{E},$$
$$\mathbf{B} = \mu_m \mathbf{H}. \qquad (2.73)$$

The material constants $\epsilon_s > 1$ and μ_m are the static dielectric constant and magnetic permeability, respectively. The dielectric constant is strictly greater than unity for all materials, reaching the value $\epsilon_s = 1$ in vacuum. Magnetic permeability is greater than unity, $\mu_m > 1$, for paramagnetic materials and less than unity, $\mu_m < 1$, for diamagnetic materials. One additionally defines the electric and magnetic susceptibilities, χ_s and χ_m, which connect polarization and magnetization to the corresponding fields

$$\mathbf{P} = \chi_s \mathbf{E},$$
$$\mathbf{M}_m = \chi_m \mathbf{H}. \qquad (2.74)$$

We use the subscript "m" for magnetization to distinguish it from the total dipole moment of the sample

$$\mathbf{M} = \int_V d\mathbf{r} \mathbf{P}, \qquad (2.75)$$

where the integral is taken over the volume V of a bulk sample. We will distinguish the volume V of bulk dielectric from the volume of dielectric Ω excluding cavities and solutes introduced to a dielectric body. Finally, $\mathbf{j}_q$ in the fourth Maxwell's equation is the current density of free charges measured as the number of positive charges crossing unit area per unit time.

2.7   Polarization of a cavity

The derivation presented in Sec. 2.5 is used here to determine the effective dipole moment assigned to a spherical cavity within the uniformly polarized dielectric (Fig. 2.11). This result is typically derived by solving the boundary-value problem for electrostatic polarization [6]. We instead present physically motivated arguments to show how to arrive at the same result. The main goal here is to demonstrate how the ideas of the surface charge density introduced in Sec. 2.5 can be productively used to incorporate self-consistency of dielectric polarization hidden behind mathematics of the boundary-value problem.

Despite being an idealized practice exercise of textbook electrostatics, the problem of cavity polarization is very essential to the formulation of dielectric theories [11]. It will be applied to the derivation of the cavity field in dielectrics in Sec. 2.16.

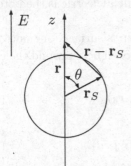

Fig. 2.11 Calculation of the potential of a spherical cavity inserted in the dielectric.

The polarization density **P** is connected to the Maxwell field **E** far from the cavity by Eqs. (2.72) and (2.73)

$$\mathbf{P} = \frac{\epsilon_s - 1}{4\pi}\mathbf{E}. \tag{2.76}$$

The goal is to determine the electrostatic potential at the point **r** placed, for simplicity, at the z-axis passing through the center of the spherical cavity (Fig. 2.11).

If one adopts an unknown polarization density $P_z = P$ at the surface of the sphere, the surface charge density is $\sigma_P = -P\cos\theta$. The negative sign accounts for the convention that the surface normal $\hat{\mathbf{n}}$ points outward from the dielectric, $\hat{\mathbf{n}} = -\hat{\mathbf{r}}$. The potential $\phi_b(\mathbf{r})$ (Eq. (2.68)) is calculated from the surface charge density given that $dS = 2\pi a^2 \sin\theta d\theta$ for the sphere with the radius a. We can drop the second term from Eq. (2.68) since it contributes only a constant term to the electrostatic potential. The surface integral over the surface charge density becomes

$$
\begin{aligned}
\phi_b &= -2\pi a^2 P \int_{-1}^{1} d\cos\theta \frac{\cos\theta}{|\mathbf{r} - \mathbf{r}_S|} \\
&= -2\pi a^2 P \int_{-1}^{1} d\cos\theta \cos\theta \frac{1}{r} \sum_{\ell=0}^{\infty} \left(\frac{a}{r}\right)^{\ell} P_\ell(\cos\theta),
\end{aligned}
\tag{2.77}
$$

where we assumed $r > a$. By using the orthogonality relation for Legendre polynomials

$$\int_{-1}^{1} dx\, x P_\ell(x) = \frac{2}{3}\delta_{\ell 1}, \tag{2.78}$$

one obtains

$$\phi_b = \frac{M^{\text{int}}}{r^2}, \quad M^{\text{int}} = -\Omega_0 P, \qquad (2.79)$$

where $\Omega_0 = (4\pi/3)a^3$ is the cavity volume. The electrostatic potential created by the polarized spherical void in the outside dielectric is the potential of the interface dipole M^{int}.

To determine the polarization P at the sphere's surface, one notes that the field of the interface dipole modifies the Maxwell field around the cavity which becomes

$$\mathbf{E}(\mathbf{r}) = \mathbf{E} + \frac{M^{\text{int}}}{r^3} \left(3\hat{\mathbf{r}}\hat{r}_z - \hat{\mathbf{r}}_z\right). \qquad (2.80)$$

One can write for the surface charge density

$$\sigma_P = -P\cos\theta = \frac{\epsilon_s - 1}{4\pi} E_n(r = a). \qquad (2.81)$$

Solving this equation for P by substituting Eq. (2.80) provides the required self-consistency for the surface polarization. One obtains

$$\boxed{M^{\text{int}} = -a^3 \frac{\epsilon_s - 1}{2\epsilon_s + 1} E.} \qquad (2.82)$$

The dipole polarized at the interface is directed opposite to the applied field. It lowers the dielectric polarization since the volume of the cavity has been removed from the polarized dielectric. However, the result is more complex than the naive assumption $M^{\text{int}} = -\Omega_0 P$. The reason is that the surface charges at the dividing surface also polarize the surrounding medium and the overall interface dipole is calculated by including this polarization in a self-consistent formalism. This self-consistency can be achieved either through the boundary-value problem of dielectric theories or through the postulate of locality of the Maxwell field used in the derivation presented here. We show in Sec. 2.14 that the negative sign of M^{int} leads to lowering of the dielectric constant for dielectrics with nonpolar impurities (dielectric mixture formula).

2.8   *Electromagnetic wave

Vector calculus:

$$\nabla \cdot (\phi \mathbf{A}) = \mathbf{A} \cdot \nabla\phi + \phi\nabla \cdot \mathbf{A}$$

$$\nabla \times (\phi \mathbf{A}) = \nabla\phi \times \mathbf{A} + \phi\nabla \times \mathbf{A}$$

$$\nabla \times (\nabla \times \mathbf{A}) = \nabla(\nabla \cdot \mathbf{A}) - \nabla^2\mathbf{A}$$

The goal of this section is to derive equations for the electromagnetic wave in terms of the vector and scalar potentials. Since the magnetic field [6] (also called the magnetic induction field) has zero divergence (second Maxwell's equation) it can be derived from the vector potential $\mathbf{A}$

$$\mathbf{B} = \nabla \times \mathbf{A}. \tag{2.83}$$

While the magnetic field is a physical field which can be measured locally at a given point, the vector potential cannot be measured locally. Only its curl, according to Eq. (2.83), has physical meaning. However, the vector potential is required for formulating Lagrangian and Hamiltonian dynamics of particles in the electromagnetic field. Since this formulation is critical for quantum mechanics of particles interacting with the electromagnetic field, this discussion is separately presented in Sec. 3.6.

Substituting Eq. (2.83) into the third Maxwell equation, one gets

$$\nabla \times \left[\mathbf{E} + c^{-1}\partial_t \mathbf{A}\right] = 0. \tag{2.84}$$

The expression in the brackets is a vector field with zero curl, and it can be represented by the gradient of a scalar potential ϕ_q. This allows one to write the electric field in the form

$$\mathbf{E} = -\nabla\phi_q - c^{-1}\partial_t \mathbf{A}. \tag{2.85}$$

When the electric field is time-independent (electrostatics), we have $\mathbf{E} = -\nabla\phi_q$ derived in Eq. (2.11).

The importance of the scalar and vector potentials is that they reduce the number of independent parameters from 6 for two vector fields $\mathbf{E}$ and $\mathbf{B}$ to 4 for ϕ and $\mathbf{A}$. The definition of the electromagnetic field in terms of $(\phi, \mathbf{A})$ is, however, not unique and can include gauge transformations. We will apply the Coulomb gauge, which requires $\nabla \cdot \mathbf{A} = 0$.

We now insert the definitions of $\mathbf{E}$ and $\mathbf{B}$ in terms of $(\phi, \mathbf{A})$ into the fourth Maxwell's equation where we assume $\mathbf{D} = \mathbf{E}$ for vacuum fields

$$\nabla \times \mathbf{B} = \frac{4\pi}{c}\mathbf{j}_q + \frac{1}{c}\partial_t \mathbf{E}. \tag{2.86}$$

One utilizes here the identity

$$\nabla \times \nabla \times \mathbf{A} = \nabla\nabla \cdot \mathbf{A} - \nabla^2 \mathbf{A}. \tag{2.87}$$

Because of the Coulomb gauge, $\nabla \cdot \mathbf{A} = 0$, it becomes

$$\nabla \times \nabla \times \mathbf{A} = -\nabla^2 \mathbf{A}. \tag{2.88}$$

One obtains the wave equation in the presence of current and potential sources

$$\nabla^2 \mathbf{A} - c^{-2}\partial_t^2 \mathbf{A} = c^{-1}\nabla\partial_t\phi_q - \frac{4\pi}{c}\mathbf{j}_q. \tag{2.89}$$

The electric current density $\mathbf{j}_q$ can be represented by the sum of longitudinal and transverse currents

$$\mathbf{j}_q = \mathbf{j}_L + \mathbf{j}_T, \tag{2.90}$$

where $\nabla \times \mathbf{j}_L = 0$ and $\nabla \cdot \mathbf{j}_T = 0$. They are also called irrotational and solenoidal currents, respectively (see Sec 1.17).

According to the Helmholtz theorem, any vector field $\mathbf{j}$ can be separated into its longitudinal and transverse components given by the following identities

$$\begin{aligned}
\mathbf{j}_L &= -\frac{1}{4\pi}\nabla \int dr' \frac{\nabla' \cdot \mathbf{j}'}{|\mathbf{r} - \mathbf{r}'|}, \\
\mathbf{j}_T &= \frac{1}{4\pi}\nabla \times \int dr' \frac{\nabla' \times \mathbf{j}'}{|\mathbf{r} - \mathbf{r}'|},
\end{aligned} \tag{2.91}$$

where $\mathbf{j}' = \mathbf{j}(\mathbf{r}')$.

One can use here an additional equation known as the continuity relation (Sec. 3.3). This equation is the differential form of the law of conservation of charge

$$\int dr\, \rho_q = \text{Const.} \tag{2.92}$$

The differential form of this equation is easily obtained from the first and the fourth Maxwell's equations. One takes the time derivative of the first equation and applies the gradient to the fourth

$$\nabla \cdot \partial_t \mathbf{D} = 4\pi \partial_t \rho_q,$$
$$\nabla \cdot \nabla \times \mathbf{H} = \frac{4\pi}{c}\nabla \cdot \mathbf{j}_q + \frac{1}{c}\nabla \cdot \partial_t \mathbf{D}. \tag{2.93}$$

Given that $\nabla \cdot \nabla \times \mathbf{H} = 0$, one obtains the continuity relation

$$\boxed{\partial_t \rho_q + \nabla \cdot \mathbf{j}_q = 0.} \tag{2.94}$$

This result can now be applied to simplify the electromagnetic wave equation. One first notes that the Coulomb law prescribes the solution for the electrostatic potential in terms of the charge density (Eq. (2.12)). By taking the time derivative and substituting the continuity relation, one obtains

$$\partial_t \phi = \int dr' \frac{\partial_t \rho_q'}{|\mathbf{r} - \mathbf{r}'|} = -\int dr' \frac{\nabla' \cdot \mathbf{j}_q'}{|\mathbf{r} - \mathbf{r}'|}. \tag{2.95}$$

From this equation and the definition of the longitudinal current in Eq. (2.91),

$$\nabla \partial_t \phi = 4\pi \mathbf{j}_L. \tag{2.96}$$

Combining this equation with the wave equation (2.89), one arrives at the wave equation for the vector potential $\mathbf{A}$ with the source term determined by the transverse current

$$\nabla^2 \mathbf{A} - c^{-2}\partial_t^2 \mathbf{A} = -\frac{4\pi}{c}\mathbf{j}_T. \qquad (2.97)$$

In the absence of solenoidal (transverse) currents, one obtains

$$\nabla^2 \mathbf{A} = c^{-2}\partial_t^2 \mathbf{A}. \qquad (2.98)$$

To be specific, assume that $\mathbf{A}(t)$ is a plane wave with the only non-zero projection along the y-axis and propagating along the z-axis

$$\mathbf{A}(z,t) = \hat{\mathbf{y}}A_0 e^{i(kz-\omega t)}. \qquad (2.99)$$

One then gets from Eq. (2.85) ($\phi_q = 0$)

$$\mathbf{E}(z,t) = (i\omega/c)\mathbf{A}(z,t). \qquad (2.100)$$

For the magnetic field, one applies Eq. (2.83)

$$\mathbf{B}(z,t) = ik\hat{\mathbf{k}} \times \mathbf{A}(z,t) = -ik\hat{\mathbf{x}}A(z,t). \qquad (2.101)$$

Here, $\hat{\mathbf{x}}$, $\hat{\mathbf{y}}$, and $\hat{\mathbf{k}} = \hat{\mathbf{z}}$ are unit vectors along the Cartesian axes.

The fields $\mathbf{E}(z,t)$ and $\mathbf{B}(z,t)$ oscillate in phase and are aligned along y-axis and x-axis, respectively. This is the plane electromagnetic wave with the circular (angular) frequency ω and the wave vector k. When the plane wave solution is substituted into the wave equation (2.98), one obtains the dispersion relation between the circular frequency and the wave vector

$$\omega = ck. \qquad (2.102)$$

2.9 Maxwell tensor

One starts with the Newton's equation of motion stating that the change of the mechanical momentum $\mathbf{P}_m$ of the material is equal to the total force integrated over the sample's volume. We consider the electromagnetic force as the sum of the electrostatic force density $\rho_q \mathbf{E}$ and the Lorentz force per unit volume $c^{-1}\mathbf{j}_q \times \mathbf{B}$. The equation of motion becomes

$$\frac{d\mathbf{P}_m}{dt} = \int d\mathbf{r}\left[\rho_q \mathbf{E} + c^{-1}\mathbf{j}_q \times \mathbf{B}\right]. \qquad (2.103)$$

By using Maxwell's equations in vacuum, one can rewrite the term in the brackets as ($\nabla \cdot \mathbf{B} = 0$)

$$\rho_q \mathbf{E} + \frac{1}{c}\mathbf{j}_q \times \mathbf{B} = \frac{1}{4\pi}\left[(\nabla \cdot \mathbf{E})\mathbf{E} + (\nabla \cdot \mathbf{B})\mathbf{B} - \mathbf{B} \times (\nabla \times \mathbf{B}) - \frac{1}{c}\partial_t\mathbf{E} \times \mathbf{B}\right].$$
$$(2.104)$$

One additionally has

$$\frac{1}{c}\partial_t \mathbf{E} \times \mathbf{B} = \frac{1}{c}\partial_t (\mathbf{E} \times \mathbf{B}) + \mathbf{E} \times (\nabla \times \mathbf{E}). \qquad (2.105)$$

Finally, one obtains

$$\frac{d\mathbf{P}_m}{dt} = -\frac{1}{4\pi c}\partial_t \int d\mathbf{r}(\mathbf{E} \times \mathbf{B})$$

$$+ \frac{1}{4\pi} \int d\mathbf{r} \left[(\nabla \cdot \mathbf{E})\mathbf{E} + (\nabla \cdot \mathbf{B})\mathbf{B} - \mathbf{B} \times (\nabla \times \mathbf{B}) - \mathbf{E} \times (\nabla \times \mathbf{E}) \right].$$

$$(2.106)$$

The first term on the right-hand-side can be related to the time derivative of the momentum of the electromagnetic field $\mathbf{P}_f = \int d\mathbf{r}\mathbf{g}$ with the momentum density $\mathbf{g}$. The latter is also related to the Poynting vector $\mathbf{S}$ (Eq. (2.136))

$$\mathbf{g} = \frac{1}{4\pi c}(\mathbf{E} \times \mathbf{B}) = \frac{1}{c^2}\mathbf{S}. \qquad (2.107)$$

The change of the total linear momentum $\mathbf{P} = \mathbf{P}_m + \mathbf{P}_f$ becomes

$$\frac{d\mathbf{P}}{dt} = \frac{1}{4\pi} \int d\mathbf{r} \left[(\nabla \cdot \mathbf{E})\mathbf{E} + (\nabla \cdot \mathbf{B})\mathbf{B} - \mathbf{B} \times (\nabla \times \mathbf{B}) - \mathbf{E} \times (\nabla \times \mathbf{E}) \right].$$

$$(2.108)$$

By calculating separate Cartesian components of the terms in the brackets [9], one can realize that they can be represented as the divergence of the second-rank tensor

$$\boxed{T_{\alpha\beta} = \frac{1}{4\pi} \left[E_\alpha E_\beta + B_\alpha B_\beta - \tfrac{1}{2}\delta_{\alpha\beta}(\mathbf{E}^2 + \mathbf{B}^2) \right].} \qquad (2.109)$$

The equation for the conservation of linear momentum can be conveniently written in terms of Cartesian components of the corresponding tensors

$$\frac{dP_\alpha}{dt} = \int d\mathbf{r}\partial_\beta T_{\alpha\beta} = \oint_S dS T_{\alpha\beta}\hat{n}_\beta. \qquad (2.110)$$

In the last part, Gauss's theorem (Sec. 1.5) transforming from the volume integral of the divergence to the surface integral is used. The tensor is projected on the Cartesian component $\hat{n}_\beta$ of the unit vector $\hat{\mathbf{n}}$ normal to the surface element dS and directed outward from the volume.

We focus more specifically on stationary fields for which Eq. (2.110) requires that the total electric force applied to a set of charges and stationary currents within a given volume is represented by the Maxwell stress integrated over the surface of this volume

$$\mathbf{f}_{el} = \oint_S dS \mathbf{T} \cdot \hat{\mathbf{n}}. \qquad (2.111)$$

Similarly to the theory of elasticity of deformable elastic bodies (Chap. 9), the force acting on a given volume can be reduced to stresses applied to the surface. Note that there is no medium to translate this stress that has been introduced so far. The stresses exist in vacuum and apply to the surface of any closed volume containing the electromagnetic field. The electromagnetic field itself is becoming a stress-transmitting entity. However, in contrast to observable propagation of electromagnetic waves considered above, transmission of stress is only an alternative mathematical description of observable volumetric forces. To demonstrate the application of the concept of the stress tensor, we consider the repulsion force between two equal charges q placed at the distance d (Fig. 2.12).

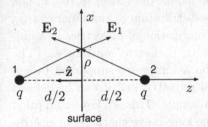

Fig. 2.12 Calculation of the force between two equal charges q, separated by the distance d, by integrating Maxwell's stress tensor over the surface perpendicular to the z-axis and dividing the distance between two charges in half.

In the absence of the magnetic field, the stress tensor simplifies to

$$T_{\alpha\beta} = \frac{1}{4\pi}\left[E_\alpha E_\beta - \tfrac{1}{2}\delta_{\alpha\beta}\mathbf{E}^2\right]. \tag{2.112}$$

The force acting on charge 2 because of the electric field created in space by charges 1 and 2 can be obtained by integrating the contraction of the Maxwell tensor with the normal to the surface perpendicular to the z-axis. Since we are calculating the force on charge 2, the normal vector should be directed outward from the volume containing charge 2. If the surface splitting the distance in half and perpendicular to the z-axis is chosen, the normal to this surface is $\hat{\mathbf{n}} = -\hat{\mathbf{z}}$, where $\hat{\mathbf{z}}$ is the unit vector specifying the direction of the z-axis (Fig. 2.12). The z-projection of the force becomes

$$f_{\text{el},z} = -\int_S dS\, T_{zz}. \tag{2.113}$$

It is clear that the sum of the fields $\mathbf{E}_1$ and $\mathbf{E}_2$ has the only nonzero projection in the dividing plane with

$$E_\alpha = \frac{2qr_\alpha}{[(d/2)^2 + \rho^2]^{3/2}}, \quad \alpha = x, y, \tag{2.114}$$

where $\rho^2 = x^2 + y^2$. One also obtains from Eq. (2.112)

$$T_{zz} = -\frac{1}{8\pi}\left(E_x^2 + E_y^2\right) \tag{2.115}$$

and

$$f_{el,z} = q^2 \int_0^\infty \frac{\rho^3 d\rho}{[(d/2)^2 + \rho^2]^3} = \frac{q^2}{d^2}. \tag{2.116}$$

The result is obviously the Coulomb repulsion force between two equal charges. This derivation is meant to illustrate that the perspective of the surface stress applied to the surface of the volume containing the electric field is equivalent to direct calculation of the Coulomb force acting on the charges within that volume.

2.10 Electrostatic free energy of dielectrics

This section describes electrostatics of polarizable dielectrics [8,10,11], i.e., materials where an external field alters the distribution of molecular charge, but does not produce a steady current. Polarization of dielectrics in a uniform electric field is described by the static dielectric constant ϵ_s. The formulation of the theory is based on the assumption, supported by experimental evidence, that ϵ_s is a material property, i.e., it can be defined independently of the shape of the material sample. This is a key assumption of the theory of dielectrics since, due to the long-range character of electrostatic interactions, a specific formulation of the theory is strongly affected by the shape of the sample polarized by the external electric field. When the boundary conditions are correctly accounted for, all such formulations should result in only one dielectric constant of the bulk material. Based on this fundamental assumption, it is often sufficient to perform the derivation for a sample of specific shape to generalize the result to an arbitrary shape of the dielectric. In order to accomplish the last step, it is important to express the result in terms of rotational invariants independent of rotations of the laboratory frame of coordinates.

Experimentally, the dielectric constant $\epsilon_s = \epsilon'(0)$ is obtained as $\omega \to 0$ (static) limit of the real part $\epsilon'(\omega)$ of the complex-valued dielectric function $\epsilon(\omega)$. It is measured in dielectric spectroscopy by applying a weak electric field oscillating with the circular frequency ω. The time-dependent response of the dielectric in a plane capacitor is discussed in Sec. 8.13; the discussion here is limited to stationary fields only.

Consider charges q_i interacting with Coulomb forces. The electrostatic potential at the charge q_i produced by the rest of charges is ϕ_i. The interaction energy becomes

$$U = \frac{1}{2} \sum_i q_i \phi_i = \frac{1}{2} \sum_{i \neq j} \frac{q_i q_j}{|\mathbf{r}_i - \mathbf{r}_j|}, \tag{2.117}$$

where the factor $1/2$ accounts for double counting of each charge [9]. This relation can be written in terms of continuous charge densities ρ_q and ρ'_q

$$U = \frac{1}{2} \int d\mathbf{r} \rho_q \phi_q = \frac{1}{2} \int d\mathbf{r} d\mathbf{r}' \frac{\rho_q \rho'_q}{|\mathbf{r} - \mathbf{r}'|}, \qquad (2.118)$$

where the electrostatic potential ϕ_q is given by Eq. (2.12). The potential energy of free charges U represents the work done to bring all charges, one by one, from infinity to their locations. When the work is done on an isolated system, it goes to increase its energy and, therefore, U is associated with the work of assembling the charge distribution.

There is no dielectric medium in this consideration. To calculate the work done to assemble charges in a dielectric, one has to allow heat flow between the dielectric body and the surroundings if temperature is maintained. The work done to assemble charges will be equal to the change in the Helmholtz free energy if, in addition, the volume of the dielectric sample is held constant. The free energy of assembling charges in the dielectric has to include both the free charges with the charge density ρ_q and the bound charges with the charge density ρ_b. Since the density ρ_b is unknown, the goal of the derivation that follows is to express the change in the free energy in terms of the density of free charges and the relevant polarization density field describing the dielectric polarization.

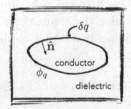

Fig. 2.13 Conductor with the electrostatic potential ϕ_q in the dielectric.

Consider a single conductor in the dielectric material with the charge q and the electrostatic potential at the surface ϕ_q [8] (Fig. 2.13). The potential is created by the charges in the dielectric and the free charges of the conductor and is constant over the surface of the conductor. If the charge of the conductor is changed by δq, the external work required is

$$\delta W_{\text{ext}} = \phi_q \delta q. \qquad (2.119)$$

If we assign $\hat{\mathbf{n}}$ as the unit vector normal to the conductor surface and pointing toward the conductor (outward from the dielectric), then the change of charge can be related to the change of the electric displacement $\delta \mathbf{D}$

$$\delta q = -\frac{1}{4\pi} \oint_S dS \delta \mathbf{D} \cdot \hat{\mathbf{n}}, \qquad (2.120)$$

where the integral is over the bounding surface and the first Maxwell's equation (Eq. (2.28)) was applied here to transform from the alteration of the charge density to the alteration of the displacement $\delta\mathbf{D}$.

Given that the potential ϕ_q is constant at the surface of the conductor (but not constant everywhere else), one can move ϕ_q under the surface integral upon substituting Eq. (2.120) to Eq. (2.119)

$$\delta W_{\text{ext}} = -\frac{1}{4\pi} \oint_S dS \phi_q \delta\mathbf{D} \cdot \hat{\mathbf{n}} = -\frac{1}{4\pi} \int_V d\mathbf{r} \nabla \cdot (\phi \delta\mathbf{D}), \qquad (2.121)$$

where the Gauss's theorem (Sec. 1.5) has been applied and the integral is taken over the volume V of the dielectric. In going from the first equation to the second, we recognized that ϕ_q is the surface value of the overall electrostatic potential ϕ created by both the free and bound charges in the medium. The following mathematical identity can be used next

$$\nabla \cdot (\phi \delta\mathbf{D}) = \nabla\phi \cdot \delta\mathbf{D} + \phi \nabla \cdot \delta\mathbf{D}. \qquad (2.122)$$

Since there are no free charges in the dielectric, one gets $\nabla \cdot \delta\mathbf{D} = 0$ and, with account for $\nabla\phi = -\mathbf{E}$, the equation for the work becomes

$$\boxed{\delta W_{\text{ext}} = \frac{1}{4\pi} \int_V d\mathbf{r} \mathbf{E} \cdot \delta\mathbf{D}.} \qquad (2.123)$$

According to the first law of thermodynamics (Sec. 5.2), the work done to polarize the dielectric adds to heat flow to determine the change in the total energy U of the sample (we use U for the total energy in this Chapter to avoid confusion with the electric field $\mathbf{E}$)

$$dU = TdS - PdV + \frac{1}{4\pi} \int_V d\mathbf{r} \mathbf{E} \cdot d\mathbf{D}. \qquad (2.124)$$

Alternatively, the change in the Helmholtz free energy of the sample F is

$$dF = -SdT - PdV + \frac{1}{4\pi} \int_V d\mathbf{r} \mathbf{E} \cdot d\mathbf{D}. \qquad (2.125)$$

If we define f and s as free energy and entropy per unit volume, then

$$df = -sdT - PdV/V + (4\pi)^{-1}\mathbf{E} \cdot d\mathbf{D}. \qquad (2.126)$$

One observes from the last two equations that the free energy is a function of the electric displacement, $F = F(\mathbf{D})$. The vector fields $\mathbf{E}$ and $\mathbf{D}$ are conjugate variables with no obvious connection between them. However, one can find the electric field as the derivative of the free energy

$$\mathbf{E} = 4\pi \left(\partial f / \partial \mathbf{D} \right)_{V,T}. \qquad (2.127)$$

The problem of polarization work becomes more tractable for linear isotropic dielectrics. In that case, one expects either local or nonlocal constitutive relations to hold. Nonlocal relations connect the electric field $\mathbf{E}(\mathbf{r})$ at point $\mathbf{r}$ to the electric displacements in all other points $\mathbf{r}'$

$$\mathbf{E}(\mathbf{r}) = \int d\mathbf{r}' \chi(\mathbf{r} - \mathbf{r}') \cdot \mathbf{D}(\mathbf{r}'). \tag{2.128}$$

Here, χ is a second-rank tensor field. The dot product in Eq. (2.128) should be understood as the tensor contraction involving summation over the common Cartesian indices

$$(\chi \cdot \mathbf{D})_\alpha = \chi_{\alpha\beta} D_\beta. \tag{2.129}$$

Any such linear relation between two conjugate fields $\mathbf{E}$ and $\mathbf{D}$ is consistent with the quadratic form for the electrostatic work

$$W_{\text{ext}} = \frac{1}{8\pi} \int_V d\mathbf{r} \mathbf{E} \cdot \mathbf{D}, \tag{2.130}$$

where non-locality of the dielectric response in terms of the non-local function $\chi(\mathbf{r} - \mathbf{r}')$ is still allowed. Equation (2.130) is proved by considering a small change

$$\delta W_{\text{ext}} = \delta \left(\frac{1}{8\pi} \int_V d\mathbf{r}' d\mathbf{r} \mathbf{D}(\mathbf{r}') \cdot \chi(\mathbf{r} - \mathbf{r}') \cdot \mathbf{D}(\mathbf{r}) \right), \tag{2.131}$$

which brings it back to Eq. (2.123).

2.11 Poynting theorem

The work required to change the energy of the electric field is given by Eq. (2.123). A similar equation can be written for the magnetic component of the electromagnetic field. Overall, a small change of both the electric and magnetic indiction fields leads to the following work

$$\delta W_{\text{ext}} = \frac{1}{4\pi} \int d\mathbf{r} \left[\mathbf{E} \cdot \delta \mathbf{D} + \mathbf{H} \cdot \delta \mathbf{B} \right]. \tag{2.132}$$

If the fields $\mathbf{E}$ and $\mathbf{H}$ are related by linear constitutive relations to the corresponding electric and magnetic induction fields $\mathbf{D}$ and $\mathbf{B}$, the rate of performing work can be written as

$$\partial_t W_{\text{ext}} = \partial_t \int \frac{d\mathbf{r}}{8\pi} \left[\mathbf{E} \cdot \mathbf{D} + \mathbf{H} \cdot \mathbf{B} \right]. \tag{2.133}$$

It is next assumed [9] that the quantity under the volume integral represents the density of the electromagnetic field energy, i.e., the difference in energy per unit volume with and without the field

$$u = \frac{1}{8\pi} \left[\mathbf{E} \cdot \mathbf{D} + \mathbf{H} \cdot \mathbf{B} \right]. \tag{2.134}$$

The reason for this assumption is that this property appears from certain manipulations of Maxwell's equations which carry the physical meaning of the conventianal continuity relation for the energy density.

One takes the dot product of the 3rd Maxwell's equation with $\mathbf{H}$ and of the 4th Maxwell's equation with $\mathbf{E}$. When the two resulting equations are subtracted, one gets

$$(\mathbf{E} \cdot \nabla \times \mathbf{H} - \mathbf{H} \cdot \nabla \times \mathbf{E}) - \frac{1}{c} (\mathbf{E} \cdot \partial_t \mathbf{D} + \mathbf{H} \cdot \partial_t \mathbf{B}) = \frac{4\pi}{c} \mathbf{j}_q \cdot \mathbf{E}. \tag{2.135}$$

One can next use the identity $\mathbf{E} \cdot \nabla \times \mathbf{H} - \mathbf{H} \cdot \nabla \times \mathbf{E} = -\nabla \cdot (\mathbf{E} \times \mathbf{H})$ and define the Poynting vector

$$\mathbf{S} = \frac{c}{4\pi} (\mathbf{E} \times \mathbf{H}). \tag{2.136}$$

One obtains

$$\frac{1}{4\pi} (\mathbf{E} \cdot \partial_t \mathbf{D} + \mathbf{H} \cdot \partial_t \mathbf{B}) = -\mathbf{j}_q \cdot \mathbf{E} - \nabla \cdot \mathbf{S}. \tag{2.137}$$

We now go back to Eq. (2.132) and write $\partial_t W_{\text{ext}} = \delta W_{\text{ext}} / \delta t$ as

$$\partial_t W_{\text{ext}} = \frac{1}{4\pi} \int d\mathbf{r} \left[\mathbf{E} \cdot \partial_t \mathbf{D} + \mathbf{H} \cdot \partial_t \mathbf{B} \right] = - \int d\mathbf{r} \left[\mathbf{j}_q \cdot \mathbf{E} + \nabla \cdot \mathbf{S} \right]. \tag{2.138}$$

Rearranging terms with account for Eq. (2.134) leads to

$$\boxed{\partial_t u + \nabla \cdot \mathbf{S} = -\mathbf{j}_q \cdot \mathbf{E}.} \tag{2.139}$$

This equation is the Poynting theorem. It states that the change of energy of electromagnetic field in a small volume is the result of the flux of energy, given by the Poynting vector, through the surface embracing the volume and the loss of energy dissipated to Joule's heat (right-hand side of the equation). If the electrical current $\mathbf{j}_q = \sigma \mathbf{E}$ is linearly connected to the electric field through the electric conductivity σ, the term $\sigma^{-1} j_q^2$ gives the power density dissipated to the Joule heat. Equation (2.139) has the form of a continuity relation arising from the requirement of conservation of energy in a closed system without dissipation. Conservation of electromagnetic energy is achieved when there are no currents and the Joule heat is zero.

2.12 Plane capacitor

An important special case of dielectric polarization used in dielectric measurements is the plane capacitor, i.e., a slab dielectric sample polarized by two parallel metal plates with their linear dimensions far exceeding the distance d between them (Fig. 2.14). The Maxwell field in the capacitor $E = \Delta\phi_q/d$ is controlled experimentally through the voltage (difference of electrostatic potentials of free charges) $\Delta\phi_q$ applied to the plates. One assumes proportionality between the uniform fields D and E in terms of the static dielectric constant ϵ_s

$$D = \epsilon_s E. \tag{2.140}$$

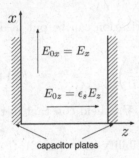

capacitor plates

Fig. 2.14 Polarization of the dielectric in a plane capacitor. $E = E_z = \Delta\phi_q/d$ is the uniform Maxwell field produced by the voltage applied to the capacitor plates. The Maxwell field is equal to the external field if applied parallel to the plates.

The electrostatic free energy (of both the free charges and of polarizing the dielectric) stored in the capacitor is equal to the reversible work done to assemble the free charges and to polarize the dielectric. This is given by Eq. (2.130), from which we have for the electrostatic free energy

$$F_{\rm el} = \frac{V}{8\pi}\epsilon_s E^2 = \frac{V}{8\pi}\epsilon_s^{-1}D^2, \tag{2.141}$$

where the volume $V = dA$ is the product of the distance between the plates d and the area of the plates A. Since the free energy of electric polarization is given in terms of capacitance C as $F_{\rm el} = \frac{1}{2}C\Delta\phi_q^2$, $\Delta\phi_q = Ed$ one gets the equation for the capacitance

$$C = \frac{\epsilon_s A}{4\pi d}. \tag{2.142}$$

Measuring the capacitance of the plane capacitor gives access to the dielectric constant ϵ_s in the dielectric experiment.

The Helmholtz free energy $F(T, V, D)$ is a function of the electric displacement D when electric polarization is concerned. Calculation of the

Table 2.2   Dielectric constants[a] and thermodynamic derivatives at 298 K.

Liquid[b]	ϵ_s	$(\partial \ln \epsilon_s / \partial \ln T)_P$	$(\partial \ln \epsilon_s / \partial \ln T)_V$[c]	$\rho(\partial \epsilon_s / \partial \rho)_T$
Water	78.5	−1.37	−1.29	81.0
Methanol	35.9	−1.29	−1.04	25.6
Acetonitrile	35.9	−1.24	−0.83	36.0
Chloroform	4.9	−1.08	−0.66	5.3

[a]Relative permittivity in SI convention. [b]From Ref. [16]. [c]From Ref. [17].

entropy of electric polarization requires keeping V and D constant in Eq. (2.141). One then takes the second equality in Eq. (2.141) to obtain

$$S_{el} = -\left(\frac{\partial F_{el}}{\partial T}\right)_{V,D} = \frac{V}{8\pi\epsilon_s^2}D^2\left(\frac{\partial \epsilon_s}{\partial T}\right)_V. \qquad (2.143)$$

The entropy and the free energy of electric polarization can be put in one equation

$$\frac{TS_{el}}{F_{el}} = \left(\frac{\partial \ln \epsilon_s}{\partial \ln T}\right)_V = \left(\frac{\partial \ln \epsilon_s}{\partial \ln T}\right)_\rho, \qquad (2.144)$$

where the number density of the material $\rho = N/V$ is held constant in the second relation.

For many polar liquids, the dielectric susceptibility $\chi_s = (\epsilon_s - 1)/(4\pi)$ scales approximately inversely with temperature as $\chi_s \propto T^{-1}$. This scaling translates to $TS_{el} \simeq -(1 - \epsilon_s^{-1})F_{el}$. From this relation, the polarization energy $U_{el} \simeq F_{el}/\epsilon_s$ becomes much smaller than the polarization free energy. For these materials, electric polarization is predominantly an entropic effect. As is seen from Table 2.2, the experimental data for the temperature derivative of the dielectric constant at constant pressure [16] and volume [17] suggest for methanol $TS_{el}/F_{el} \simeq -1.29$ at $P = $ Const and $TS_{el}/F_{el} \simeq -1.04$ at $V = $ Const. The latter result implies $U_{el} \simeq -0.04F_{el}$ from Eq. (2.144). However, the polarization energy $U_{el} \simeq 0.34F_{el}$ is more substantial for chloroform which has a lower dielectric constant.

2.13   Free energy of polarizing the dielectric

The total free energy of a polarized dielectric (Eq. (2.130)) includes both the free energy of free charges and the free energy of orienting the dipole moments in the material

$$F_{el} = \frac{1}{8\pi}\int d\mathbf{r}\, \mathbf{E} \cdot \mathbf{D}, \qquad (2.145)$$

where the integral can be extended to the entire space since there is no field in the conductor constraining the volume in deriving Eq. (2.130). The free energy of free charges is the work to assemble a given charge distribution ρ_q. One can alternatively ask the following question: What is the free energy of polarizing the dielectric for a given distribution of free charge? For instance, consider a collection of ions with the charge density ρ_q. One would like to determine the change in the free energy (reversible work) of placing a polar liquid in the field of these fixed charges.

Given that Eq. (2.145) yields the combined vacuum and polarization free energies, one needs to subtract the free energy of the vacuum field $\mathbf{E}_0$ from the total free energy

$$\Delta F = \frac{1}{8\pi} \int d\mathbf{r} \left[\mathbf{E} \cdot \mathbf{D} - \mathbf{E}_0 \cdot \mathbf{E}_0 \right]. \tag{2.146}$$

We can add and subtract $\mathbf{E} \cdot \mathbf{E}_0$ in the brackets in the integral and first consider the term

$$\int d\mathbf{r}\mathbf{E} \cdot (\mathbf{D} - \mathbf{E}_0) = - \int d\mathbf{r} \nabla \phi \cdot (\mathbf{D} - \mathbf{E}_0)$$
$$= \int d\mathbf{r} \phi \nabla \cdot (\mathbf{D} - \mathbf{E}_0). \tag{2.147}$$

Since $\nabla \cdot \mathbf{D} = 4\pi\rho_q$ and $\nabla \cdot \mathbf{E}_0 = 4\pi\rho_q$, this term is identically zero and one gets

$$\Delta F = \frac{1}{8\pi} \int d\mathbf{r} \left[\mathbf{E} - \mathbf{E}_0 \right] \cdot \mathbf{E}_0. \tag{2.148}$$

The difference of the fields $\mathbf{E} - \mathbf{E}_0$ is the field $\mathbf{E}_b$ created by the bound charges

$$\mathbf{E}_b = \nabla \int d\mathbf{r}' \frac{\nabla' \cdot \mathbf{P}'}{|\mathbf{r} - \mathbf{r}'|}. \tag{2.149}$$

According to Eq. (1.44), $\mathbf{E}_b$ is directly related to the longitudinal component of the polarization field

$$\mathbf{E}_b = -4\pi \mathbf{P}_L. \tag{2.150}$$

One therefore obtains for the free energy of polarizing the dielectric

$$\boxed{\Delta F = -\tfrac{1}{2} \int_V d\mathbf{r} \mathbf{P}_L \cdot \mathbf{E}_0.} \tag{2.151}$$

The integral here is extended to the volume V of the dielectric only since $\mathbf{P}_L$ vanishes outside of it.

This result is especially useful for developing theories of electrostatic solvation (solvation thermodynamics is considered in Sec. 12.2). Assume that the polarization of the dielectric is related to the field of external charge by a nonlocal linear convolution equation

$$\mathbf{P}_L(\mathbf{r}) = \int_V d\mathbf{r}' \chi(\mathbf{r} - \mathbf{r}') \cdot \mathbf{E}_0(\mathbf{r}'), \tag{2.152}$$

where, similarly to Eq. (2.128), $\chi(\mathbf{r} - \mathbf{r}')$ is the tensor field of the second rank. The free energy of polarizing the dielectric medium becomes

$$\Delta F = -\tfrac{1}{2} \int_V d\mathbf{r} d\mathbf{r}' \mathbf{E}_0(\mathbf{r}) \cdot \chi(\mathbf{r} - \mathbf{r}') \cdot \mathbf{E}_0(\mathbf{r}'). \tag{2.153}$$

This relation becomes particularly useful when the direct-space field $\mathbf{E}_0(\mathbf{r})$ is Fourier transformed to reciprocal space to produce the field

$$\tilde{\mathbf{E}}_0(\mathbf{k}) = \int_V d\mathbf{r} \mathbf{E}_0(\mathbf{r}) e^{i\mathbf{k}\cdot\mathbf{r}}. \tag{2.154}$$

Note that the integral is taken over the volume of the dielectric, thus excluding the volume occupied by free charges (e.g., the solvated molecule). The advantage of this representation is that two direct-space integrals in Eq. (2.153) are replaced by one integral in reciprocal space (convolution theorem, Eq. (1.54))

$$\Delta F = -\tfrac{1}{2} \int \frac{d\mathbf{k}}{(2\pi)^3} \tilde{\mathbf{E}}_0(\mathbf{k}) \cdot \tilde{\chi}(\mathbf{k}) \cdot \tilde{\mathbf{E}}_0(\mathbf{k}). \tag{2.155}$$

Here, the Fourier transform of the nonlocal susceptibility

$$\tilde{\chi}(\mathbf{k}) = \int d\mathbf{r} \chi(\mathbf{r}) e^{i\mathbf{k}\cdot\mathbf{r}} \tag{2.156}$$

does not include the excluded volume due to the source of the external field since it has been incorporated into $\tilde{\mathbf{E}}_0(\mathbf{k})$.

2.14   Dielectric with impurities

The general equations for the free energy of dielectric polarization derived above are applied here to the problem of calculating the decrement (decrease) of the dielectric constant of a polar material to which N_0 nonpolar impurities have been introduced (Fig. 2.15). To connect this problem with dielectric polarization in the plane capacitor, we assume that the polar material is placed in such a capacitor with plates carrying free charges to produce the electric displacement D. The electric field of free charges E_0

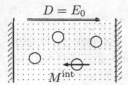

Fig. 2.15 Nonpolar impurities in the polarized dielectric.

considered in the previous section is equal to D for this geometry. The charges on the plates are held constant and $D = \text{Const}$.

One first notes that the empty capacitor carries the free energy $VD^2/(8\pi)$. When the dielectric is added, the free energy becomes (Eqs. (2.141) and (2.151))

$$F_{\text{el}} = \frac{V}{8\pi}D^2 - \tfrac{1}{2}MD. \tag{2.157}$$

Given that $M = (1 - \epsilon_s^{-1})VD/(4\pi)$, one gets

$$F_{\text{el}} = \frac{V}{8\pi\epsilon_s}D^2. \tag{2.158}$$

This is the result already listed in Eq. (2.141).

When nonpolar impurities are introduced, dielectric discontinuities become polarized by the field $E_0 = D$ and each impurity with the volume Ω_0 gains the interface dipole M^{int} (Eq. (2.82))

$$M^{\text{int}} = -\frac{3\Omega_0}{4\pi\epsilon_s}\frac{\epsilon_s - 1}{2\epsilon_s + 1}D. \tag{2.159}$$

The electrostatic free energy of polarizing the interfaces of N_0 impurities becomes

$$\Delta F = -\tfrac{1}{2}N_0 M^{\text{int}}D. \tag{2.160}$$

On the other hand, this free energy change can be characterized by the dielectric constant of the mixture ϵ_{mix}

$$\Delta F = \frac{V}{8\pi}\left(\frac{1}{\epsilon_{\text{mix}}} - \frac{1}{\epsilon_s}\right)D^2. \tag{2.161}$$

Equating these two relations and assuming that $\Delta\epsilon = \epsilon_{\text{mix}} - \epsilon_s \ll \epsilon_s$, one obtains

$$\boxed{\frac{\Delta\epsilon}{\epsilon_s} = -3\eta_0\frac{\epsilon_s - 1}{2\epsilon_s + 1},} \tag{2.162}$$

where $\eta_0 = N_0\Omega_0/V$ is the volume fraction of nonpolar impurities. This result is the Maxwell-Wagner mixture formula [18]. The dielectric constant is decreased because the polar material is expelled from the volume of nonpolar impurities.

2.15 Lorentz field

The arguments leading to the definition of the electric displacement $\mathbf{D}$ (Eq. (2.27)) made it clear that some type of statistical averaging, followed by coarse graining of small-scale oscillations, is required to introduce smooth functions making calculations with Maxwell's equations practical. It is also clear that the microscopic field acting on individual molecules in dielectrics can be very different from the average electric displacement arising from Maxwell's boundary value problem. Development of theories of dielectrics required answering the question of what is the local field forcing a molecular dipole to orient along the field of external charges. Answering this question has greatly affected the development of theories of dielectrics [10, 11]. The first solution to this problem was offered by Lorentz in terms of the field inside a virtual cavity created in the dielectric. This solution is presented here, followed, in the next section, by the calculation of the field inside a physical cavity carved in the dielectric.

The Lorentz recipe for finding the local field E_{loc} starts with the idea of a virtual cavity. This mathematical construct assumes that a void can be created in the material by eliminating from consideration a spherical volume with the radius a. The notion of a virtual cavity implies that no physical interface is created by taking the volume $(4\pi/3)a^3$ out of consideration. Assuming that the dielectric is polarized along the z-axis in the laboratory frame, the field at the center of the cavity can be viewed as the result of adding the field of external charges E_0 and the field of the uniformly polarized dielectric occupying the volume Ω surrounding the virtual cavity

$$
\begin{aligned}
E_{\text{loc}} &= E_0 + \int_\Omega d\mathbf{r}' T_{z\alpha}(\mathbf{r}') P_\alpha(\mathbf{r}') \\
&= E_0 + \int \frac{d\mathbf{k}}{(2\pi)^3} \tilde{T}_{z\alpha}(\mathbf{k}) \tilde{P}_\alpha(\mathbf{k}).
\end{aligned}
\tag{2.163}
$$

Here,

$$
T_{\alpha\beta} = \nabla_\alpha \nabla_\beta \frac{1}{r} = (3\hat{r}_\alpha \hat{r}_\beta - \delta_{\alpha\beta}) \frac{1}{r^3}
\tag{2.164}
$$

is the second-rank dipolar tensor (Eq. (1.67)). The term $T_{z\alpha}(\mathbf{r}) P_\alpha(\mathbf{r}) d\mathbf{r}$ describes the electric field along the z-axis produced at the center of the virtual cavity by the dipoles of the dielectric occupying the small volume $d\mathbf{r}$ and distributed with the polarization density $\mathbf{P}(\mathbf{r})$. The second line in Eq. (2.163) shows the transformation of the real-space integral to reciprocal space of Fourier-transformed fields. This is done to avoid diverging integrals typical for long-range Coulomb interactions, which are easier to handle in reciprocal space.

When there is no physical interface between the cavity and the bulk polarized by the external field, the polarization $\mathbf{P} = \hat{\mathbf{z}}P$ is constant everywhere in the dielectric. Therefore, only the z-component needs to be taken in Eq. (2.163) and one needs to calculate $\tilde{T}_{zz}(\mathbf{k})$ (Sec. 1.9).

The convenience of transforming to reciprocal space is that one can shift the constraint $r > a$, imposed on the polarization density, to the dipolar tensor. In other words, one can rewrite the integral over the dielectric volume as

$$\int_\Omega d\mathbf{r}' T_{zz}(\mathbf{r}') P_z(\mathbf{r}') = \int d\mathbf{r}' h_\Omega(\mathbf{r}') T_{zz}(\mathbf{r}') P_z(\mathbf{r}'), \tag{2.165}$$

where $h_\Omega(\mathbf{r})$ is the step function equal to zero inside the virtual cavity and equal to unity outside of it. The Fourier transform of $h_\Omega(\mathbf{r})T_{z\alpha}(\mathbf{r})$ then involves the space integration outside of the virtual cavity (Eq. (1.79))

$$T_{zz}(\mathbf{k}) = \int_\Omega d\mathbf{r} T_{zz}(\mathbf{r}) e^{i\mathbf{k}\cdot\mathbf{r}}$$

$$= -4\pi \frac{j_1(ka)}{ka} \hat{\mathbf{z}} \cdot \left(3\hat{\mathbf{k}}\hat{\mathbf{k}} - 1\right) \cdot \hat{\mathbf{z}}. \tag{2.166}$$

Here, $\hat{\mathbf{k}} = \mathbf{k}/k$ and

$$j_1(x) = \frac{\sin x}{x^2} - \frac{\cos x}{x} \tag{2.167}$$

is the first-order spherical Bessel function [7] (Sec. 1.9).

Since the restriction of the virtual cavity has been shifted to the dipolar tensor, the Fourier transform of the uniform polarization is particularly simple: $P_z(\mathbf{k}) = P(2\pi)^3 \delta(\mathbf{k})$. The integral with the delta-function $\delta(\mathbf{k})$ means taking the limit $\mathbf{k} \to 0$ in $T_{zz}(\mathbf{k})$. This transition requires putting $\hat{\mathbf{k}} = \mathbf{k}/k$ along the z-axis to avoid altering the symmetry of the problem. An external field along the z-direction establishes axial symmetry to the problem in direct space. To avoid switching to biaxial symmetry, z-axis and the direction of $\mathbf{k}$ in reciprocal space, one needs to put $\hat{\mathbf{k}} = \hat{\mathbf{z}}$, with the result

$$T_{zz}(\mathbf{k})\Big|_{k\to 0} = -\frac{8\pi}{3}. \tag{2.168}$$

The equation for the local field becomes

$$E_{\text{loc}} = E_0 - \frac{8\pi}{3}P. \tag{2.169}$$

Table 2.3 Molecular polarizability α and refractive index n_D.

Molecule	α, Å³	$8\alpha^{*a}$	ρ^{*b}	n_D^2	σ_s, Å^c
Water	1.47	0.50	0.79	1.78	2.87
Methanol	1.31	0.49	0.78	1.76	3.77
Acetonitrile	4.48	0.50	0.81	1.80	4.14
Benzene	10.39	0.57	0.98	2.24	5.27

$^a\alpha^* = \alpha/\sigma_s^3$. $^b\rho^* = \rho\sigma_s^3$ ^{c}All data are from Ref. [14].

With $P = (\epsilon_s - 1)/(4\pi\epsilon_s)E_0$ and $E = E_0/\epsilon_s$ for a slab sample consistent with the anticipated axial symmetry, one gets the Lorentz local field

$$E_{\text{loc}}^L = \frac{\epsilon_s + 2}{3\epsilon_s}E_0 = \frac{\epsilon_s + 2}{3}E. \tag{2.170}$$

The Lorentz field is used to derive the dielectric constant of a liquid of polarizable molecules. If each molecule carries the polarizability α (Sec. 4.14), the total dipole of the virtual sphere including N polarizable molecules is

$$M_z = N\alpha E_{\text{loc}}^L. \tag{2.171}$$

If this dipole is identified with the dielectric response

$$M_z = \Omega E\frac{\epsilon_s - 1}{4\pi}, \tag{2.172}$$

one obtains

$$\frac{\epsilon_s - 1}{\epsilon_s + 2} = \frac{4\pi}{3}\rho\alpha, \tag{2.173}$$

where $\rho = N/\Omega$ is the number density of polarizable molecules. This result is the Clausius-Mossotti equation for nonpolar liquids [10, 19].

If the molecule is viewed as a conducting sphere with the radius a_s, its polarizability is equal to a_s^3. The right-hand side of the Clausius-Mossotti equation can then be viewed as the packing fraction $\eta_s = (4\pi/3)a_s^3\rho$ of molecules viewed as conducting spheres.

When the molecular polarizability α is known, one can view the right-hand side of Eq. (2.173) as the unitless density of induced dipoles

$$y_e = (4\pi/3)\rho\alpha. \tag{2.174}$$

The dielectric constant can be obtained in terms of this parameter

$$\epsilon_s = \frac{1 + 2y_e}{1 - y_e}. \tag{2.175}$$

It is easy to appreciate that ϵ_s can be calculated from this equation only at $y_e < 1$: ϵ_s shows a ferroelectric divergence at $y_e \rightarrow 1$. The Clausius-Mossotti equation cannot be used for polar liquids with the density of dipoles exceeding this limit. The issue was resolved by the Onsager equation and, more precisely, by the Kirkwood-Onsager equation considered below. The Clausius-Mossotti equation gives good estimates for the refractive index $n_D^2 = \epsilon_s$ due to electronically induced dipoles. The parameter y_e can be scaled with the molecular diameter σ_s: $y_e = (4\pi/3)\rho^*\alpha^*$, where the scaled dimensionless parameters are $\rho^* = \rho\sigma_s^3$ and $\alpha^* = \alpha/\sigma_s^3$. For most molecular systems $8\alpha^* \simeq 0.5$ and $\rho^* \simeq 0.9$ in the liquid phase (Table 2.3). One gets $y_e \simeq 0.24$ and $\epsilon_s \simeq 1.9$. A separate set of equations is required for polar liquids carrying permanent dipoles. This problem is fully resolved in the Kirkwood-Onsager framework.

2.16 Cavity field

One step from the virtual Lorentz field is the introduction of the physical cavity carved from the dielectric. The field produced by a uniformly polarized dielectric inside such a cavity is called the cavity field.

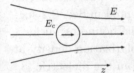

Fig. 2.16 Illustration of a thought experiment to measure the Maxwell field E through the cavity field E_c.

The concept of the cavity field was initiated by Thompson and Maxwell in an attempt to formulate a conceptual framework for measuring inhomogeneous fields inside dielectrics. As mentioned in Sec. 2.12, the homogeneous Maxwell field inside a plane capacitor is controlled by the voltage and is directly measured experimentally. In contrast, an inhomogeneous Maxwell field cannot be measured since only the line integral of the Maxwell field, that is the voltage between two points, is accessible experimentally (Eq. (2.31) and Fig. 2.4). The resolution of this problem suggested by Thompson was to create a tiny cavity in the dielectric and measure the field inside it with some device or molecular probe (Fig. 2.16). One obviously needs a formalism connecting the measured cavity field to the Maxwell field in the dielectric. This connection turned out to be very difficult to establish in a general case. Here, an approximate solution, leading to the first successful theory of dielectrics, is discussed.

The field inside the cavity follows from the arguments similar to the ones applied to derive the Lorentz field. One starts with Eq. (2.163) to write the cavity field as a sum of the field of external charges and the field created by the surrounding polarized dielectric. We again assume that the external field is directed along the z-axis of the laboratory frame and consider the projection of the cavity field on the same axis

$$E_c = E_0 + \int_\Omega dr' T_{z\alpha}(\mathbf{r}')P_\alpha(\mathbf{r}'),$$ (2.176)

where the integral is over the volume Ω outside of the cavity.

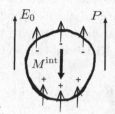

Fig. 2.17   Cavity in a uniformly polarized dielectric.

The difference between the physical cavity considered here and the Lorentz cavity considered in the previous section is in the presence of a physical interface at the surface bounding the dielectric. Since the cavity surface is polarized nonuniformly, the arguments presented in Sec. 2.2 suggest that bound surface charge should be present at the bounding surface. This bound charge is given by the projection of the polarization density at the surface on the unit vector $\hat{\mathbf{n}}_S$ pointing outward from the dielectric [8]. The surface charge density caused by dielectric discontinuity becomes

$$\sigma_P = \hat{\mathbf{n}}_S \cdot \mathbf{P}(\mathbf{r}_S).$$ (2.177)

This charge density creates an overall interface dipole moment found by multiplying the charge element $\sigma_P(\mathbf{r}_S)dS$ with the radius-vector $\mathbf{r}_S$ pointing to this charge and integrating over the surface

$$\mathbf{M}^{\text{int}} = \oint_S dS\sigma_P(\mathbf{r}_S)\mathbf{r}_S.$$ (2.178)

The physics of this problem is illustrated in Fig. 2.17. Two charged lobes at the opposite sides of the void can be visualized as oriented dipoles of the medium cut through by the dividing surface. The corresponding charges at the surface can be represented by the ends of polarized dipoles carrying opposite charges. There is no requirement of physical cutting through the dipoles by the dividing surface. As explained in Fig. 2.10, these arrows represent slight orientational preferences of mostly randomly oriented dipoles arriving to the interface in the process of thermal motion. The plus and minus charges of the corresponding interfacial regions represent specific orientational preferences of the medium dipoles pointing either toward of outward from the interface. The overall interface dipole resulting from all such microscopic preferences points from the negative surface lobe to the positive lobe. It is aligned oppositely to the external field, thus effectively screening the external field inside the cavity. From this physical picture, one anticipates the field E_c to be reduced compared to the external field E_0.

The interface dipole $\mathbf{M}^{\text{int}}$ modifies the Maxwell field around the physical cavity, which now becomes

$$\mathbf{E} = \epsilon_s^{-1}\mathbf{E}_0 + \mathbf{T} \cdot \mathbf{M}^{\text{int}}, \tag{2.179}$$

where $\mathbf{T}$ is the dipolar tensor given by Eq. (2.164). One can now use the locality of the Maxwell field to substitute $\mathbf{P} = \mathbf{E}(\epsilon_s-1)/(4\pi)$ to Eq. (2.176). If the cavity carries axial symmetry, $\mathbf{M}^{\text{int}}$ is aligned along the z-axis and one gets

$$E_c = E_0\left(1 + \frac{\epsilon_s - 1}{4\pi\epsilon_s}\int_\Omega d\mathbf{r}T_{zz}\right) + \frac{\epsilon_s - 1}{4\pi}M^{\text{int}}\int_\Omega d\mathbf{r}T_{z\alpha}T_{\alpha z}. \tag{2.180}$$

The integral in the brackets can be Fourier transformed and then evaluated according to Eq. (2.168). For the contraction of two dipolar tensors in the last term in Eq. (2.180), one obtains for the spherical cavity with the radius a

$$\int_\Omega d\mathbf{r}T_{z\alpha}T_{\alpha z} = 8\pi\int_a^\infty \frac{dr}{r^4} = \frac{8\pi}{3a^3}. \tag{2.181}$$

Taking these two results together, one arrives at the cavity field

$$\boxed{E_c = E_{\text{loc}}^L + \frac{2(\epsilon_s - 1)}{3a^3}M^{\text{int}}.} \tag{2.182}$$

The first term here is the local Lorentz field from Eq. (2.170). One, therefore, recovers the result for the virtual Lorentz cavity when no physical interface exists and $M^{\text{int}} = 0$.

There is no general prescription for the interface dipole M^{int}. It is clear that, given the definition in terms of the surface charge density in Eq. (2.178), its magnitude is strongly affected by specific properties of the physical interface between the cavity and the medium. Nevertheless, Eq. (2.182) is important since it establishes a general framework seamlessly connecting between the Lorentz and Maxwell scenarios for the cavity field. This latter limit deserves special mentioning.

The bulk polarization **P** and its value at the surface of the cavity can be calculated from the solution of the electrostatic boundary value problem. This is a standard problem of electrostatics covered in a number of textbooks [6]. Equation (2.82) presents an alternative derivation of this result based on self-consistent formulation establishing the surface charge density σ_P consistent with the macroscopic constitutive relation of Maxwell's electrostatics. When Eq. (2.82) is substituted to Eq. (2.182), one obtains the field inside a spherical cavity consistent with the Maxwell boundary value problem

$$E_c = \frac{3\epsilon_s}{2\epsilon_s + 1}E = \frac{3}{2\epsilon_s + 1}E_0. \tag{2.183}$$

As expected, the external field is strongly screened by the surface charge to the value $E_c \simeq 3E_0/(2\epsilon_s)$ at $\epsilon_s \gg 1$. However, this result also indicates that screening of the external field by the curved dividing surface of a spherical cavity is less efficient than screening by the plane surface of the dielectric slab in the plane capacitor. One gets the reduction factor of $3/(2\epsilon_s)$ in the former case and $1/\epsilon_s$ in the latter.

2.17 Onsager equation

The concept of the cavity field was applied by Onsager to develop the first successful theory of the dielectric constant of polar liquids [11]. Here, one shrinks the cavity to the size of a single molecule carrying the dipole moment **m**. The interaction of the molecular dipole with the external field is given in terms of the local field for which the cavity field is used. The interaction Hamiltonian becomes

$$H' = -m\cos\theta E_c, \tag{2.184}$$

where θ is the polar angle between the direction of the dipole moment and the z-axis along which the external field is applied. The average dipole of a single molecule along the external field is found as the Boltzmann average over the possible angles

$$\langle m_z \rangle = (m/Q)\int_0^\pi d\theta \sin\theta \cos\theta e^{\cos\theta\beta mE_c}, \tag{2.185}$$

where $\beta = (k_BT)^{-1}$ is the inverse temperature and Q is the configuration integral (see Sec. 5.5)

$$Q = \int_0^\pi d\theta \sin\theta e^{\cos\theta\beta mE_c} = \frac{2\sinh\beta mE_c}{\beta mE_c}. \tag{2.186}$$

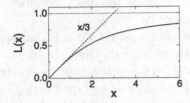

Fig. 2.18   Langevin function.

The average dipole is found as

$$\langle m_z \rangle = m \frac{\partial}{\partial(\beta m E_c)} \ln Q. \tag{2.187}$$

The result is

$$\langle m_z \rangle = m L(\beta m E_c), \tag{2.188}$$

where

$$L(x) = \coth(x) - \frac{1}{x} \tag{2.189}$$

is the Langevin function (Fig. 2.18). It describes the average orientation of a single dipole in the external field. The maximum value of the dipole fully aligned along the field is $\langle m_z \rangle = m$.

Saturation of dipolar orientations described by the Langevin function is not reached by most polar materials and expansion at $x = \beta m E_c \ll 1$ is a very good approximation. In this limit $L(x) \simeq x/3$ and one obtains

$$\langle m_z \rangle = (\beta m^2/3) E_c. \tag{2.190}$$

A slab of the dielectric material with N dipoles placed in the external field will produce the dipole moment

$$M_z = N \langle m_z \rangle = \Omega E \frac{\epsilon_s - 1}{4\pi}. \tag{2.191}$$

Substituting Eq. (2.190) for the average dipole, one arrives at the Onsager equation

$$\frac{(\epsilon_s - 1)(2\epsilon_s + 1)}{9\epsilon_s} = y. \tag{2.192}$$

Here, y is the effective (unitless) density of dipoles in the liquid

$$y = (4\pi/9)\beta m^2 \rho, \tag{2.193}$$

which is distinct from the density of induced dipoles y_e given by Eq. (2.174). A finite solution for the dielectric constant is found for any value of $y > 0$

from Eqs. (2.192) and (2.193). The Onsager equation resolves the limitation $y_e < 1$ present in the Clausius-Mossotti equation.

The derivation leading to the Onsager equation assumes that each dipole responds to the external field independently of other dipoles. The effect of the rest of the liquid on the dipole's orientation is replaced with the cavity field E_c, which becomes a mean field of the surrounding liquid acting on a chosen target dipole. The field E_c is taken from the Maxwell solution for an empty cavity discussed in Eqs. (2.182) and (2.183). Therefore, the possibility that a dipole in the liquid might modify the surrounding liquid to lead to M^{int} distinct from the Maxwell's solution is not considered by the theory. The need for such a modification, which would be an extension of the mean-field theory, is bypassed by direct account of correlations between the dipolar orientations leading to the Kirkwood-Onsager equation considered next.

2.18 Reaction field

Consider a spherical cavity with the radius a carved from the surrounding polar medium with a point dipole at its center (Fig. 2.19). This is an idealized model for many practical situations when an impurity (solute) carrying the dipole moment $\mathbf{m}_0$ is placed in a polarizable medium, usually a polar liquid. The solute dipole polarizes the surrounding medium inducing a nonuniform polarization density $\mathbf{P}(\mathbf{r})$. This induced polarization is responsible for the electric field on its own. The field by the polarized medium calculated at the position of the solute dipole is called the reaction field. The interaction of this electric field with the solute dipole can be recorded as the solvent-induced shift of the energy of optical excitation of the solute. This is known as spectral solvatochromism discussed in more detail in Sec. 11.5.

For a spherical shape of the repulsive core of the solute expelling the polar medium from its volume, the reaction field is aligned along the solute dipole, which we direct along the z-axis of the laboratory frame (Fig. 2.19). The reaction field becomes the z-projection of the electric field of the polarized medium.

The standard solution for this problem is sought from the boundary-value formalism of continuum electrostatics [11]. We adopt a more intuitive approach following the arguments already applied in considering the problem of polarizing a spherical cavity by the uniform electric field (Sec. 2.7). Specifically, based on the symmetry of the problem, we assume that

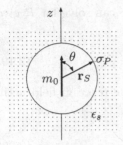

Fig. 2.19 Dipolar spherical solute in a polarizable medium.

the surface charge at the dividing surface changes with the polar angle θ according to the first Legendre polynomial

$$\sigma_P = P \cos\theta. \tag{2.194}$$

As already found in Eq. (2.79), this surface charge distribution creates the interfacial dipole $M^{\text{int}} = P\Omega_0$, where $\Omega_0 = (4\pi/3)a^3$ is the solute volume.

The z-projection of the field created by the surface charge is

$$R = R_z = -\nabla_z \oint_S dS \frac{\sigma_P(\mathbf{r}_S)}{|\mathbf{r} - \mathbf{r}_S|}\bigg|_{r=0} = -\frac{4\pi}{3}P. \tag{2.195}$$

The missing parameter here is the surface charge density P. It is found by using the locality of the Maxwell field composed of the external field of the solute dipole $\mathbf{E}_0$ and the field of bound charge $\mathbf{E}_b$. One needs to find the normal projection of the medium polarization at the dividing surface, which is equal to the surface charge density σ_P. Since the normal is drawn from the medium toward the solute, the normal projection is the negative of the radial projection. One can write

$$\sigma_P = P \cos\theta = -\frac{\epsilon_s - 1}{4\pi}\left(E_{0r} + E_{br}\right), \tag{2.196}$$

where E_{0r} and E_{br} are the radial projections of the corresponding fields taken at the dividing surface $r = a$ (Fig. 2.19).

The field of the solute dipole at the dividing surface is

$$\mathbf{E}_0 = (m_0/a^3)\left[3\hat{\mathbf{r}}\cos\theta - \hat{\mathbf{z}}\right], \tag{2.197}$$

where $\hat{\mathbf{r}} = \mathbf{r}/r$. The field of the surface charge is given by the same expression with the replacement of m_0 with the surface dipole moment M^{int}. One obtains

$$E_{0r} + E_{br} = 2(m_0 + M^{\text{int}})\cos\theta/a^3. \tag{2.198}$$

Substituting this equation back to Eq. (2.196), one can solve for P given that $M^{\text{int}} = P\Omega_0$

$$P = -\frac{3}{4\pi} \frac{2(\epsilon_s - 1)}{2\epsilon_s + 1} \frac{m_0}{a^3}. \qquad (2.199)$$

Plugging this result to Eq. (2.195), one obtains the reaction field acting back on the solute dipole

$$\boxed{R = \frac{2(\epsilon_s - 1)}{2\epsilon_s + 1} \frac{m_0}{a^3}.} \qquad (2.200)$$

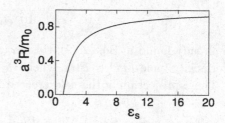

Fig. 2.20   Reaction field R vs the solvent dielectric constant ϵ_s.

The reaction field is zero at $\epsilon_s = 1$, which is the condition of vacuum, and saturates to its maximum value for typical polar liquids with $\epsilon_s \gg 1$ (Fig. 2.20). The free energy of the solute dipole in the polarizable medium is given as the electrostatic component of the solute's chemical potential (see Eq. (2.151) and a more extensive discussion in Chap. 12)

$$\mu_{\text{el}} = -\tfrac{1}{2} m_0 R = -\frac{\epsilon_s - 1}{2\epsilon_s + 1} \frac{m_0^2}{a^3}. \qquad (2.201)$$

A significant result of the derivation for the reaction field in a spherical void is the value of the interface dipole polarized by the solute dipole m_0 (Fig. 2.21). It is found by multiplying the surface charge density P with the void volume Ω_0

$$M^{\text{int}} = -\frac{2(\epsilon_s - 1)}{2\epsilon_s + 1} m_0. \qquad (2.202)$$

For polar media with $\epsilon_s \gg 1$, the interface dipole nearly compensates the dipole in the void m_0 being oriented opposite to it. This result can be contrasted with Eq. (2.82) where the interface dipole induced at a spherical void by an external uniform electric field turned out to carry the same functionality involving the dielectric constant of the medium, but was found to be proportional to the void volume and the field in the medium.

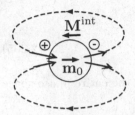

Fig. 2.21 Interface dipole created in the polarized dielectric by a dipole in the spherical void. The dashed lines sketch the field lines of the dipole m_0.

The total dipole that is responsible for the field outside the sphere is the sum of m_0 and the interface dipole

$$m_0 + M^{\text{int}} = \frac{3}{2\epsilon_s + 1} m_0. \tag{2.203}$$

It is reduced by the cavity field factor (Eq. (2.183)) compared to m_0.

2.19 Kirkwood-Onsager equation

The Kirkwood-Onsager equation [10,11] connects the linear static dielectric constant ϵ_s to the variance of the dipole moment of the dielectric sample $\mathbf{M}$. Since the variance of a macroscopic variable scales linearly with the number of molecules N (Sec. 5.9), one can anticipate that an intensive material property, such as the dielectric constant, should depend on $N^{-1}\langle \mathbf{M}^2 \rangle$. This indeed turns out to be correct. Note that the average is taken in the absence of the field when the average dipole $\langle \mathbf{M} \rangle$ is zero for a paraelectric isotropic material.

Consider a slab sample of dielectric placed between the plates of a plane capacitor (Fig. 2.14). One can first apply the electric field perpendicular to the plates, as is typically done in the dielectric experiment. If the Hamiltonian of the dielectric without the external field is H_0, the perturbed dielectric is described by the Hamiltonian

$$H = H_0 - \mathbf{M} \cdot \mathbf{E}_0, \tag{2.204}$$

where $\mathbf{E}_0$ is the electric field of the external charges on the plates of the capacitor. This equation for the energy is different from the free energy of polarizing the dielectric in Eq. (2.151) by a factor of $1/2$. The external carriers are distributed on the plates with the charge density σ_0 (charge per unit area) and produce the electric field of external charges $E_{0z} = 4\pi\sigma_0$. This field is also equal to the electric displacement in Eq. (2.140) and thus $E_{0z} = D = \epsilon_s E$.

The dipole moment is zero before the field is switched on. The average dipole moment after the application of the field can be found by calculating the statistical average in the canonical ensemble (Sec. 6.4)

$$\langle M_z \rangle_E = Q^{-1} \int d\Gamma M_z e^{-\beta H}, \quad Q = \int d\Gamma e^{-\beta H}, \tag{2.205}$$

where $\beta = (k_B T)^{-1}$ and $d\Gamma$ denotes integration over the entire phase space of the molecules of the sample.

In most cases of practical interest the applied fields are so weak that $\beta \mathbf{M} \cdot \mathbf{E}_0 \ll 1$ and one can expand both the nominator and denominator in Eq. (2.205). Given that the average dipole without field, $\langle M_z \rangle$, is zero, one obtains

$$\langle M_\alpha \rangle_E = \beta \langle M_\alpha^2 \rangle E_{0\alpha}, \tag{2.206}$$

where $\alpha = x, z$ is now introduced to reflect two possibilities of the external field applied either along the z-axis or along the x-axis (Fig. 2.14).

For the slab geometry, the connection between the field of external charges E_0 and the Maxwell field E depends on whether the external field is applied perpendicular or parallel to the slab plane [6]: $E_{0z} = \epsilon_s E_z$ and $E_{0x} = E_x$ (Fig. 2.14). On the other hand, the dipole moment in response to the Maxwell field is given in terms of the dielectric constant

$$4\pi \langle M_\alpha \rangle_E = V(\epsilon_s - 1) E_\alpha. \tag{2.207}$$

Since the dielectric constant is a material property, the relation between the dipole moment and the Maxwell field is invariant in respect to the direction of the field in an isotropic dielectric

$$4\pi \beta \langle M_\alpha^2 \rangle E_{0\alpha} = V(\epsilon_s - 1) E_\alpha. \tag{2.208}$$

One can further write the total variance of the dipole moment of the slab sample $\langle \mathbf{M}^2 \rangle = \langle M_z^2 \rangle + 2 \langle M_x^2 \rangle$ to arrive at the Kirkwood-Onsager equation

$$\boxed{\frac{(\epsilon_s - 1)(2\epsilon_s + 1)}{9\epsilon_s} = \frac{4\pi \rho}{9N} \beta \langle \mathbf{M}^2 \rangle.} \tag{2.209}$$

As anticipated, the variance of the macroscopic dipole moment enters with the factor N^{-1} to produce the dielectric constant (an intensive parameter). The variance of the total dipole in the absence of an external field is an invariant of rotations of the laboratory frame. It is therefore independent of the shape of a macroscopic sample and the Kirkwood-Onsager equation can be applied to any macroscopic material.

One can make one step further and connect the variance of the macroscopic dipole moment to microscopic correlations between molecular dipoles. Neglecting the induced dipoles, the dipole $\mathbf{M}$ can be written as a sum of N molecular permanent dipoles $\mathbf{m}_j$

$$\mathbf{M} = \sum_{j=1}^{N} \mathbf{m}_j = m \sum_{j=1}^{N} \hat{\mathbf{u}}_j, \qquad (2.210)$$

where $\hat{\mathbf{u}}_j$ is the unit vector along the direction of molecular dipole $\mathbf{m}_j$. The right-hand side of Eq. (2.209) becomes

$$(4\pi\rho/9N)\beta\langle \mathbf{M}^2 \rangle = y g_K. \qquad (2.211)$$

Here, the unitless density of dipoles is from Eq. (2.193) and g_K is the Kirkwood correlation factor [10, 11]

$$g_K = N^{-1} \sum_{i,j} \langle \hat{\mathbf{u}}_i \cdot \hat{\mathbf{u}}_j \rangle. \qquad (2.212)$$

This measure of orientational correlations of dipoles in the material is alternatively defined as the average cosine of the angles between a given target dipole i and all dipoles in the liquid surrounding it

$$\boxed{g_K = 1 + \sum_{j \neq i} \langle \hat{\mathbf{u}}_i \cdot \hat{\mathbf{u}}_j \rangle.} \qquad (2.213)$$

Note that the first term in this equation is the self term arising from $i = j$; the sum runs over all $j \neq i$ in the second term. Therefore, if dipoles are fully uncorrelated, one gets $g_K = 1$. The interpretation of the Kirkwood factor as the average cosine between the dipolar orientations is often reduced to only a few neighboring dipoles, but this interpretation is too restrictive. As an illustration, the Kirkwood factor formally turns to infinity at the transition to the ferroelectric phase with $\langle \mathbf{M} \rangle \neq 0$. Therefore, interactions between all dipoles within a ferroelectric domain, and not only between the neighbors, should be included in Eq. (2.213).

By combining Eqs. (2.209) and (2.211), one arrives at the standard form of the Kirkwood-Onsager equation

$$\frac{(\epsilon_s - 1)(2\epsilon_s + 1)}{9\epsilon_s} = y g_K, \qquad (2.214)$$

which differs from the Onsager equation (2.192) by the Kirkwood factor on the right-hand side. This equation is limited to permanent dipoles in the liquid, while the more general Eq. (2.209) does not specify the origin of the dipole, which can include induced dipoles. If this extension is allowed, the dipolar density y in Eq. (2.214) is replaced with the density of both permanent and induced dipoles assuming that they carry the same Kirkwood factor

$$y_{\text{eff}} = (4\pi/9)\beta(m')^2\rho + y_e, \qquad (2.215)$$

where m' is the dipole moment in the condensed phase [19] and the density of induced dipoles y_e has already appeared in the Clausius-Mossotti equation (2.173). The assumption of equality of the Kirkwood factor for permanent and induced dipoles is based on the observation that the reaction field induced by the permanent dipole at a given molecule is the strongest local field in a polar medium. It is directed along the permanent dipole making the permanent dipole and the induced dipole polarized by the reaction field parallel on average.

For many polar liquids, $\epsilon_s \gg 1$ and Eq. (2.214) can be simplified to

$$\epsilon_s \simeq (9/2)yg_K \propto (\rho/T), \qquad (2.216)$$

where the second scaling relation neglects the density and temperature variations of the Kirkwood factor. How well this approximation is justified can be seen from $(\partial \ln \epsilon_s/\partial \ln T)_V$ and $\rho(\partial \epsilon_s/\partial \rho)_T$ listed in Table 2.2. Equation (2.216) suggests the value of -1 for the former and ϵ_s for the latter. We used different formats for these two parameters since $\rho(\partial \epsilon_s/\partial \rho)_T$ appears in the Maxwell tensor considered in Sec. 9.9. It is seen from Table 2.2 that $\epsilon_s \propto \rho$ agrees better with the empirical data than $\epsilon_s \propto T^{-1}$. This implies that g_K is approximately a constant when density is varied at constant temperature, but changes when temperature is varied either at constant density or constant pressure.

The Onsager equation (2.192) was the first successful theory for the dielectric constant of the liquid. Historically, it was preceded by Debye's formula. It was based on the Lorentz local field acting on each dipole in the liquid, instead of the cavity field used in Onsager's formulation. This derivation follows the steps of Sec. 2.15 with the result

$$\epsilon_s = \frac{1+2y}{1-y}. \qquad (2.217)$$

It is clear that this formula predicts a ferroelectric transition $\epsilon \to \infty$ at $y \to 1$. Figure 2.22 shows values of ϵ_s vs y for 56 molecular liquids with their molecular dipoles exceeding 0.5 D [14]. Most of them fall between the Onsager and Debye predictions indicating that the Kirkwood factor $g_K > 1$ needs to be involved for a correct estimate. The Kirkwood factor is in turn affected by details of local intermolecular interactions, which, among other things, include strong influence of the molecular quadrupole on the short-range orientational order. Accounting for these effects requires microscopic theories directly addressing the local liquid structure and polarizability effects already included in Eq. (2.215).

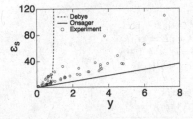

Fig. 2.22 Dielectric constant ϵ_s vs y for 56 molecular liquids with molecular dipoles exceeding 0.5 D. The lines show the Debye (Eq. (2.217)) and Onsager (Eq. (2.192)) predictions.

2.20   SI units primer

$$e^2/(4\pi\epsilon_0\text{Å}) \simeq 14.4 \text{ eV}, \ (1 \text{ D})^2/(4\pi\epsilon_0\text{Å}^3) \simeq 0.6242 \text{ eV}$$
$$1 \text{ D} \simeq 3.33 \times 10^{-30} \text{ C m}, \ \text{e} \times \text{Å} = 4.8032 \text{ D}$$

The main focus of this Chapter is on electrostatics for which SI units are widely used. Some simple recipes on how to connect formulas presented above to those given in SI units are presented here. The discussion involves only the electric field and electric polarization density and does not touch on magnetic field and magnetization.

The main distinction between the Gaussian and SI units arise from the adopted form of the Coulomb law (Eq. (2.6)). It is written in the SI units by introducing the numerical factor of 4π multiplied with the vacuum permittivity $\epsilon_0 \simeq 8.854 \times 10^{-12}$ (F/m) in the denominator

$$F = \frac{q_1 q_2}{4\pi\epsilon_0 r^2}. \tag{2.218}$$

This change in the numerical form of the basic law propagates through many equations. Since the Coulomb law is brought back to the Gaussian units by putting $4\pi\epsilon_0 = 1$, this convention provides a simple recipe to convert the final expressions to the SI units. In many cases, the appearance of the factor 4π in equations derived in Gaussian units implies that it needs to be replaced with ϵ_0^{-1} in the SI units. There is no need to perform this replacement in all mathematical manipulations and it suffices to do it in the final expressions.

Table 2.4 Conversion of electrostatic properties between Gaussian and SI systems of units.

Gaussian	SI
$D = E + 4\pi P$	$D = \epsilon_0 E + P$
$P = \chi E$	$P = \chi \epsilon_0 E$
$4\pi\chi$	χ
$P = \frac{\epsilon-1}{4\pi}E$	$P = \epsilon_0(\varepsilon - 1)E$
$D = \epsilon E$	$D = \epsilon E = \varepsilon\epsilon_0 E$
$\epsilon = 1 + 4\pi\chi$	$\varepsilon = \epsilon/\epsilon_0 = 1 + \chi$
$U = -mE$	$U = -mE$
$U = q\phi$	$U = q\phi$
$p = \alpha E^a$	$p = \alpha E$
α^b	$\alpha/(4\pi\epsilon_0)$

[a]Induced dipole. [b]α is the electronic polarizability expressed in m^3 in the Gaussian units. If the polarizability is given in cm^3, the conversion factor becomes $10^6/(4\pi\epsilon_0)$.

The main distinction between two sets of units when electrostatics is concerned comes in the connection between the electric displacement and the Maxwell field and the corresponding definition of the dielectric constant. Consider dipolar polarization density P induced in a dielectric by the Maxwell field E. P and E are related by the dielectric susceptibility χ such that $P = \chi E$ in the Gaussian units and $P = \chi\epsilon_0 E$ in the SI units. Correspondingly, the electric displacement D is related to E as

$$D = (1 + 4\pi\chi)E = \epsilon E \quad \text{(Gaussian)},$$
$$D = (\epsilon_0 + \epsilon_0\chi)E = \epsilon E \quad \text{(SI)}. \tag{2.219}$$

Therefore, the dielectric constant $\epsilon = 1+4\pi\chi$ in the Gaussian units becomes the permittivity relative to vacuum $\varepsilon = \epsilon/\epsilon_0 = 1 + \chi$ in the SI units. The dielectric constants ϵ and $\varepsilon = \epsilon/\epsilon_0$ are dimensionless in both systems of units and have equal values. A short list of conversions between electrostatic parameters is displayed in Table 2.4. The subscript "s" used in the previous sections for the static dielectric constant ϵ_s and dielectric susceptibility χ_s is dropped here for brevity.

Many of the final results of the theory of polarized dielectrics come in terms of dimensionless parameters and require only minor modification. The density of electronic induced dipoles y_e in the Clausius-Mossotti equation is dimensionless. Given by Eq. (2.174), it requires the replacement of

$4\pi\alpha$ with $\alpha\epsilon_0^{-1}$ to be converted to SI units

$$y_e = \rho\alpha/(3\epsilon_0).\qquad(2.220)$$

Likewise, the unitless density of permanent dipoles in the Kirkwood-Onsager equation is a dimensionless parameter which reads in SI units as $(4\pi \to \epsilon_0^{-1})$

$$y = \beta m^2 \rho/(9\epsilon_0).\qquad(2.221)$$

A simple formula for calculating y follows when m is in debye units

$$y \simeq 34(T_0/T)(m^2/v),\qquad(2.222)$$

where $T_0 = 300$ K and v is the volume occupied in the material by a single molecule and expressed in Å^3.

The most challenging situations appear when applying properties measured in mixed units in one equation. Electrostatic calculations involving the interaction of the electric field with the dipole moment are often challenging because molecular dipoles are given in debye units and the field is often measured in either V/m or in V/cm. Converting V to eV, the energy unit, often simplifies calculations. As an example, let's assume that the field is measured in V/cm and the dipole is given in debye units (D). We apply the identity $e \times \text{Å} = 4.8032$ D, where e is the unit charge

$$M \times E = M\ \cancel{D}\frac{e \times \cancel{\text{Å}}}{4.8032\ \cancel{D}}E\frac{\text{V}}{10^8\cancel{\text{Å}}} = 2.08 \times 10^{-9} ME\ \text{eV}.\qquad(2.223)$$

This estimate shows that only microscopic electric fields of the order of V/Å$= 10^8$ V/cm can produce the interaction energy of the order of an eV when the typical molecular dipole moment of the order of a few debye units is adopted.

Chapter 3

Classical Mechanics

This Chapter briefly covers the laws of *classical mechanics* [20–22]. It starts with the principle of least action from which the Euler-Lagrange equation is derived, followed with the transition to the Hamiltonian mechanics. The mechanical action is further generalized when path integrals in quantum mechanics are considered in Sec. 4.17. The Hamilton equations of motion are used to formulate Liouville's theorem for the density of trajectories in the phase space of conjugate coordinates and momenta. This theorem is essential for formulating the dynamics of statistical ensembles covered in Sec. 8.1.

3.1 Principle of least action

Classical mechanics is derived from the principle of least action which states that from all possible trajectories a classical system can potentially travel between points $x_0 = x(0)$ and $x = x(t)$ the path actually taken is the one that minimizes the mechanical action [20, 23]

$$S[x] = \int_0^t d\tau L(x(\tau), \dot{x}(\tau), \tau). \tag{3.1}$$

Here, the square brackets indicate that $S[x]$ is a functional depending on the function $x(t)$, i.e., this is a rule producing a number for each function given as input. This definition distinguishes a functional from a function, which is a rule specifying a number corresponding to a given variable. We will encounter a number of occasions of using functionals in this book and, for instance, the density functional $F[\rho]$ discussed in Sec. 6.11 establishes a rule producing a number for each given scalar function of the number density $\rho(\mathbf{r})$ defined in 3D space.

The function $L(x, \dot{x}, t)$ in Eq. (3.1) is the Lagrangian depending on the coordinate x and its derivative (velocity) $\dot{x}$. The Lagrangian does not depend on higher derivatives of the coordinate (e.g., $\ddot{x}$) and depends at most quadratically on $\dot{x}$. Given these limitations, one can produce the differential equation of motion for the trajectory minimizing the action.

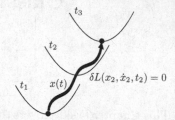

Fig. 3.1   Trajectory minimizing the action.

The equation of motion is produced by finding the trajectory corresponding to the lowest value of $S[x]$ along it. To understand how to proceed, one can represent the integral in Eq. (3.1) with a discretized sum over N small steps $\epsilon = t/N$

$$S[x] = \epsilon \sum_{n=0}^{N-1} L(x(t_n), \dot{x}(t_n), t_n), \tag{3.2}$$

where $t_n = \epsilon n$. The requirement of minimizing the trajectory implies that at each time t_n intermediate between the end points, the Lagrangian $L(x_n, \dot{x}_n, t_n)$ should be at minimum in respect to $x_n = x(t_n)$ and $\dot{x}_n = \dot{x}(t_n)$ (Fig. 3.1)

$$\delta L(x_n, \dot{x}_n, t_n) = \frac{\partial L}{\partial x_n} \delta x_n + \frac{\partial L}{\partial \dot{x}_n} \delta \dot{x}_n = 0. \tag{3.3}$$

One can now extend this equation from all internal points to all points along the trajectory by requiring $\delta x_0 = \delta x(0) = \delta x_N = \delta x(t) = 0$. The variation of the action becomes

$$\delta S[x] = \epsilon \sum_{n=0}^{N-1} \delta L(x(t_n), \dot{x}(t_n), t_n) = 0. \tag{3.4}$$

We now can return from the discretized sum back to the continuous integral to write

$$\begin{aligned}
\delta S &= \int_0^t d\tau \left[\frac{\partial L}{\partial x} \delta x + \frac{\partial L}{\partial \dot{x}} \delta \dot{x} \right] \\
&= \int_0^t d\tau \left[\frac{\partial L}{\partial x} - \frac{d}{dt} \frac{\partial L}{\partial \dot{x}} \right] \delta x + \frac{\partial L}{\partial \dot{x}} \delta x \Big|_0^t .
\end{aligned} \tag{3.5}$$

In the second line, we applied the identity

$$\frac{\partial L}{\partial \dot{x}}\delta \dot{x} = \frac{d}{dt}\left(\frac{\partial L}{\partial \dot{x}}\delta x\right) - \frac{d}{dt}\frac{\partial L}{\partial \dot{x}}\delta x. \tag{3.6}$$

The last term in Eq. (3.5) vanishes because the boundary conditions for the trajectory are fixed and $\delta x(0) = \delta x(t) = 0$. To satisfy the condition of action's minimum on the trajectory, $\delta S = 0$, the terms in the brackets in Eq. (3.5) must be zero. This condition is the Euler-Lagrange equation

$$\boxed{\frac{d}{dt}\frac{\partial L}{\partial \dot{x}} = \frac{\partial L}{\partial x}.} \tag{3.7}$$

For a particle with the mass m and velocity $\dot{x}$, its kinetic energy is defined as $\frac{1}{2}m\dot{x}^2$. If the particle interacts with other particles or the external field with the potential energy $U(x)$, which depends on the coordinate x and not on the velocity, the Lagrangian becomes

$$L = \tfrac{1}{2}m\dot{x}^2 - U(x) \tag{3.8}$$

i.e., the Lagrangian is the kinetic energy minus the potential energy. When substituted to the Euler-Lagrange equation, this form leads to Newton's equation of motion

$$\boxed{m\ddot{x} = f = -\partial U/\partial x.} \tag{3.9}$$

The assumption that the potential energy depends on the coordinate only implies an instantaneously propagated interaction. This is a limitation of classical mechanics rooted in the absolute nature of time in Galileo's relativity principle.

3.2 Classical Hamiltonian mechanics

An alternative formulation of classical mechanics is based on the function called the Hamiltonian, which is obtained from the Lagrangian by Legendre transformation. This transformation, which is widely used in statistical mechanics (Sec. 5.11), amounts to subtracting the product of conjugate variables from the function transformed. In the case of the Hamiltonian function H, it is the Lagrangian that is subtracted from the product of the velocity $\dot{x}$ and the conjugate canonical momentum p

$$H = p\dot{x} - L. \tag{3.10}$$

The conjugate canonical momentum is given by the derivative of the Lagrangian over the velocity

$$p = \partial L/\partial \dot{x}. \tag{3.11}$$

While the Lagrangian is a function of x and $\dot{x}$, the Hamiltonian is a function of x and p. This means that Eq. (3.7) needs to be solved to provide the function $\dot{x}(x, p)$. This solution is straightforward for a common situation of $\dot{x}$ entering the Lagrangian as kinetic energy: $L = m\dot{x}^2/2 - \ldots$. In this case, $p = m\dot{x}$ is the standard linear momentum. Sec. 3.6 discussing a particle in the electromagnetic field provides a more complex example for the calculation of the canonical momentum.

From the Euler-Lagrange equation and Eq. (3.11) one obtains

$$\dot{p} = \partial L/\partial x. \tag{3.12}$$

One, therefore, can write the differential of the Lagrangian as

$$dL = \frac{\partial L}{\partial x}dx + \frac{\partial L}{\partial \dot{x}}d\dot{x} = \dot{p}dx + pd\dot{x} = d(p\dot{x}) - \dot{x}dp + \dot{p}dx. \tag{3.13}$$

By rearranging the terms one obtains

$$d(p\dot{x} - L) = dH = \dot{x}dp - \dot{p}dx. \tag{3.14}$$

This differential means that $H = H(x, p)$ is a function of the coordinate and the canonical momentum with the following derivatives

$$\boxed{\dot{p} = -\frac{\partial H}{\partial x}, \quad \dot{x} = \frac{\partial H}{\partial p}.} \tag{3.15}$$

These are Hamilton equations of motion.

It is instructive to illustrate the application of Hamilton equations of motion to a harmonic oscillator since harmonic motion sets foundation for many developments discussed below. According to Hooke's law, the force pulling back on the spring stretched by the displacement x is defined by the spring constant k as $f = -kx = -dU/dx$. This yields the potential energy of the stretched spring $U = \frac{1}{2}kx^2$ and the Lagrangian

$$L = \frac{1}{2}m\dot{x}^2 - \frac{1}{2}kx^2. \tag{3.16}$$

The canonical momentum is equal to the linear momentum and $\dot{x} = p/m$. With this transformation from velocity to canonical variables, one obtains for the Hamiltonian from Eq. (3.10)

$$H = \frac{p^2}{2m} + \frac{1}{2}kx^2. \tag{3.17}$$

The second Hamiltonian equation gives the expected result $\dot{x} = p/m$ and the first equation is the Newton's equation

$$\dot{p} = m\ddot{x} = -kx. \tag{3.18}$$

Its solution is the harmonic oscillatory motion with the frequency
$$\omega_0 = \sqrt{k/m}. \tag{3.19}$$
The solution depends on two constants: initial position $x(0)$ and the initial velocity $\dot{x}(0)$. Adopting $\dot{x}(0) = 0$ and $x(0) = x_0$, one obtains
$$x(t) = x_0 \cos \omega_0 t. \tag{3.20}$$
This solution corresponds to vibrations of the oscillator stretched to the displacement x_0 at $t = 0$. As it oscillates, the total energy is conserved, $E = H(x,p)$, and is equal to the energy stored in the initial stretch
$$\frac{p(t)^2}{2m} + \frac{1}{2}kx(t)^2 = E = \frac{1}{2}kx_0^2. \tag{3.21}$$

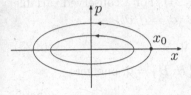

Fig. 3.2 Phase space of harmonic oscillators with $x(0) = x_0$.

The set of points $\{x(t), p(t)\}$ allowed by the Hamilton equations of motion is called the phase space. Since Eq. (3.21) is the equation of an ellipse in the coordinates x and p, the phase space of a single harmonic oscillator is an elliptical orbit, which it circles passing through the initial point $\{x_0, 0\}$ with the time period $T = 2\pi/\omega_0$ (Fig. 3.2). One can imagine a number of oscillators, each starting with the initial stretch x_0 and propagating over the its own trajectory with the same period T. This set of particles with different initial positions and, generally, with different initial velocities is known as an ensemble. If the initial coordinates and velocities are distributed according to the Boltzmann distribution, one arrives at a canonical ensemble considered in statistical mechanics (Chap. 5).

Each trajectory $\{x(t), p(t)\}$ in the phase space is fully specified by the initial conditions $\{x(0), p(0)\}$ and the Hamiltonian describing the dynamics of the system. Therefore, trajectories in the phase space never cross, as is clearly the case for two harmonic oscillators with different initial conditions shown in Fig. 3.2.

3.3 Poisson brackets

A function of coordinates, momenta, and time $A(x, p, t)$ has the following time derivative
$$\frac{dA}{dt} = \frac{\partial A}{\partial t} + \dot{x}\frac{\partial A}{\partial x} + \dot{p}\frac{\partial A}{\partial p}. \tag{3.22}$$

If the path of the variables $x(t)$ and $p(t)$ is given by a classical trajectory, it must satisfy the Hamilton equations of motion. By applying Eq. (3.15), one obtains

$$\frac{dA}{dt} = \frac{\partial A}{\partial t} + \{A, H\}. \tag{3.23}$$

Here, $\{A, H\}$ denotes Poisson brackets

$$\boxed{\{A, H\} = \frac{\partial A}{\partial x}\frac{\partial H}{\partial p} - \frac{\partial A}{\partial p}\frac{\partial H}{\partial x}.} \tag{3.24}$$

One can similarly define Poisson brackets for any pair of dynamic variables $\{A, B\}$, in which case B replaces H in Eq. (3.24). For a conserved variable (integral of motion), $dA/dt = 0$ and one gets ($\partial_t = \partial/\partial t$)

$$\partial_t A + \{A, H\} = 0. \tag{3.25}$$

An important special case is $A = H$. Since $\{H, H\} = 0$, one gets

$$dH/dt = 0 \tag{3.26}$$

as long as the energy function is not affected by time-changing external fields and $\partial_t H = 0$. The energy is an integral of motion for trajectories satisfying the Hamilton equations of motion.

As an example, one can consider the variable of the center of mass

$$\mathbf{r}_c = M^{-1}\sum_i m_i\mathbf{r}_i, \quad M = \sum_i m_i. \tag{3.27}$$

The center of mass is not a physical body and it moves only because parts of the body with masses m_i and positions $\mathbf{r}_i$ move. Therefore, the time derivative of $\mathbf{r}_c(t)$ is the combined effect of time derivatives of $\mathbf{r}_i(t)$

$$\dot{\mathbf{r}}_c = \sum_i \left(\dot{\mathbf{r}}_i \frac{\partial \mathbf{r}_c}{\partial \mathbf{r}_i} + \dot{\mathbf{p}}_i \frac{\partial \mathbf{r}_c}{\partial \mathbf{p}_i} \right) = \{\mathbf{r}_c, H\} = M^{-1}\mathbf{P}, \tag{3.28}$$

where $\mathbf{P} = \sum_i \mathbf{p}_i$ is the total momentum of the body. This example illustrates the use of Poisson brackets: one can choose any function of the canonical coordinates and momenta and its dynamics is described by Eq. (3.23).

Another useful result easily derived with Poisson brackets and used multiple times in subsequent Chapters is the continuity relation. Consider a collective variable of the number density of a macroscopic body

$$\rho = \sum_j \delta(\mathbf{r} - \mathbf{r}_j), \tag{3.29}$$

where $\delta(\mathbf{r} - \mathbf{r}_j)$ is the three-dimensional delta function establishing the fact that there are no particles but those located at the positions $\mathbf{r}_j$. As these particle move according to the rules of Hamiltonian dynamics, the density changes in time.. The dynamic equation for the number density is

$$\dot{\rho} = \{\rho, H\}. \tag{3.30}$$

One can how assume that the Hamiltonian of the system is the sum of the kinetic energies of all the particles plus the potential energy U depending on their coordinates

$$H = \frac{1}{2m} \sum_j \mathbf{p}_j^2 + U \tag{3.31}$$

The Poisson brackets become

$$\dot{\rho} = \sum_j \frac{\partial}{\partial \mathbf{r}_j} \delta(\mathbf{r} - \mathbf{r}_j) \frac{\partial H}{\partial \mathbf{p}_j}. \tag{3.32}$$

For the delta function, one can apply the following identity

$$\frac{\partial}{\partial \mathbf{r}_j} \delta(\mathbf{r} - \mathbf{r}_j) = -\frac{\partial}{\partial \mathbf{r}} \delta(\mathbf{r} - \mathbf{r}_j). \tag{3.33}$$

With this identity applied, the equation for the density evolution becomes

$$\boxed{\dot{\rho} + \nabla \cdot \mathbf{j} = 0,} \tag{3.34}$$

where

$$\mathbf{j} = \sum_j \mathbf{v}_j \delta(\mathbf{r} - \mathbf{r}_j) \tag{3.35}$$

is the current density given in terms of particle velocities $\mathbf{v}_j = \mathbf{p}_j/m$. This is the continuity relation stating that the change of the number of particles in a small volume is caused by the overall flux of particles outward and inward of the volume. The particles entering the volume are counted with the positive flux and the particle leaving the volume produce the negative flux. If the total flux is zero, the number of particles does not change.

There is an important distinction between Newtonian and Hamiltonian dynamics. The Newton's equation of motion

$$\boxed{m\ddot{x} = f} \tag{3.36}$$

remains valid if the force f includes some friction force leading to dissipation of energy. This possibility is explored in Chap. 7 dealing with diffusion and the Langevin equation incorporating frictional forces into Newton's equation of motion. It might seem that the Hamiltonian dynamics are less general than Newtonian dynamics. This is not true since the introduction of friction is the recognition of heat exchange between a system and a thermal bath. The mechanical energy is still conserved, but because accounting for its transfer to a large bath with many degrees of freedom becomes increasingly difficult, a coarse-grained view of the system dynamics involves the loss of mechanical energy as friction or heat.

3.4   Liouville's theorem

The trajectories following Hamilton dynamics never cross, but one can ask whether they become denser as an ensemble of systems with different initial conditions evolves from the initial time $t = 0$. Liouville's theorem states that the density of trajectories remains constant. More specifically, it states that if the phase space volume $d\Omega_0 = dx_0 dp_0$ is chosen at $t = 0$, the same phase element will be found at time t

$$\boxed{d\Omega_t = d\Omega_0.} \tag{3.37}$$

To prove this theorem, one can view the transition from $d\Omega_0$ to $d\Omega_t$ as a transformation in the phase space with the Jacobian determinant $J(t)$

$$d\Omega_t = J(t)d\Omega_0, \quad J(t) = \left| \frac{\partial(x(t), p(t))}{\partial(x_0, p_0)} \right|, \tag{3.38}$$

where

$$J(t) = \begin{vmatrix} \dfrac{\partial x(t)}{\partial x_0} & \dfrac{\partial x(t)}{\partial p_0} \\[2mm] \dfrac{\partial p(t)}{\partial x_0} & \dfrac{\partial p(t)}{\partial p_0} \end{vmatrix}. \tag{3.39}$$

It is clear that $J(0) = 1$ $(\partial x_0/\partial p_0 = \partial p_0/\partial x_0 = 0)$. One can make one small step dt forward and calculate $x(dt)$ and $p(dt)$ keeping the terms linear in dt

$$x(dt) = x_0 + \frac{\partial H}{\partial p_0}dt,$$
$$p(dt) = p_0 - \frac{\partial H}{\partial x_0}dt. \tag{3.40}$$

In the next step, one obtains

$$\frac{\partial x(dt)}{\partial x_0} = 1 + \frac{\partial^2 H}{\partial p_0 \partial x_0}dt,$$
$$\frac{\partial p(dt)}{\partial p_0} = 1 - \frac{\partial^2 H}{\partial x_0 \partial p_0}dt. \tag{3.41}$$

The off-diagonal derivatives are not required since they give terms quadratic in dt in the Jacobian determinant. One obtains in terms linear in dt

$$J(dt) = 1. \tag{3.42}$$

However, the total Jacobian determinant $J(t)$ can be calculated as a sequence of n steps dt such that $t = ndt$: $J(t) = J(t, t - dt) \ldots J(dt) = 1$. This proves $J(t) = 1$. One therefore concludes that the volume element in the phase space does not change along the classical trajectories satisfying the Hamiltonian dynamics.

One can illustrate Liouville's theorem by considering the harmonic oscillator with $x(0) = x_0$ and $p(0) = p_0$. The solution for $x(t)$ and $p(t)$ is

$$x(t) = x_0 \cos \omega_0 t + \frac{p_0}{m\omega_0} \sin \omega_0 t,$$

$$p(t) = -m\omega_0 x_0 \sin \omega_0 t + p_0 \cos \omega_0 t. \tag{3.43}$$

One obtains the Jacobian determinant

$$J(t) = \begin{vmatrix} \cos \omega_0 t & (m\omega_0)^{-1} \sin \omega_0 t \\ -(m\omega_0) \sin \omega_0 t & \cos \omega_0 t \end{vmatrix} = 1. \tag{3.44}$$

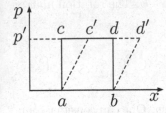

Fig. 3.3 Phase-space element of a free particle.

One can additionally appreciate the meaning of Liuoville's theorem by looking at the simplest case of a free particle and the rectangular phase-space element shown in Fig. 3.3. The two points at the bottom, "a" and "b", have zero momenta and they do not move with time. Two other points, "c" and "d", have the momentum p' and will change to "c'" and "d'" with the coordinates

$$x_c(t) = x_c + (p'/m)t, \quad x_d(t) = x_d + (p'/m)t. \tag{3.45}$$

The Jacobian determinant is

$$J(t) = \begin{vmatrix} 1 & t/m \\ 0 & 1 \end{vmatrix} = 1. \tag{3.46}$$

This result can be illustrated by looking at the area of the phase-space element: the square at $t = 0$ becomes a rhombus at time t, but with the same overall area (Fig. 3.3).

3.5   Rigid body rotations

Rotations of a rigid body are conveniently described by applying Euler's theorem that states that the displacement of a body with one fixed point can be represented by a rotation about some axis. One also realizes that an active rotation of a body can be replaced by an equivalent rotation of

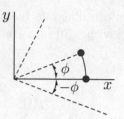

Fig. 3.4   An active rotation of a particle in the x, y-plane through the angle ϕ is equivalent to the rotation of the laboratory frame through the angle $-\phi$.

the laboratory coordinate frame (Sec. 1.2). This is illustrated in Fig. 3.4 where the active rotation of the particle in the x, y-plane through the angle ϕ can be represented by the rotation of the coordinate system about the z-axis in the opposite direction through the angle $-\phi$. The rotation matrix for representing the active rotation becomes

$$\mathbf{R} = \begin{pmatrix} \cos\phi & -\sin\phi & 0 \\ \sin\phi & \cos\phi & 0 \\ 0 & 0 & 1 \end{pmatrix}. \tag{3.47}$$

This view can be extended to rotations in space. One can consider some vector property $\mathbf{A}$ in the body-frame and laboratory-frame coordinates. One can think, for instance, of an atomic position within the molecule which can change through molecular flexibility. The change $d\mathbf{A}'$ in the body frame can be related to $d\mathbf{A}$ in the laboratory frame through the vector $d\boldsymbol{\Omega}$ of rotation transformation of the laboratory frame. One writes

$$d\mathbf{A}' = d\mathbf{A} + \mathbf{A} \times d\boldsymbol{\Omega}. \tag{3.48}$$

The vector $d\boldsymbol{\Omega}$ specifies the direction of the rotation axis; its magnitude is equal to the angle of rotation around this axis. If no change of $\mathbf{A}'$ occurs in the body frame, $d\mathbf{A}' = 0$ (e.g., the molecule is rigid), the vector $d\boldsymbol{\Omega}$ can be viewed as the rotation of the laboratory frame representing the active rotation of the rigid body. Figure 3.5 illustrates this infinitesimal rotation in the case when $\mathbf{A} = \mathbf{r}$ represents a point within a rigid body.

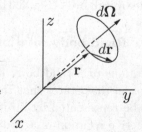

Fig. 3.5   Change in the coordinate $\mathbf{A} = \mathbf{r}$ in the rigid body caused by an infinitesimal rotation about the rotation axis $d\boldsymbol{\Omega}$: $d\mathbf{r} = d\boldsymbol{\Omega} \times \mathbf{r}$.

For the time derivatives one obtains

$$\frac{d\mathbf{A}}{dt} = \frac{d\mathbf{A}'}{dt} + \boldsymbol{\omega} \times \mathbf{A}, \tag{3.49}$$

where $\boldsymbol{\omega} = d\boldsymbol{\Omega}/dt$ is the angular velocity of the body. If the position $\mathbf{A}' = \mathbf{r}'$ does not change within the rigid body, one obtains the kinematic equation for the rigid body rotation

$$\boxed{\dot{\mathbf{r}} = \boldsymbol{\omega} \times \mathbf{r}.} \tag{3.50}$$

The angular frequency $\boldsymbol{\omega}$ enters the angular momentum $\mathbf{L}$ of rigid body rotation

$$\mathbf{L} = \sum_i m_i \left(\mathbf{r}_i \times \mathbf{v}_i \right), \tag{3.51}$$

where the sum runs over all mass elements m_i of the rigid body with linear velocities $\mathbf{v}_i$ and positions $\mathbf{r}_i$ (Fig. 3.6). One can now use Eq. (3.49) with $\mathbf{A} = \mathbf{r}_i$

$$\mathbf{L} = \sum_i m_i \left(\mathbf{r}_i \times (\boldsymbol{\omega} \times \mathbf{r}_i) \right). \tag{3.52}$$

Expanding the double cross product leads to

$$\mathbf{L} = \sum_i m_i \left(\boldsymbol{\omega} r_i^2 - \mathbf{r}_i (\boldsymbol{\omega} \cdot \mathbf{r}_i) \right). \tag{3.53}$$

This result can be alternatively written as the dot product of the angular frequency $\boldsymbol{\omega}$ with the second-rank tensor (Sec. 1.3) $\mathbf{I}_0$ of the moment of inertia

$$\boxed{\mathbf{L} = \mathbf{I}_0 \cdot \boldsymbol{\omega},} \tag{3.54}$$

where

$$I_0^{\alpha\beta} = \sum_i m_i \left(r_i^2 \delta_{\alpha\beta} - r_{i\alpha} r_{i\beta} \right). \tag{3.55}$$

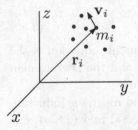

Fig. 3.6 A rigid body represented by elements $\mathbf{r}_i$ carrying masses m_i and possessing velocities $\mathbf{v}_i$.

The moment of inertia tensor is diagonal in its principal axes associated with the body frame: $I_0^{\alpha\beta} = I_0^{\alpha} \delta_{\alpha\beta}$. As an example, one can consider a linear diatomic molecule with two atoms separated by the distance d and carrying equal masses m_0. If the z'-axis of the body frame is taken to coincide with the symmetry axis of the molecule, the moment of inertia becomes a diagonal matrix defined by a single value $I_\perp = m_0 d^2/2$

$$\mathbf{I}_0 = \begin{pmatrix} I_\perp & 0 & 0 \\ 0 & I_\perp & 0 \\ 0 & 0 & 0 \end{pmatrix}. \tag{3.56}$$

By taking the time derivative of the angular momentum in Eq. (3.51), one obtains

$$\frac{d\mathbf{L}}{dt} = \sum_i m_i \left(\mathbf{v}_i \times \mathbf{v}_i + \mathbf{r}_i \times \dot{\mathbf{v}}_i \right). \tag{3.57}$$

The first term in the bracket is obviously zero and the equation of motion for the angular momentum becomes

$$\boxed{\frac{d\mathbf{L}}{dt} = \mathbf{I}_0 \cdot \dot{\boldsymbol{\omega}} = \mathbf{T} = \sum_i \mathbf{r}_i \times \mathbf{f}_i,} \tag{3.58}$$

where $\mathbf{f}_i$ is the force acting on the element i carrying the mass m_i. In this equation, $\mathbf{T}$ is the torque applied to the body.

3.6   *Charged particle in the field

$$\boxed{\begin{aligned} \mathbf{B} &= \nabla \times \mathbf{A}, \\ \mathbf{E} &= -\nabla \phi_q - c^{-1} \partial_t \mathbf{A} \end{aligned}}$$

The motion of a charged particle with the mass m and the charge q in electric and magnetic fields is considered next as an example of applying the Lagrangian and Hamiltonian dynamics. One has to postulate the law for the force acting on the particle. When magnetic field $\mathbf{B}$ is added to the electric field $\mathbf{E}$, one has to combine the Coulomb law with the Lorentz law in the total force $\mathbf{f}$

$$\mathbf{f} = q\mathbf{E} + \frac{q}{c}\mathbf{v} \times \mathbf{B}, \tag{3.59}$$

where $\mathbf{v}$ is the particle velocity. The second term in this equation is the Lorentz force. The goal is to derive the Lagrangian and the Hamiltonian of the particle from this expression for the force.

Substituting the equations for the electric field and magnetic induction in terms of the scalar and vector potentials (Eqs. (2.83) and (2.85)), one obtains

$$\mathbf{f} = -q\nabla\phi - \frac{q}{c}\partial_t \mathbf{A} + \frac{q}{c}\mathbf{v} \times \nabla \times \mathbf{A}. \tag{3.60}$$

Here, one can apply the identity from differential vector calculus

$$\mathbf{v} \times \nabla \times \mathbf{A} = \nabla(\mathbf{v} \cdot \mathbf{A}) - (\mathbf{v} \cdot \nabla)\mathbf{A}. \tag{3.61}$$

Additionally, the full time derivative of the vector field $\mathbf{A}(\mathbf{r})$ includes the partial time derivative and an additional derivative over $\mathbf{r}$ caused by the fact that the position of the particle changes, $\mathbf{r} = \mathbf{r}(t)$,

$$\frac{d\mathbf{A}}{dt} = \partial_t \mathbf{A} + (\mathbf{v} \cdot \nabla)\mathbf{A}. \tag{3.62}$$

Combining these two equations, the force becomes

$$\mathbf{f} = -q\nabla\phi - \frac{q}{c}\frac{d\mathbf{A}}{dt} + \frac{q}{c}\nabla(\mathbf{v} \cdot \mathbf{A}). \tag{3.63}$$

The equation of motion, $m\dot{\mathbf{v}} = \mathbf{f}$, with this force is derived from the Euler-Lagrange equation with the Lagrangian in the form

$$\boxed{L = \tfrac{1}{2}mv^2 - q\phi + \frac{q}{c}(\mathbf{v} \cdot \mathbf{A}).} \tag{3.64}$$

In order to prove that, we first re-write the Euler-Lagrange equation in the vectorial form

$$\frac{d}{dt}\nabla_{\mathbf{v}}L - \nabla_{\mathbf{r}}L = 0. \tag{3.65}$$

The gradient over the velocity field is

$$\nabla_{\mathbf{v}}L = m\mathbf{v} + \frac{q}{c}\mathbf{A}. \tag{3.66}$$

The gradient over the position of the particle $\nabla_{\mathbf{r}} = \nabla$ is

$$\nabla_{\mathbf{r}}L = -q\nabla\phi + \frac{q}{c}\nabla(\mathbf{v} \cdot \mathbf{A}). \tag{3.67}$$

Combining these two equations in the Euler-Lagrange equation, one obtains Eq. (3.63) for the Coulomb-Lorentz force.

One can now derive the Hamiltonian of the particle in the electromagnetic field. The canonical momentum in classical mechanics is $\mathbf{p} = \nabla_{\mathbf{v}}L$

$$\mathbf{p} = m\mathbf{v} + \frac{q}{c}\mathbf{A}. \tag{3.68}$$

Correspondingly, the Hamiltonian is $H = \mathbf{p} \cdot \mathbf{v} - L$ with the result

$$H = mv^2/2 + q\phi. \tag{3.69}$$

This result is simply the sum of the kinetic and potential energies. However, the Hamiltonian is a function of the canonical momentum and coordinate and the velocity has to be expressed in terms of the momentum. The result is

$$H = \frac{1}{2m}\left(\mathbf{p} - \frac{q}{c}\mathbf{A}\right)^2 + q\phi. \tag{3.70}$$

As an example of applying the Lagrangian to derive equations of motion, we consider here the motion of charge in a uniform magnetic field aligned along the z-axis of the laboratory frame [24]. Since $\mathbf{B} = \nabla \times \mathbf{A}$, one can represent $\mathbf{B} = (0, 0, B)$ as $\mathbf{A} = (0, xB, 0)$. However, this is not the only possible representation. Since $\nabla \times \nabla g$ is identically zero, one can add ∇g of an arbitrary function g to $\mathbf{A}$ without altering the experimentally measured magnetic field. This is known as gauge transformation

$$\mathbf{A}' = \mathbf{A} + \nabla g. \tag{3.71}$$

If $g = -Bxy$, one obtains $\mathbf{A} = (-By, 0, 0)$, which is an equivalent representation of the magnetic field along the z-axis.

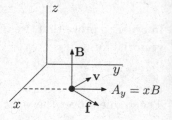

Fig. 3.7   Particle in the magnetic field.

When $\mathbf{A} = (0, xB, 0)$ (Fig. 3.7) and $\phi = 0$, the Lagrangian in Eq. (3.64) becomes

$$L = \tfrac{1}{2}mv^2 + \frac{q}{c}v_y xB. \tag{3.72}$$

Since L depends only on the coordinate x, p_y and p_z are conserved. This follows from Eq. (3.65) in which $\nabla_\alpha L = 0$ for $\alpha = y, z$. One, therefore, gets $\dot{p}_\alpha = 0$, which is the meaning of conservation of p_y and p_z. We therefore obtain $\dot{v}_z = 0$ and, by taking the derivative over v_y in Eq. (3.72),

$$p_y = mv_y + (q/c)xB. \tag{3.73}$$

and since $\dot{p}_y = 0$

$$m\dot{v}_y = -(q/c)v_x B. \tag{3.74}$$

One can next use an alternative representation of the vector potential $\mathbf{A} = (-By, 0, 0)$ to obtain

$$L = \tfrac{1}{2}mv^2 - \frac{q}{c}v_x yB. \tag{3.75}$$

This equation implies that p_x and p_z are now conserved. The change of the conservation law reflects the fact that the canonical momentum is not gauge invariant. By taking the derivative over v_x and then applying $\dot{p}_x = 0$ we get additionally

$$m\dot{v}_x = (q/c)v_y B. \tag{3.76}$$

One next can form the vector $\mathbf{v} = v_x\hat{\mathbf{i}} + v_y\hat{\mathbf{j}}$, which, by combining Eqs. (3.74) and (3.76), satisfies the equation

$$m\dot{\mathbf{v}} = (q/c)\mathbf{v} \times \mathbf{B}. \tag{3.77}$$

This is Newton's equation for a particle moving under the action of the Lorentz force. The conservation of p_z adds an additional relation $v_z =$ Const.

Chapter 4

Quantum Mechanics

It is not the goal of this Chapter to provide a comprehensive coverage of *quantum mechanics*. Familiarity with basic concepts of the subject as described in standard textbooks [25–28] is anticipated. The focus of this Chapter is on clarifying the fundamental ideas with minimal formalism and on discussing specific subjects not sufficiently covered by introductory courses. The choice of these subjects is driven by the topics presented in the following Chapters.

4.1 Early ideas of quantum mechanics

Black-Body radiation. Planck derived his equation for black body radiation by assuming that electromagnetic waves, when interacting with materials, transfer energy is small chunks of $h\nu$, where h is the Planck constant and ν is the frequency of oscillations in the electromagnetic wave, $\nu = c/\lambda$. Einstein further suggested that light itself transfers energy in discrete amounts associated with photons

$$E = h\nu = \hbar\omega. \tag{4.1}$$

Deriving Planck's equation requires counting classical oscillators of electromagnetic radiation in a cavity. One typically assumes that a cubic box with the side length L should fit integer number of half wavelength λ: $L = n_a(\lambda/2)$, where $a = x, y, z$. These oscillators produce standing electromagnetic waves inside the cube, with the magnitude of the wave vector $k = 2\pi/\lambda$ given as

$$k^2 = k_x^2 + k_y^2 + k_z^2 = (\pi/L)^2(n_x^2 + n_y^2 + n_z^2) = (2\pi/\lambda)^2. \tag{4.2}$$

This is the equation of a sphere in terms of n_x, n_y, n_z coordinates: $n_x^2 + n_y^2 + n_z^2 = (2L/\lambda)^2$. One finds the total number of oscillators by taking $1/8$

of its volume $V = 4\pi/3(2L/\lambda)^3$ (to account for $n_a > 0$ only), multiplied by a factor of 2 to account for two directions of light polarization. This gives the number of oscillators with the wavelength λ in the box

$$N(\lambda) = (8\pi/3)(L/\lambda)^3. \tag{4.3}$$

The spectral density is the amount of radiative energy per unit volume and per frequency increment $d\nu$

$$\frac{d\rho(\nu, T)}{d\nu} = \rho_\nu(T) = \frac{\epsilon(T)}{L^3}\left|\frac{dN(\lambda)}{d\lambda}\frac{d\lambda}{d\nu}\right|, \tag{4.4}$$

where $\epsilon(T)$ is the energy of a single oscillator depending on temperature T. This transforms to

$$\rho_\nu(T) = \epsilon(T)(8\pi\nu^2/c^3). \tag{4.5}$$

The Rayleigh-Jeans theory assumed classical equipartition theorem and $\epsilon(T) = k_B T$ for a classical harmonic oscillator ($(1/2)k_B T$ for the kinetic and $(1/2)k_B T$ for the potential energy). Plank's equation replaces $k_B T$ with the energy of the photon $h\nu$ multiplied with the Bose-Einstein population $n_B(T)$ for the photons in thermal equilibrium

$$\epsilon(T) = h\nu n_B(T), \quad n_B(T) = \left[e^{\beta h\nu} - 1\right]^{-1}. \tag{4.6}$$

The result is the Planck's equation

$$\boxed{\rho_\nu(T) = \frac{8\pi h\nu^3}{c^3}n_B(T).} \tag{4.7}$$

Bohr atom. Based on the structure of atomic spectra, Bohr suggested that energy states in the atoms are discrete such that the energy of the absorbed or emitted photon must match the energy difference between the atomic states

$$E_m - E_n = \hbar\omega_{mn} = \hbar\omega. \tag{4.8}$$

Heisenberg formulated matrix quantum mechanics based on the Bohr postulate of discrete energy levels.

The Bohr orbital model seeks the radius of the circular orbit of the electron. The total mechanical energy E of the electron on such an orbit is the sum of the potential energy $U = -Ze^2/r$ and the kinetic energy K, where Z is the nuclear charge. The kinetic energy, which is positive, is equal to the negative of the half of the potential energy: $K = -U/2$ (Fig. 4.1). The total energy is therefore

$$E = -Ze^2/(2r). \tag{4.9}$$

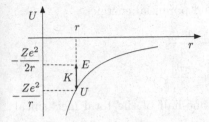

Fig. 4.1 Potential energy $U(r)$ of the electron on the circular orbit. The total energy, $E = K + U$, is one half of the potential energy.

The radius of the orbit r_n characterized by the principal quantum number $n = 1, 2, 3, \ldots$ is found from Bohr's hypothesis of quantization of the angular momentum (now obsolete) combined with the condition of mechanical stability of the electron on the circular orbit (second equation)

$$m_e v_n r_n = n\hbar, \quad m_e v_n^2 = Ze^2/r_n, \tag{4.10}$$

where m_e is the electron mass. Combining these two equations, one obtains

$$r_n = (n^2/Z)a_0, \quad a_0 = \hbar^2/(m_e e^2). \tag{4.11}$$

Here, $a_0 \simeq 0.53$ Å is the Bohr radius. Substituting the orbital radius to the total energy (4.9), one arrives at the Born solution for the energy levels E_n in the hydrogen atom

$$\boxed{E_n = -(Ze)^2/(2n^2 a_0).} \tag{4.12}$$

4.2 Harmonic oscillator and uncertainty

Classical harmonic oscillator was the starting point for Heisenberg to derive his uncertainty relation. The displacement $x(t)$ of the harmonic oscillator satisfies the dynamic equation

$$\ddot{x}(t) + \omega_0^2 x(t) = 0, \tag{4.13}$$

where ω_0 is the circular frequency in Eq. (3.19). The oscillatory solution of this equation satisfying the initial conditions $x(0) = x_0$ and $\dot{x}(0) = v_0$ is given by Eq. (3.43).

One defines the average over the oscillation period

$$\langle x \rangle = \frac{1}{T} \int_0^T x(t)dt \tag{4.14}$$

from which

$$\langle x \rangle = 0. \tag{4.15}$$

One next calculates the average kinetic and potential energies

$$\langle K \rangle = \frac{m\omega_0^2 x_0^2}{4} + \frac{mv_0^2}{4},$$

$$\langle V \rangle = \frac{m\omega_0^2 x_0^2}{4} + \frac{mv_0^2}{4}. \tag{4.16}$$

The averages are equal and amount to one-half of the total mechanical energy E

$$\langle K \rangle = \langle V \rangle = E/2. \tag{4.17}$$

The average kinetic and potential energies can be related to standard deviations of the position and momentum of the oscillator

$$\langle \delta x^2 \rangle = E/(m\omega_0^2), \quad \langle \delta p^2 \rangle = mE, \tag{4.18}$$

where in terms of standard deviations $\sigma_{x,p}$ one has $\langle \delta x^2 \rangle = \sigma_x^2$ and $\langle \delta p^2 \rangle = \sigma_p^2$. Combining these standard deviations into a product, one obtains

$$\sigma_x \sigma_p = E/\omega_0. \tag{4.19}$$

At this point one can apply the Bohr principle and assume that any energy of a quantum system should be associated with a photon carrying the energy $E = \hbar\omega_0 \times \text{Const}$. One arrives at the Heisenberg uncertainty relation

$$\boxed{\sigma_x \sigma_p = \text{Const} \times \hbar.} \tag{4.20}$$

This is the starting point of the Heisenberg representation of quantum mechanics. It is driven by the realization that quantization of energy levels consistent with absorption and emission of photons must lead to fundamental uncertainties in the ability to measure the position and momentum of a quantum particles. Deterministic equations of motions for dynamic variables $x(t)$ and $p(t)$ must then be replaced with equations allowing such uncertainties. This is the language of operators, instead of functions. When specific basis sets of functions are used to represent operators, one arrives at the matrix representation of operators and Heisenberg's matrix quantum mechanics.

The formulation for the dynamic variables that we seek should allow standard deviations for the dynamic variables x and p, but provide sharp spectral lines, for which well-defined energy states are required. A possible mathematical way to realize this requirement is to represent x and p by matrices

$$x_{nm} = \langle n|x|m \rangle, \tag{4.21}$$

where we use Dirac bra-ket notation (see below) for building matrix elements between quantum sates. We, therefore, can produce the potential energy matrix

$$V_{nm} = \frac{m\omega_0^2}{2} \sum_k x_{nk}x_{km} = E_{nm}/2. \tag{4.22}$$

However, in order to avoid standard deviation (uncertainty) in energy, the resulting matrix of energy must be diagonal,

$$E_{nm} = \delta_{nm}E_n, \tag{4.23}$$

where δ_{nm} is the Kronecker delta (Eq. (1.9)). Mathematically it means that the states consistent with spectroscopic experiments should be eigenstates of the total energy, which then becomes the Hamiltonian operator

$$\boxed{\hat{H}|n\rangle = E_n|n\rangle, \quad \hat{H} = \hat{K} + \hat{V}.} \tag{4.24}$$

The question addressed by the Heisenberg representation is how to construct such operators in order to produce energy states E_n consistent with well defined (without uncertainty) energy states.

Before we proceed to the next step, it is useful to ask what do we need to know about the quantum states $|n\rangle$. Textbook problems of quantum mechanics are devoted to solving the eigenvalue problem for very few representative cases when the full set of quantum states $|n\rangle$ can indeed be calculated. For most practical calculations of quantum systems this is a unrealistic program and, instead, the eigenstates $|n\rangle$ are sought as linear combinations of the "basis set" functions $|\psi_m\rangle$

$$|n\rangle = \sum_{m=1}^{M} c_{nm}|\psi_m\rangle. \tag{4.25}$$

The ability to expand an arbitrary quantum state in a complete set of basis functions $|\psi_m\rangle$ is known as the expansion postulate. For a full basis set, the sum extends to infinity, but is truncated in practical calculations at some large value M allowing sufficiently accurate solution for the eigenenergies E_n. Those are calculated by linear algebra by solving the equation for the expansion coefficients

$$\sum_{k=1}^{M} H_{nk}c_{km} = E_n c_{nm}, \tag{4.26}$$

where $H_{nk} = \langle n|\hat{H}|k\rangle$ are the matrix elements of the Hamiltonian operator. In this matrix formulation of the eigenvalue problem, the main requirement for the states $|\psi_m\rangle$ becomes the ability to quickly evaluate massive numbers of matrix elements H_{nk}.

4.3 Heisenberg formulation

The goal of the Heisenberg picture is to replace the classical equations of motion for the dynamic variables with the corresponding quantum equations. A classical dynamic variable A is replaced in this formulation by an operator $\hat{A}$ which acts on a quantum state $|\psi\rangle$ (represented by a function, vector, etc). The operator represents a measurement. One recognizes that the measurement can change the quantum state, which is why an operator altering the state is required: $\hat{A}|\psi\rangle = |\phi\rangle$. Special attention is paid to quantum states $|\psi_n\rangle$ that are not altered by the measurement

$$\hat{A}|\psi_n\rangle = a_n|\psi_n\rangle. \tag{4.27}$$

Those are eigenstates, the measurements performed on eigenstates produce eigenvalues. No uncertainties of measurements are possible if the quantum state is prepared in one of the eigenstates before the measurement is performed.

The equations of motion for classical dynamic variables are established by Poisson brackets (Sec. 3.3). One puts $\partial A/\partial t = 0$ in Eq. (3.23) and notes that $\{H, A\} = -\{A, H\}$ to obtain

$$\dot{A} = \{A, H\}. \tag{4.28}$$

The Heisenberg dynamics suggest replacing the Poisson brackets with the commutator of the operators or the commutator of matrices representing the operators

$$\{A, H\} \rightarrow \frac{1}{i\hbar}[\hat{A}, \hat{H}], \tag{4.29}$$

where $[\hat{A}, \hat{H}] = \hat{A}\hat{H} - \hat{H}\hat{A}$. Specifically, the equation of motion now reads

$$\frac{d\hat{A}}{dt} = \left(\frac{\partial A}{\partial t}\right)_H + \frac{1}{i\hbar}[\hat{A}, \hat{H}]. \tag{4.30}$$

Here, $(\partial A/\partial t)_H$ means the derivative taken at constant H. If the alteration of A is driven only by the changes of dynamic variables in the Hamiltonian, this term is dropped. The formal solution of the Heisenberg equation in this case is

$$\hat{A}(t) = e^{i\hat{H}t/\hbar}\hat{A}e^{-i\hat{H}t/\hbar}. \tag{4.31}$$

One can check if the results obtained for the harmonic oscillator are consistent with these new equations. We can write the equation of motion for the matrix element by taking the Dirac bracket of the Heisenberg equation $\langle n|\dot{x}|m \rangle$. When using the eigenstates of the Hamiltonian operator $|n\rangle$ the result is

$$\dot{x}_{nm} = i\omega_{nm}x_{nm}, \tag{4.32}$$

where $\omega_{nm} = (E_n - E_m)/\hbar$. The differential equation is solved to yield

$$x_{nm}(t) = x_{nm}e^{i\omega_{nm}t}. \tag{4.33}$$

The average potential energy

$$\langle \hat{V} \rangle = \sum_n V_{nn} = \frac{m\omega_0^2}{2} \sum_{n,m} x_{nm}(t)x_{mn}^*(t) = \frac{m\omega_0^2}{2}\langle x^2 \rangle, \tag{4.34}$$

where $\langle x^2 \rangle = \sum_n \langle n|\hat{x}^2|n \rangle = \sum_{nm} x_{nm}x_{mn}^*$. We have arrived at a constant average potential energy for a quantum harmonic oscillator, consistent with the result obtained from the Bohr principle.

One can also expect that the transition to the classical limit should lead to the anticipated connection between $\dot{x}(t)$ and the classical momentum

$$m\dot{x}(t) = p(t). \tag{4.35}$$

If we require that this relation holds for operators $\hat{x}(t)$ and $\hat{p}(t)$, then we have to require a certain commutation relation between the position and momentum operators

$$[\hat{x}(t), \hat{p}(t)] = i\hbar. \tag{4.36}$$

It is easy to see that the same relation holds for the initial values at $t = 0$:

$$\hat{x}(t)\hat{p}(t) = e^{i\hat{H}t/\hbar}\hat{x}e^{-i\hat{H}t/\hbar}e^{i\hat{H}t/\hbar}\hat{p}e^{-i\hat{H}t/\hbar} = e^{i\hat{H}t/\hbar}\hat{x}\hat{p}e^{-i\hat{H}t/\hbar} \tag{4.37}$$

and

$$\hat{p}(t)\hat{x}(t) = e^{i\hat{H}t/\hbar}px e^{-i\hat{H}t/\hbar}. \tag{4.38}$$

Therefore one obtains

$$\boxed{[\hat{x}(t), \hat{p}(t)] = i\hbar = [\hat{x}, \hat{p}].} \tag{4.39}$$

We can now use this rule in the Heisenberg equation of motion with the Hamiltonian

$$\hat{H} = K(\hat{p}) + V(\hat{x}), \quad K(\hat{p}) = \hat{p}^2/(2m). \tag{4.40}$$

We need in addition the rule that any operator commutes with a function of itself. By this rule,

$$[\hat{x}, V(\hat{x})] = 0 \tag{4.41}$$

and $[\hat{x}, \hat{H}] = [\hat{x}, K(\hat{p})]$. We obtain next from the commutation relation

$$[\hat{x}, K(\hat{p})] = (i\hbar/m)\hat{p}. \tag{4.42}$$

From the Heisenberg equation of motion, we obtain

$$\dot{\hat{x}}(t) = \hat{p}(t)/m, \tag{4.43}$$

which is consistent with the standard equation (4.35) of classical mechanics. The standard relation between the time derivative of the coordinate and the linear momentum is preserved when the commutation relations given by Eq. (4.39) are imposed on these operators.

4.4   Quantum wave mechanics

The wave representation of quantum mechanics due to Schrödinger puts emphasis on the evolution of the "quantum state" in contrast to evolution of quantum variables in the Heisenberg picture. We can write the average of $\hat{x}(t)$ in two alternative ways, either as the average of a dynamic variable $\hat{x}(t)$ or as the average of the static variable $\hat{x}$ taken on the propagating quantum state $|\psi_n(t)\rangle$

$$\langle x \rangle = \sum_n \langle n|\hat{x}(t)|n\rangle = \sum_n \langle\psi_n(t)|\hat{x}|\psi_n(t)\rangle, \tag{4.44}$$

where

$$|\psi_n(t)\rangle = \hat{U}(t)|n\rangle \tag{4.45}$$

and

$$\hat{U}(t) = e^{-i\hat{H}t/\hbar} \tag{4.46}$$

is the evolution operator propagating the initial state $|n\rangle$ in time to $|\psi_n(t)\rangle$.

The dynamic quantum state satisfies the Schrödinger equation

$$\boxed{i\hbar\partial_t|\psi_n(t)\rangle = \hat{H}|\psi_n(t)\rangle.} \tag{4.47}$$

If one chooses the eigenstates $\hat{H}|n\rangle = E_n|n\rangle$ for the initial conditions, one obtains

$$|\psi_n(t)\rangle = e^{-i\hat{E}_n t/\hbar}|n\rangle \tag{4.48}$$

and

$$x_{nm}(t) = x_{nm}e^{i\omega_{nm}t}, \tag{4.49}$$

which is consistent with the result of the Heisenberg picture (Eq. (4.33)).

The advantage offered by the Schrödinger picture is that quantum dynamics is cast in terms of the eigenvalue problem, which is a well-established mathematical tool in the theory of wave propagation. The distinction between the Heisenberg and Schrödinger pictures is merely in casting the problem in terms of different mathematical formalisms, linear algebra in Heisenberg's case and differential equations for wave mechanics in the Schrödinger picture. The present-day interest in solving multi-particle problems of quantum mechanics and the speed of matrix computations have made the Heisenberg formalism a more attractive choice for the development of computational algorithms.

The equation for eigenvalue functions is

$$\hat{H}|n\rangle = E_n|n\rangle \tag{4.50}$$

and one needs to specify the meaning of the operator $\hat{H}$ when representing the states $|n\rangle$ in terms of some functions. Since the form of the kinetic energy is preserved for all mechanical problems and the form of the potential energy is problem-specific, representation in terms of functions of the coordinate x is the preferred way for constructing the wave mechanics. We discuss x- and p-representations of wave functions after a short overview of Dirac's bra-ket formalism.

4.5   Dirac bra-kets

Here we give a quick list of rules for the Dirac bra-ket formalism (also bracket notation) as already used in the arguments presented above. The bra-ket formalism is the realization that every quantum state can be viewed as a point in the infinite-dimension Hilbert space. The variable x used to specify a function is just a continuous index allowing a specific representation of a given state. "Representation" means that for each x there is a rule specifying a numerical value, which is what a function really is. Multiple representations can be used for a given quantum state by using different indexes; x and p representations are most common (Sec. 4.6).

An integral between two functions $\phi(x)$ and $\psi(x)$ can be viewed as summation over the common index. If the functions are given in terms of the variable x, one applies the x-representation of two states in the Hilbert space, bra $\langle\phi|$ and ket $|\psi\rangle$. The integral of two functions becomes a sum over the common index values

$$\int dx\phi^*(x)\psi(x) = \langle\phi|\psi\rangle = \sum_x \langle\phi|x\rangle\langle x|\psi\rangle. \tag{4.51}$$

This representation requires the completeness relation

$$\sum_x |x\rangle\langle x| = 1. \tag{4.52}$$

Two vectors, bra and ket, are required to perform calculus involving complex-valued functions since every complex-valued function requires two real-valued functions, the real and imaginary parts. Taking complex conjugate of a function involves switching between ket and bra

$$\psi(x) = \langle x|\psi\rangle, \quad \psi(x)^* = \langle\psi|x\rangle. \tag{4.53}$$

The ket and bra vectors are also acted upon by, correspondingly, the operator $\hat{A}$ and its Hermitian conjugate (adjoint) operator $\hat{A}^\dagger$

$$|\phi\rangle = \hat{A}|\psi\rangle, \quad \langle\phi| = \langle\psi|\hat{A}^\dagger. \tag{4.54}$$

All physical operators are self-adjoint,

$$\boxed{\hat{A} = \hat{A}^\dagger.} \tag{4.55}$$

One also obtains the rule

$$\langle\phi|\hat{A}|\psi\rangle^* = \langle\psi|\hat{A}^\dagger|\phi\rangle = \langle\psi|\hat{A}|\phi\rangle. \tag{4.56}$$

The completeness relation (4.52) leads to the expansion in which a quantum state $|\psi\rangle$ is expanded in projections $\psi(x)$ on each basis state $|x\rangle$ of the infinite-dimension Hilbert space

$$|\psi\rangle = \sum_x |x\rangle\langle x|\psi\rangle. \tag{4.57}$$

In the special case $|\psi\rangle = |x'\rangle$ one has

$$|\psi\rangle = |x'\rangle = \sum_x |x\rangle\langle x|x'\rangle = \int dx |x\rangle\langle x|x'\rangle. \tag{4.58}$$

This condition is satisfied if the bra-ket of two x-states is given by the Dirac delta-function

$$\langle x|x'\rangle = \delta(x - x'). \tag{4.59}$$

4.6 x- and p-representation

The completeness relation is based on the quantum-mechanical postulate that any physical observable can be replaced by a quantum-mechanical operator producing a complete set of eigenstates. Since position x is an observable property, there must exist a hermitian position operator producing the complete set of $|x\rangle$-kets

$$\hat{x}|x\rangle = x|x\rangle. \tag{4.60}$$

The same statement applies to the momentum operator which should produce the complete set of p-kets

$$\hat{p}|p\rangle = p|p\rangle. \tag{4.61}$$

They also satisfy the completeness relation

$$\sum_p |p\rangle\langle p| = 1. \tag{4.62}$$

The standard Schrödinger representation of the $\hat{x}$ and $\hat{p}$ operators anticipates that the action of $\hat{x}$ on $\psi(x)$ is replaced with $x\psi(x)$ and the action $\hat{p}$ on $\psi(x)$ becomes the derivative, $\hat{p}\psi(x) = -i\hbar\partial_x\psi(x)$. This prescription is consistent with the following rule

$$\begin{aligned}\langle x|\hat{p}|\psi\rangle &= -i\hbar\partial_x\langle x|\psi\rangle, \\ \langle x|\hat{x}|\psi\rangle &= x\langle x|\psi\rangle.\end{aligned} \tag{4.63}$$

Replacing $|\psi\rangle$ with $|p\rangle$, we arrive at a differential equation

$$\langle x|\hat{p}|p\rangle = -i\hbar\partial_x\langle x|p\rangle = p\langle x|p\rangle, \tag{4.64}$$

where Eq. (4.61) was used in the second step. The solution of the differential equation is

$$\langle x|p\rangle = Ce^{ipx/\hbar}. \tag{4.65}$$

To determine the normalization constant C, one applies the completeness relation (4.62) for the momentum kets

$$\psi(x) = \langle x|\psi\rangle = \sum_p \langle x|p\rangle\langle p|\psi\rangle = C\int dpe^{ipx/\hbar}\psi(p). \tag{4.66}$$

One the other hand, one can write for $\psi(p)$

$$\psi(p) = \langle p|\psi\rangle = \sum_x \langle p|x\rangle\langle x|\psi\rangle = C\int dxe^{-ipx/\hbar}\psi(x). \tag{4.67}$$

By substituting Eq. (4.67) to (4.66), one gets $(2\pi)\hbar C^2 = 1$ and

$$\langle x|p\rangle = \frac{1}{\sqrt{2\pi\hbar}}e^{ipx/\hbar}. \tag{4.68}$$

4.7   Interaction representation

Heisenberg:

$$\hat{A}(t) = e^{i\hat{H}t/\hbar}\hat{A}e^{-i\hat{H}t/\hbar}$$

Schrödinger:

$$|\psi(t)\rangle = \hat{U}(t)|\psi(0)\rangle, \quad \hat{U}(t) = e^{-i\hat{H}t/\hbar}$$

In order to simplify the algebraic manipulations, we will replace $\hat{H}/\hbar$ with H in this section. The correct units can be restored in the final expressions. We will also suppress the carets over the operators $\hat{H}$ and $\hat{U}$ and will keep them only for the operators in the interaction representation. In this notation, the propagation of the wave function in the Schrödinger picture is given by the evolution operator

$$|\psi(t)\rangle = U(t)|\psi(0)\rangle, \quad U(t) = e^{-iHt}. \tag{4.69}$$

The time evolution of a quantum system is given through the wave function and the operators, such as the Hamiltonian operator H, are independent of time in this representation.

Formally equivalent representation of quantum mechanics is the Heisenberg picture, in which the wave functions form the time-independent basis states to calculate the observables and time evolution is assigned to operators

$$A(t) = e^{iHt}Ae^{-iHt}. \tag{4.70}$$

A representation intermediate between Schrödinger and Heisenberg pictures is to assign time dependence to both operators and wave functions based on the separation of the Hamiltonian H into the part H_0 which forms the basis of wave functions and the remaining part V. The assumption is that H_0 is a part of the Hamiltonian for which the eigenvalue problem can be solved and the states $|\psi_n\rangle$ produced

$$H_0|\psi_n\rangle = \omega_n|\psi_n\rangle, \tag{4.71}$$

where, according to our convention adopted here, $\omega_n = E_n/\hbar$ and E_n is the eigenenergy.

This is the interaction representation in which the dynamics of operators is given by the component H_0 (we use caret to distinguish this type of operators)

$$\hat{A}(t) = e^{iH_0t}Ae^{-iH_0t}. \tag{4.72}$$

The dynamics of the wave function is given by the operator

$$\hat{U}(t) = e^{iH_0 t} e^{-iHt} \tag{4.73}$$

such that

$$|\hat{\psi}(t)\rangle = \hat{U}(t)|\psi(0)\rangle. \tag{4.74}$$

One can obtain for the derivative

$$\partial_t \hat{U}(t) = i e^{iH_0 t}(H_0 - H)e^{-iHt}$$
$$= -i e^{iH_0 t} V e^{-iH_0 t + iH_0 t} e^{-iHt} = -i\hat{V}(t)\hat{U}(t). \tag{4.75}$$

The Schrödinger equation is replaced in the interaction representation by the equation based on the interaction operator $\hat{V}(t)$

$$i\partial_t |\hat{\psi}(t)\rangle = \hat{V}(t)|\hat{\psi}(t)\rangle. \tag{4.76}$$

The first-order differential equation for $\hat{U}(t)$ can be formally integrated. Given $\hat{U}(0) = 1$, one obtains

$$\hat{U}(t) = 1 - i \int_0^t dt_1 \hat{V}(t_1)\hat{U}(t_1). \tag{4.77}$$

This can be iterated by substituting $\hat{U}(t_1)$ from the left to the integral in the right

$$\hat{U}(t) = 1 - i \int_0^t dt_1 \hat{V}(t_1) + (-i)^2 \int_0^t dt_1 \int_0^{t_1} dt_2 \hat{V}(t_1)\hat{V}(t_2) + \ldots. \tag{4.78}$$

Here, one wants to replace the series expansion with a more compact relation converting the series into an exponent. This is achieved by introducing the time-ordering operator $\hat{T}$. This operator orders the product of operators with the earliest time to the right

$$\hat{T}[\hat{V}(t_1)\hat{V}(t_2)] = \hat{V}(t_1)\hat{V}(t_2), \quad t_1 > t_2. \tag{4.79}$$

The use of the time-ordering operator allows one to convert each time-nested integral in the series in Eq. (4.78) into the product of integrals all extending from $t_i = 0$ to $t_i = t$. For each term in the series with n operators and n time integrals, there are $n!$ ways one can obtain the time-nested integral from the one without time nesting. Therefore, each such term is divided by $n!$ and one obtains

$$\hat{U}(t) = \sum_{n=0}^{\infty} \frac{(-i)^n}{n!} \int_0^t dt_1 \cdots \int_0^t dt_n \hat{T}\left[\hat{V}(t_1)\ldots\hat{V}(t_n)\right]. \tag{4.80}$$

This is obviously the series expansion of the exponential function leading to a compact form for the evolution operator

$$\boxed{\hat{U}(t) = \hat{T}\exp\left[-i\int_0^t d\tau \hat{V}(\tau)\right].} \tag{4.81}$$

4.8 Fermi's golden rule

We want to calculate the probability of a quantum system to arrive to the eigenstate $|\psi_2\rangle$ of the operator H_0 at time t given that the system was in the state $|\psi_1\rangle$ at $t = 0$. The typical application of this problem is to calculate the rate of transitions between eigenstates of H_0 under the action of a weak perturbation V.

The probability P_{12} to find the system at $|\psi_2\rangle$ at time t is found in terms of the projection of the evolving wave function $|\hat{\psi}(t)\rangle$ on the state $|\psi_2\rangle$

$$P_{12}(t) = \left| \langle \psi_2 | \hat{\psi}(t) \rangle \right|^2. \tag{4.82}$$

In terms of the evolution operator $\hat{U}(t)$, one finds

$$\langle \psi_2 | \hat{\psi}(t) \rangle = \langle \psi_2 | \hat{T} \exp \left[-i \int_0^t d\tau \hat{V}(\tau) \right] | \psi_1 \rangle. \tag{4.83}$$

Since $|\psi_n\rangle$ are eigenstates of H_0, they are orthogonal and $\langle \psi_2 | \psi_1 \rangle = 0$. Therefore, assuming that the perturbation $\hat{V}(t)$ is small, one can approximate the above bra-ket with the first terms in the series expansion

$$\langle \psi_2 | \hat{\psi}(t) \rangle \simeq -i \int_0^t d\tau \langle \psi_2 | \hat{V}(\tau) | \psi_1 \rangle. \tag{4.84}$$

This expansion can be used in the probability P_{12} or in the rate of transitions $w_{12} = dP_{12}/dt$. For the latter one gets

$$\begin{aligned} w_{12}(t) = &\int_0^t d\tau \langle \psi_2 | \hat{V}(t) | \psi_1 \rangle \langle \psi_1 | \hat{V}(\tau) | \psi_2 \rangle \\ + &\int_0^t d\tau \langle \psi_2 | \hat{V}(\tau) | \psi_1 \rangle \langle \psi_1 | \hat{V}(t) | \psi_2 \rangle. \end{aligned} \tag{4.85}$$

Since $|\psi_n\rangle$ are eigenstates of H_0, the off-diagonal matrix element of $\hat{V}(t)$ becomes

$$\langle \psi_2 | \hat{V}(t) | \psi_1 \rangle = e^{i\Delta\omega t} V_{21}, \tag{4.86}$$

where $V_{21} = \langle \psi_2 | V | \psi_1 \rangle$ and $\Delta\omega = \omega_2 - \omega_1$ (Eq. (4.72)). Equation (4.85) becomes

$$w_{12}(t) = |V_{12}|^2 \int_{-t}^t d\tau e^{i\Delta\omega\tau}. \tag{4.87}$$

By assuming that $t\Delta\omega \gg 1$, one can put $t \to \infty$, with the integral over t converting to delta-function. The result is

$$w_{12} = 2\pi |V_{12}|^2 \delta(\Delta\omega). \tag{4.88}$$

Returning back to the energy units $V \to V/\hbar$, one obtains the standard form of the Fermi's golden rule

$$w_{12} = \frac{2\pi}{\hbar}|V_{12}|^2\delta(\Delta E). \tag{4.89}$$

This derivation was performed for the perturbation independent of time and can be extended to time-dependent perturbations. The result given by the golden rule derived here indicates that there are no transitions violating the energy conservation $\Delta E = 0$. In practice, the energy levels E_i are affected by interactions with the fluctuating medium, which involve both static and dynamic effects (Kubo's line shape discussed in Sec. 11.8). If only static effects are considered, the transition rate is averaged over the statistical distribution of the energy levels E_i and one arrives at the equation often used to describe radiationless (not involving absorption or emission of radiation photons) transitions in condensed materials

$$w_{12} = \frac{2\pi}{\hbar}|V_{12}|^2\langle\delta(\Delta E)\rangle. \tag{4.90}$$

Here, angular brackets denote an ensemble average.

When the interaction energy is an oscillatory function, $V(t) = V e^{-i\omega t}$, the same sequence of steps leads to the result

$$w_{12} = \hbar^{-2}\int_{-\infty}^{\infty} d\tau \langle V_{12}V_{21}\rangle e^{i(\Delta\omega-\omega)\tau}. \tag{4.91}$$

This equation is a starting point for calculations of spectral lines for transitions produced by the interaction of a molecule with the electromagnetic field oscillating with the circular frequency ω (Chap. 11). This equation applies to $\Delta\omega > 0$ and thus to absorption of radiation energy. One has to replace $\omega \to -\omega$ when $\Delta\omega < 0$ for the emission transition (rotating wave approximation).

4.9 Raising and lowering operators

The Hamiltonian of a harmonic oscillator is obtained from the classical Hamiltonian in Eq. (3.17) by replacing the dynamic variable of linear momentum with the differential operator

$$p \to -i\hbar\partial_x. \tag{4.92}$$

The Hamiltonian is thus given by

$$\hat{H} = -\frac{\hbar^2}{2m}\partial_x^2 + \frac{1}{2}k\hat{x}^2. \tag{4.93}$$

One can introduce the dimensionless variable

$$\hat{\xi} = \hat{x}\sqrt{\frac{m\omega}{\hbar}}. \tag{4.94}$$

The second derivative becomes

$$\partial_x^2 = \frac{m\omega}{\hbar}\partial_\xi^2 \tag{4.95}$$

and one gets for the Hamiltonian

$$\hat{H} = \frac{\hbar\omega}{2}\left(-\partial_\xi^2 + \hat{\xi}^2\right). \tag{4.96}$$

In addition, the commutation relation between $\hat{x}$ and $\hat{p}$ (Eq. (4.39)) becomes

$$[\hat{\xi}, \hat{p}_\xi] = i, \quad \hat{p}_\xi = -i\partial_\xi. \tag{4.97}$$

One can write the expression in the brackets in Eq. (4.96) as

$$-\partial_\xi^2 + \hat{\xi}^2 = \hat{p}_\xi^2 + \hat{\xi}^2 = (\hat{\xi} + i\hat{p}_\xi)(\hat{\xi} - i\hat{p}_\xi) + i[\hat{\xi}, \hat{p}_\xi]. \tag{4.98}$$

By using the commutation relation from Eq. (4.97), one obtains for the Hamiltonian

$$\hat{H} = \frac{\hbar\omega}{2}(\hat{\xi} + i\hat{p}_\xi)(\hat{\xi} - i\hat{p}_\xi) - \frac{\hbar\omega}{2}. \tag{4.99}$$

The linear combinations of $\hat{\xi}$ and $\hat{p}_\xi$ in the brackets are defined as raising and lowering operators

$$\boxed{\hat{a} = \frac{1}{\sqrt{2}}\left(\hat{\xi} + i\hat{p}_\xi\right), \quad \hat{a}^\dagger = \frac{1}{\sqrt{2}}\left(\hat{\xi} - i\hat{p}_\xi\right).} \tag{4.100}$$

The Hamiltonian becomes

$$\hat{H} = \hbar\omega\left[\hat{a}\hat{a}^\dagger - \tfrac{1}{2}\right]. \tag{4.101}$$

One can also notice that

$$[\hat{a}, \hat{a}^\dagger] = -i[\hat{\xi}, \hat{p}_\xi] = 1. \tag{4.102}$$

Replacing $\hat{a}\hat{a}^\dagger$ with $1 + \hat{a}^\dagger\hat{a}$ in Eq. (4.101), one arrives at the commonly used form for the Hamiltonian of the harmonic oscillator expressed in raising and lowering operators

$$\boxed{\hat{H} = \hbar\omega\left[\hat{a}^\dagger\hat{a} + \tfrac{1}{2}\right].} \tag{4.103}$$

The eigenstates $|n\rangle$ of this Hamiltonian provide the eigenenergies

$$E_n = \hbar\omega(n + \tfrac{1}{2}). \tag{4.104}$$

To arrive to this result, one has to require the following rules for the raising and lowering operators acting on $|n\rangle$

$$\boxed{\hat{a}^\dagger|n\rangle = \sqrt{n+1}|n+1\rangle, \quad \hat{a}|n\rangle = \sqrt{n}|n-1\rangle.} \tag{4.105}$$

It is easy to prove that $\hat{a}^\dagger\hat{a}|n\rangle = \sqrt{n}\hat{a}^\dagger|n-1\rangle = n|n\rangle$ thus leading to Eq. (4.104).

4.10 Angular momentum

Angular momentum appears for any kind of mechanical motion involving the projection of the particle velocity on the direction perpendicular to the radial motion. One can generally write the velocity in three-dimensional spherical coordinates as as the sum of the radial component $\mathbf{v}_r = \dot{r}\hat{\mathbf{r}}$ and the perpendicular projection $\mathbf{v}_\perp$

$$\mathbf{v} = \dot{r}\hat{\mathbf{r}} + \mathbf{v}_\perp. \tag{4.106}$$

Here, $\hat{\mathbf{r}} = \mathbf{r}/r$ is the unit vector along $\mathbf{r}$. The angular momentum is defined as the vector product of $\mathbf{r}$ and the linear particle momentum $\mathbf{p} = m\mathbf{v}$

$$\mathbf{L} = \mathbf{r} \times \mathbf{p} = m\mathbf{r} \times \mathbf{v}. \tag{4.107}$$

The radial component $\mathbf{v}_r$ produces identically zero in the vector product and only the perpendicular component of the velocity determines the magnitude of the angular momentum, $L = mrv_\perp$ (Fig. 4.2).

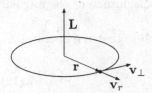

Fig. 4.2 Angular momentum.

In quantum mechanics, the differential operator replaces the classical linear momentum and the angular momentum operator becomes

$$\hat{\mathbf{L}} = \hat{\mathbf{r}} \times \hat{\mathbf{p}} = -i\hbar\mathbf{r} \times \nabla. \tag{4.108}$$

Because of the known commutation relations between $\hat{r}_\alpha$, $\alpha = x, y, z$ and $\hat{p}_\alpha$ (Eq. (4.39)), commutations between different projections $\hat{L}_\alpha$ can be established. They do not vanished and can be summarized by the formula [29]

$$\boxed{\hat{\mathbf{L}} \times \hat{\mathbf{L}} = i\hat{\mathbf{L}}.} \tag{4.109}$$

Note that the cross product of a classical vector with itself is always zero. From this equation,

$$i\hat{L}_z = \hat{L}_x\hat{L}_y - \hat{L}_y\hat{L}_x = [\hat{L}_x, \hat{L}_y]. \tag{4.110}$$

The rest of the commutation relations follow by considering other Cartesian projections.

The lack of commutations between the angular momentum projections means that quantum states cannot include among their quantum numbers those specifying the values of any two components of the angular momentum. For a closed dynamic system, a meaningful set of quantum numbers can be obtained for integrals of motion: the Hamiltonian and any operators commuting with the Hamiltonian and among each other. The squared angular momentum $\hat{\mathbf{L}}^2$ is conserved for systems with rotational isotropy and commutes with the Hamiltonian. One should be able to set a quantum number characterizing the total angular momentum of the system.

As an operator in spherical coordinates, $\hat{\mathbf{L}}^2$ is given in terms of the angular part of the Laplacian operator $\nabla^2_{\theta\phi}$ (Eq. (1.36))

$$\hat{\mathbf{L}}^2 = -\hbar^2\nabla^2_{\theta\phi}. \tag{4.111}$$

It is clear that the operator $\hat{\mathbf{L}}^2$ depends only on the angular coordinates $\{\theta, \phi\}$ and, therefore, its eigenfunctions should be represented in the space of these two variables. Since the angles do not carry any significant units, this relation also sets the physical scale for the magnitude of the angular momentum, which should be multiples of $\hbar$.

In a similar fashion, one can find from Eq. (4.108) that the operator $\hat{L}_z$ is given as

$$\hat{L}_z = -i\hbar\partial_\phi. \tag{4.112}$$

Since the part of $\nabla^2_{\theta\phi}$ involving operation on ϕ is $\sin^{-2}\theta\partial^2_\phi$ (Eq. (1.36)), this operator commutes with the first derivative in ϕ specifying $\hat{L}_z$. One gets the commutation relation

$$[\hat{L}_z, \hat{\mathbf{L}}^2] = 0. \tag{4.113}$$

This equation implies that $\hat{\mathbf{L}}^2$ and $\hat{L}_z$ should share a set of eigenfunctions.

The set of eigenfunctions of shared by $\hat{\mathbf{L}}^2$ and $\hat{L}_z$ are important for rotational motion in general, extending far beyond quantum-mechanical applications (applications to rotational diffusion are discussed in Secs. 7.7 and 11.11). Those are spherical harmonics specified by two quantum number, ℓ and m. The quantum number ℓ yields the eigenvalues of $\hat{\mathbf{L}}^2$ and $m = -\ell, -\ell + 1, \ldots, \ell$ specifies $(2\ell + 1)$ quantum numbers of $\hat{L}_z$ for each given ℓ

$$\boxed{\hat{\mathbf{L}}^2|\ell m\rangle = \hbar^2\ell(\ell + 1)|\ell m\rangle, \quad \hat{L}_z|\ell m\rangle = \hbar m|\ell m\rangle.} \tag{4.114}$$

Spherical harmonics $Y_{\ell m}(\theta, \phi)$ are representations of Hilbert-space quantum eigenstates $|\ell m\rangle$ in coordinates of spherical angles.

In 3D space, the one-dimensional x-representation of quantum states (Sec. 4.6) is replaced with the coordinate representation $|\mathbf{r}\rangle = |x, y, z\rangle$, which becomes the representation in terms of the radial coordinate r and two polar angles θ, ϕ in the spherical coordinates. Since the eigenstates of $\hat{\mathbf{L}}^2$ and $\hat{L}_z$ should depend only on θ, ϕ, we can define the spherical harmonic function

$$Y_{\ell m}(\theta, \phi) = \langle \theta\phi | \ell m \rangle. \tag{4.115}$$

The angles θ, ϕ are coordinates on the surface of a unit sphere and any physical problem fully defined by motions on the unit sphere can be represented in terms of bra and ket states, $\langle \theta\phi|$ and $|\theta\phi\rangle$. For all such problems, the completeness relation holds

$$\sum_{\theta,\phi} |\theta\phi\rangle\langle\theta\phi| = \int d\Omega |\theta\phi\rangle\langle\theta\phi| = \int_0^{2\pi} d\phi \int_0^\pi d\theta \sin\theta |\theta\phi\rangle\langle\theta\phi| = 1. \tag{4.116}$$

By multiplying with $\langle \theta'\phi'|$ on the left, one can prove

$$\sum_{\theta,\phi} \langle\theta'\phi'|\theta\phi\rangle\langle\theta\phi| = \int_0^{2\pi} d\phi \int_0^\pi d\theta \sin\theta \langle\theta'\phi'|\theta\phi\rangle\langle\theta\phi|, \tag{4.117}$$

which requires

$$\langle\theta'\phi'|\theta\phi\rangle = \frac{1}{\sin\theta}\delta(\theta - \theta')\delta(\phi - \phi'). \tag{4.118}$$

Any quantum state fully determined by coordinates on a unit sphere should be expandable in $|\ell m\rangle$

$$|\psi\rangle = \sum_{\ell=0}^\infty \sum_{m=-\ell}^\ell |\ell m\rangle\langle\ell m|\psi\rangle. \tag{4.119}$$

The completeness relation then becomes

$$\sum_{\ell=0}^\infty \sum_{m=-\ell}^\ell |\ell m\rangle\langle\ell m| = 1. \tag{4.120}$$

If we multiply this equation by $\langle\theta'\phi'|$ on the left and by $|\theta\phi\rangle$ on the right, we obtain, by using Eq. (4.118), a sum rule for spherical harmonics

$$\sum_{\ell=0}^\infty \sum_{m=-\ell}^\ell Y_{\ell m}(\theta', \phi')Y_{\ell m}^*(\theta, \phi) = \frac{1}{\sin\theta}\delta(\theta - \theta')\delta(\phi - \phi'). \tag{4.121}$$

4.11   Spin

Spin of the electron is considered here. It is the intrinsic angular momentum of the electron. The eigenvalue characterizing the total spin $\hat{S}^2$ is $s = 1/2$, which allows only two quantum numbers for the z-projection: $m_s = -1/2$ and $m_s = +1/2$. Since only one value of $\ell = s$ is allowed, the quantum number $\ell = 1/2$ can be dropped from $|\ell m\rangle$ for brevity. The eigenvalue problem becomes

$$
\begin{aligned}
\hat{S}^2|m_s\rangle &= \tfrac{3}{4}\hbar^2|m_s\rangle, \\
\hat{S}_z|m_s\rangle &= m_s\hbar|m_s\rangle.
\end{aligned}
\tag{4.122}
$$

There are no internal coordinates associated with the spin operator and no coordinate representation, like with the spherical harmonics for the orbital motion, is possible. Since there are only two states allowed, there is no need in a continuous index to distinguish measurement outcomes utilized in ordinary functions. An alternative representation in terms of matrices is more convenient. One starts by introducing the dimensionless matrix $\hat{\boldsymbol{\sigma}}$

$$
\hat{\mathbf{S}} = \frac{\hbar}{2}\hat{\boldsymbol{\sigma}}.
\tag{4.123}
$$

Like any other operator of quantum mechanics, $\hat{\boldsymbol{\sigma}}$ can change the state of a quantum system. Given that $\hat{\mathbf{S}}$ is a vector, $\hat{\boldsymbol{\sigma}}$ must have three components, $\hat{\sigma}_\alpha$, $\alpha = x, y, z$. These are Pauli matrices.

There are only two states shared by $\hat{S}^2$ and $\hat{S}_z$, which differ by m_s. This simplification, and the absence of continuous intrinsic variables, allows a vector representation in terms of eigenspinors $|+\rangle$ ($m_s = +1/2$) and $|-\rangle$ ($m_s = -1/2$)

$$
|+\rangle = \begin{pmatrix} 1 \\ 0 \end{pmatrix}, \quad |-\rangle = \begin{pmatrix} 0 \\ 1 \end{pmatrix}.
\tag{4.124}
$$

The orthonormality relations hold

$$
\langle +|+\rangle = 1, \quad \langle -|-\rangle = 1, \quad \langle +|-\rangle = 0,
\tag{4.125}
$$

where the bra vectors are found by transposing the ket spinors

$$
\langle +| = \begin{pmatrix} 1 & 0 \end{pmatrix}, \quad \langle -| = \begin{pmatrix} 0 & 1 \end{pmatrix}.
\tag{4.126}
$$

One next needs to find the matrix $\hat{\sigma}_z$ satisfying the following eigenvalue problem

$$
\hat{\sigma}_z|\pm\rangle = \pm|\pm\rangle.
\tag{4.127}
$$

The matrix is found by forming matrix elements, e.g., $\langle -|\sigma_z|-\rangle = -1$. The result is

$$\hat{\sigma}_z = \begin{pmatrix} 1 & 0 \\ 0 & -1 \end{pmatrix}. \tag{4.128}$$

From Eqs. (4.122) and (4.123) one also finds that

$$(\hat{\sigma}_x^2 + \hat{\sigma}_y^2 + \hat{\sigma}_z^2)|\pm\rangle = 3|\pm\rangle. \tag{4.129}$$

Since $\hat{\sigma}_z^2$ is the identity matrix $\mathbf{I}$ (Eq. (1.9)) and there is no reason to distinguish between $\hat{\sigma}_x^2$ and $\hat{\sigma}_y^2$, one obtains

$$\hat{\sigma}_x^2 = \hat{\sigma}_y^2 = \hat{\sigma}_z^2 = \mathbf{I}. \tag{4.130}$$

The remaining Pauli matrices can be found by solving linear equations between matrix elements arising from the commutation relations

$$\hat{\sigma}_x = \begin{pmatrix} 0 & 1 \\ 1 & 0 \end{pmatrix}, \quad \hat{\sigma}_y = \begin{pmatrix} 0 & -i \\ i & 0 \end{pmatrix}. \tag{4.131}$$

4.12 *Spin-orbit coupling

Spin-orbit coupling appears as the interaction between the magnetic moment of the orbital motion of the electron and the magnetic dipole of the electron. The intrinsic magnetic dipole of the electron is related to its intrinsic angular momentum by the relation

$$\boldsymbol{\mu}_s = -g\frac{e}{2cm_e}\mathbf{S} \approx -\frac{e}{cm_e}\mathbf{S}, \tag{4.132}$$

where g is the electron g-factor, which is very close to exactly the factor of two, $g \simeq 2.0023$. The orbital motion of the electron with the angular momentum $\mathbf{L}$ produces the orbital magnetic dipole

$$\boldsymbol{\mu}_L = -\frac{e}{2cm_e}\mathbf{L}. \tag{4.133}$$

The orbital and electron intrinsic magnetic dipoles interact, and one anticipates that the interaction should scale as r^{-3} with the distance r between the magnetic dipoles. In order to obtain the precise form of the interaction potential, one can consider the magnetic field (or magnetic induction field) $\mathbf{B}$ produced by the orbital motion of the charge. It interacts with the magnetic dipole of the electron $\boldsymbol{\mu}_s$ with the energy

$$V_{SO} = -\tfrac{1}{2}\boldsymbol{\mu}_s \cdot \mathbf{B}, \tag{4.134}$$

where the one-half factor is a relativistic effect (Thomas precession).

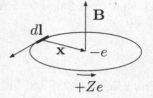

Fig. 4.3 Magnetic field **B** created by the positive nucleus with the charge $+Ze$ circling around the negative electron $-e$ in the system of coordinates associated with the electron.

The magnetic field **B** can be calculated from classical electromagnetism by assigning the system of coordinate to the electron. In this system of coordinates, the nucleus carrying the positive charge Z is circling around the electron with the angular frequency ω (Fig. 4.3). The current of the positive charge $+Ze$ circling the orbit with the period $T = 2\pi/\omega$ is equal to

$$I = Ze/T = Ze\omega/(2\pi). \tag{4.135}$$

We can next use the Biot and Savart law [6] which states that the element of the length $d\mathbf{l}$ of the wire carrying the current I produces the increment of the magnetic field equal to

$$d\mathbf{B} = \frac{I}{c}\frac{(d\mathbf{l} \times \mathbf{x})}{|\mathbf{x}|^3}, \tag{4.136}$$

where **x** is the vector pointing from the element of the wire $d\mathbf{l}$ to the observation point where the field is measured. This field is perpendicular to the plane of the circular orbit as shown in Fig. 4.3. Integration over the length of the spherical orbit leads to the total magnetic induction

$$B = \frac{2\pi I}{cr} = \frac{Ze\omega}{cr} = \frac{Ze}{m_e cr^3}L, \tag{4.137}$$

where $|\mathbf{x}| = r$ is the radius of the circular orbit. From the last relation, connecting the magnetic field to the angular momentum, we obtain for the operator of the interaction energy

$$\hat{V}_{\text{SO}} = \frac{Ze^2}{2(m_e c)^2 r^3}\hat{\mathbf{S}} \cdot \hat{\mathbf{L}} = \frac{1}{2r(m_e c)^2}\frac{\partial V_{\text{C}}}{\partial r}\hat{\mathbf{S}} \cdot \hat{\mathbf{L}}. \tag{4.138}$$

As expected, the scaling of the spin-orbit interaction energy with the distance r between the electron and the nucleus is r^{-3}. In the last relation, r^{-3} is represented in terms of the radial derivative of Coulomb attraction, $V_{\text{C}} = -Ze^2/r$. The physical origin of the spin-orbit interaction is in the magnetic field acting on a particle moving in the electric field. Because of spin-orbit interaction, electric field acts on a moving magnetic moment (spin).

The scalar product of the spin $\hat{\mathbf{S}}$ and orbital $\hat{\mathbf{L}}$ angular momenta can be expressed through the total angular momentum $\hat{\mathbf{J}} = \hat{\mathbf{L}} + \hat{\mathbf{S}}$ as

$$2\,\hat{\mathbf{S}} \cdot \hat{\mathbf{L}} = \hat{\mathbf{J}}^2 - \hat{\mathbf{L}}^2 - \hat{\mathbf{S}}^2. \tag{4.139}$$

In this representation, the spin-orbit coupling is expressed as

$$\hat{V}_{SO} = \frac{1}{4r(m_e c)^2} \frac{\partial V_C}{\partial r} \left[\hat{\mathbf{J}}^2 - \hat{\mathbf{L}}^2 - \hat{\mathbf{S}}^2 \right]. \tag{4.140}$$

One can estimate the change in the energy due to spin-orbit coupling in terms of the first-order perturbation theory by taking the expectation value of $\hat{V}_{SO}$. For the angular momentum operator, one expects $\langle \hat{\mathbf{J}}^2 \rangle = j(j+1)\hbar^2$, where j is the quantum number of the total angular momentum operator. The energy change becomes

$$\langle \hat{V}_{SO} \rangle = E_{SO} \left[j(j+1) - l(l+1) - s(s+1) \right], \tag{4.141}$$

where E_{SO} is obtained by taking the expectation value of $\hat{V}_{SO}$ with the radial wave function. It establishes the energy scale of spin-orbit coupling.

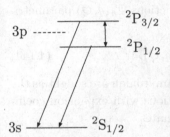

Fig. 4.4 Spin-orbit splitting of the 3p atomic orbital of sodium leading to the doublet in the emission of light in the transition from 3p to 3s electronic states. The state with $j = 3/2$ is higher in energy than the state with $j = 1/2$. The splitting between $j = 3/2$ and $j = 1/2$ states is $3E_{SO}$.

The allowed values of j are obtained from the rules of summation of angular momenta [27–29]. For the electron spin one has $s = 1/2$, and l specifies the orbital angular momentum, $\langle \hat{\mathbf{L}}^2 \rangle = \hbar^2 l(l+1)$. For $l > 0$, the allowed values of j are either $j = l + 1/2$ or $j = l - 1/2$. These two values of j will produce two possible values of the spin-orbit energy. As an example, consider $l = 1$. In this case, one finds either $j = 3/2$ with $\langle \hat{V}_{SO} \rangle = E_{SO}$ or $j = 1/2$ with $\langle \hat{V}_{SO} \rangle = -2E_{SO}$. The splitting of the $l = 1$ state is responsible for the famous doublet of the sodium line corresponding to 3p→3s light emission (Fig. 4.4).

4.13 Born-Oppenheimer approximation

The Hamiltonian of a molecule depends on coordinates of both nuclei, Q, and electrons, q. Here, Q and q are used as just labels to specify all Cartesian coordinates of N_e electrons and N_n nuclei. The total Hamiltonian is a sum of kinetic, T, and potential, U, energies of all electrons and all nuclei

$$H(q,Q) = T(q) + T(Q) + U(q,Q) + U_n(Q). \qquad (4.142)$$

The potential energy $U(q,Q)$ includes all attractions between the electrons and the nuclei and the electron-electron repulsions. The separate term, $U_n(Q)$, is the repulsion Coulomb energy between the nuclei.

The Schrödinger equation with the eigenfunction $\Phi_i(q,Q)$ and eigenenergy ϵ_i cannot be obtained without an approximation allowing separation of electronic q and nuclear Q variables. Two major approaches to achieve such an approximate separation are the Born-Oppenheimer (adiabatic) and Herzberg-Teller (crude adiabatic) approximations [30].

The adiabatic solution is sought by separating the Hamiltonian into the electronic and nuclear parts. The electronic Hamiltonian is

$$H_e = T(q) + U(q,Q). \qquad (4.143)$$

It produces the eigenvalues $E_n(Q)$ and eigenfunctions $\Psi_n(q,Q)$ parametrically depending on the nuclear coordinates

$$H_e\Psi_n(q,Q) = E_i(Q)\Psi_n(q,Q). \qquad (4.144)$$

The eigenfunctions $\Psi_i(q,Q)$ of the molecular Hamiltonian are sought as the linear combination of the electronic wave-functions with expansion coefficients $\chi_{i,n}(Q)$ depending on the nuclear coordinates

$$\Phi_i(q,Q) = \sum_n \chi_{i,n}(Q)\Psi_n(q,Q). \qquad (4.145)$$

This approximate solution is now substituted in the Schrödinger equation for the molecular system

$$[H_e(q,Q) + T(Q) + U_n(Q)] \sum_n \chi_{i,n}(Q)\Psi_n(q,Q)$$
$$= \epsilon_i \sum_n \chi_{i,n}(Q)\Psi_n(q,Q). \qquad (4.146)$$

The coupling between q and Q comes from the kinetic energy operator for the nuclear motion

$$T(Q)\chi_{i,n}(Q)\Psi_n(q,Q) = \Psi_n(q,Q)[T(Q)\chi_{i,n}(Q)] + \chi_{i,n}(Q)[T(Q)\Psi_n(q,Q)]$$
$$- \hbar^2 \sum_k \partial\chi_{i,n}(Q)/\partial Q_k \partial\Psi_n(q,Q)/\partial Q_k.$$

$$(4.147)$$

By closing the bra-ket with $\langle \Psi_n| = \langle n|$, one obtains

$$\begin{aligned}
&\left[T(Q) + U_n(Q) + E_n(Q) + \langle n|T(Q)|n\rangle - \epsilon_i\right]\chi_{i,n} \\
&= \sum_{m \neq n} \left[\hbar^2 \sum_k \langle n|\partial/\partial Q_k|m\rangle \partial/\partial Q_k - \langle n|T(Q)|m\rangle\right]\chi_{i,m}.
\end{aligned} \quad (4.148)$$

The adiabatic approximation neglects all terms with $m \neq n$ on the right-hand side of this equation. The left-hand side then determines the vibrational wave function of the nuclear motion $\chi_{i,n}(Q)$ associated with the electronic state n. The Born-Oppenheimer approximation makes the next step and neglects also the expectation value of the nuclear kinetic energy, $\langle n|T(Q)|n\rangle$, in the equation for $\chi_{i,n}(Q)$.

In the crude adiabatic approximation proposed by Herzberg and Teller, the electronic wave function is evaluated at a fixed nuclear configuration Q_0

$$\left[T(q) + U(q, Q_0)\right]\Psi_n(q, Q_0) = E_n(Q_0)\Psi_n(q, Q_0). \quad (4.149)$$

The dependence on the nuclear coordinates is fully allocated in the vibrational wave function $\chi_{i,n}(Q)$ such that the wave function for the molecular system is sought in the form

$$\Phi_i(q, Q) = \sum_n \Psi_n(q, Q_0)\chi_{i,n}(Q). \quad (4.150)$$

The vibrational wave function is defined by the following equation

$$\left[T(Q) + U_n(Q) + E_n(Q_0) + \langle n|\Delta U(q, Q)|n\rangle - \epsilon_i\right]\chi_{i,n} = 0, \quad (4.151)$$

where $\Delta U(q, Q) = U(q, Q) - U(q, Q_0)$.

Both adiabatic and crude adiabatic approximations are formally complete solutions of the problem of vibronic coupling between the electronic and nuclear molecular coordinates in the range of parameters where they are applicable (the splitting of vibrational states is much smaller than the splitting of the electronic sates). However, the adiabatic approximation has a broader range of applications. The energy surfaces $E_n(Q)$ can be directly calculated from solving the electronic Born-Oppenheimer equation at each Q. In contrast, $\langle n|\Delta U(q, Q)|n\rangle$, required for the vibrational wave function in the crude adiabatic approximation, is often unknown.

4.14 Polarizability

Dispersion (or London) interactions apply to all atoms and molecules and require quantum mechanics for their derivation. All atoms and molecules undergo virtual quantum transitions in which they virtually occupy their excited quantum states returning back to the ground state (Fig. 4.5). These transitions are responsible for the appearance of the induced dipole moment in an external electric field, which is related to molecular and atomic polarizabilities. The linear dipolar polarizability relates the observable induced dipole moment p_α to the external electric field $E_{0\beta}$

$$p_\alpha = \alpha_{\alpha\beta} E_{0\beta}, \qquad (4.152)$$

where α, β denote the Cartesian components x, y, z of the vectors $\mathbf{p}$ and $\mathbf{E}_0$ and summation is over the common indices.

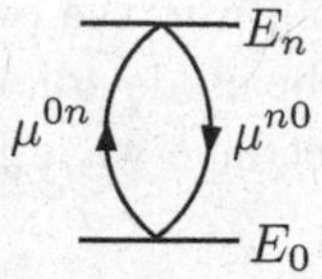

Fig. 4.5 Virtual quantum transitions responsible for atomic and molecular polarizability.

The molecular polarizability $\alpha_{\alpha\beta}$ is generally a second-rank tensor, which means that the dipole along the Cartesian axis α can be induced by the a field applied along the Cartesian axis β. For instance, an induced dipole p_x can be measured if the field is applied along the axis z: the polarizability component α_{xz} is responsible for this induced dipole moment. One often pictures the appearance of the induced dipole in classical terms as the deformation of the electronic clouds of atoms and molecules by the external field. This representation is only a mnemonic picture of a truly quantum phenomenon, which cannot be described by classical electrodynamics.

In quantum mechanics, the polarizability tensor is given by the following equation

$$\alpha_{\alpha\beta} = 2 \sum_{n>0} \frac{\langle 0|\hat{\mu}_\alpha|n\rangle\langle n|\hat{\mu}_\beta|0\rangle}{E_n - E_0}. \qquad (4.153)$$

Here, $\hat{\mu}$ is the quantum mechanical operator of the dipole moment and $|n\rangle$ is the quantum state of the molecule or atom with the quantum energy E_n. The state $|0\rangle$ is the ground state with the lowest energy and the sum runs over all excited states with $E_n > E_0$. The off-diagonal matrix element of the dipole moment operator $\mu_\alpha^{0n} = \langle 0|\hat{\mu}_\alpha|n\rangle$ is the transition dipole between states 0 and n. The intensity of absorption of a photon with the energy $\hbar\omega = E_n - E_0$ is proportional to $|\mu_\alpha^{0n}|^2$ when the radiation is polarized along axis α (discussed in more detail in Sec. 11.1).

The induced dipole at the molecule can be created by an oscillatory external field $\mathbf{E}_0 e^{i\omega t}$. In that case, the polarizability tensor becomes the polarizability function depending on the field frequency

$$\alpha_{\alpha\beta}(\omega) = \sum_{n>0} \left[\frac{\langle 0|\hat{\mu}_\alpha|n\rangle\langle n|\hat{\mu}_\beta|0\rangle}{\Delta E_n - \hbar\omega} + \frac{\langle 0|\hat{\mu}_\alpha|n\rangle^*\langle n|\hat{\mu}_\beta|0\rangle^*}{\Delta E_n + \hbar\omega} \right], \qquad (4.154)$$

where $\Delta E_n = E_n - E_0$ is the excitation energy.

4.15 *Dispersion forces

The dispersion interaction related to dipolar polarizabilities of atoms and molecules decays as r^{-6} with the separation r. It is always an attractive interaction, and it arises as the second-order quantum mechanical perturbation in terms of the interaction between the dipole moment operators $\hat{\mu}_1$ and $\hat{\mu}_2$ of molecules 1 and 2

$$\hat{V} = -\hat{\mu}_1 \cdot \mathbf{T} \cdot \hat{\mu}_2, \qquad (4.155)$$

where $\mathbf{T}$ is dipolar tensor (Eq. (1.66)) representing the effect of mutual orientations and separation of two dipoles on the interaction energy. Since the dipole moments are vectors and the interaction energy $\hat{V}$ is the scalar, the dipolar tensor $\mathbf{T}$ must be a second-rank tensor to produce the scalar from contraction over the common Cartesian indexes: $\hat{\mu}_1 \cdot \mathbf{T} \cdot \hat{\mu}_2 = \hat{\mu}_{1\alpha} T_{\alpha\beta} \hat{\mu}_{2\beta}$

$$T_{\alpha\beta} = r^{-3}(3\hat{r}_\alpha \hat{r}_\beta - \delta_{\alpha\beta}), \qquad (4.156)$$

where $\hat{r}_\alpha$ are the directional cosines of the unit radial vector (Fig. 4.6).

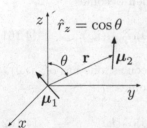

Fig. 4.6 Geometry of interacting point dipoles.

The dispersion interaction arises from considering quantum virtual excitations in which both molecules are virtually excited to states n_1 and n_2. The ground state of two molecules is therefore $|00\rangle$ and the excited state is $|n_1 n_2\rangle$, where in both cases the first index indicates the first molecule and the second index indicates the second molecule. The second-order quantum-mechanical perturbation reads

$$E^{(2)} = -\sum_{n1,n2>0} \frac{\langle 00|\hat{V}|n_1 n_2\rangle\langle n_1 n_2|\hat{V}|00\rangle}{\Delta E_{n1} + \Delta E_{n2}}, \qquad (4.157)$$

where $\Delta E_{n1} = E_{n1} - E_{01}$ and $\Delta E_{n2} = E_{n2} - E_{02}$ are the excitation energies for each molecule. In terms of the dipole-dipole interaction energy from Eq. (4.155), one obtains

$$E^{(2)} = -T_{\alpha\beta}T_{\gamma\delta} \sum_{n1,n2>0} \frac{\langle 00|\hat{\mu}_{1\alpha}\hat{\mu}_{2\beta}|n_1 n_2\rangle\langle n_1 n_2|\hat{\mu}_{1\gamma}\hat{\mu}_{2\delta}|00\rangle}{\Delta E_{n1} + \Delta E_{n2}}. \qquad (4.158)$$

The quantum states on different molecules are assumed not to overlap and one has $\langle 00|\hat{\mu}_{1\alpha}\hat{\mu}_{2\beta}|n_1 n_2\rangle = \mu_{1\alpha}^{0n_1}\mu_{2\beta}^{0n_2}$.

Further simplification of the dispersion energy requires separation of the properties of the molecules. The main difficulty here is the denominator where the excitation energies of two molecules add up. This can be avoided by using the mathematical identity

$$\frac{1}{x+y} = \frac{2}{\pi}\int_0^\infty d\omega \frac{xy}{(x^2+\omega^2)(y^2+\omega^2)}, \quad x,y > 0. \qquad (4.159)$$

In order to use this equation, one notes that the real part of the polarizability function $\alpha'_{\alpha\beta}(\omega)$ estimated at the imaginary frequency $i\omega$ is given according to Eq. (4.154) by the relation

$$\alpha'_{\alpha\gamma}(i\omega) = 2\sum_{n>0} \frac{\Delta E_n \mu_\alpha^{0n}\mu_\gamma^{n0}}{\Delta E_n^2 + (\hbar\omega)^2}. \qquad (4.160)$$

The relation for the second-order perturbation then becomes

$$E^{(2)} = -\frac{\hbar}{2\pi}T_{\alpha\beta}T_{\gamma\delta}\int_0^\infty d\omega\alpha'_{1,\alpha\gamma}(i\omega)\alpha'_{2,\beta\delta}(i\omega). \qquad (4.161)$$

In the case of isotropic polarizabilities of the molecules, one has $\alpha'_{\alpha\gamma}(i\omega) = \alpha'(i\omega)\delta_{\alpha\gamma}$. The dispersion interaction energy becomes

$$E^{(2)} = -\frac{\hbar}{2\pi}T_{\alpha\beta}T_{\alpha\beta}\int_0^\infty d\omega\alpha'_1(i\omega)\alpha'_2(i\omega). \qquad (4.162)$$

One can further obtain

$$T_{\alpha\beta}T_{\alpha\beta} = \frac{1}{r^6}(3\hat{r}_\alpha\hat{r}_\beta - \delta_{\alpha\beta})(3\hat{r}_\alpha\hat{r}_\beta - \delta_{\alpha\beta}) = \frac{6}{r^6}. \qquad (4.163)$$

This relation follows from the scalar product of the unit vector $\hat{r}_\alpha\hat{r}_\alpha = 1$. One finally obtains

$$\boxed{E^{(2)} = -\frac{3\hbar}{\pi r^6}\int_0^\infty d\omega\alpha'_1(i\omega)\alpha'_2(i\omega) \propto \frac{1}{r^6}.} \qquad (4.164)$$

4.16 Drude oscillator

The Drude oscillator model assumes that a polarizable particle can be modeled by the charge q_D on a harmonic string with the force constant k_D. The potential energy as a function of displacement $\mathbf{d}$ is

$$U(\mathbf{d}) = \tfrac{1}{2} k_D \mathbf{d}^2. \tag{4.165}$$

If an oscillatory external force $\mathbf{F}(t) = \mathbf{F}_0 e^{i\omega t}$ is applied to the particle, the stationary solution for the displacement can be sought in terms of oscillations with the frequency of the external field ω: $\mathbf{d}(t) = \mathbf{d}_0 e^{i\omega t}$. The Newton's equation of motion then becomes

$$- m_D \omega^2 \mathbf{d}(t) = -k_D \mathbf{d}(t) + \mathbf{F}(t), \tag{4.166}$$

where m_D is the mass of the Drude particle. The self-frequency of particle's oscillations is $\omega_0^2 = k_D/m_D$. One therefore obtains

$$\mathbf{d}(t) = \frac{\mathbf{F}(t)}{m_D} \frac{1}{\omega_0^2 - \omega^2}. \tag{4.167}$$

One can use the displacement to calculate the dipole moment $p_\alpha(t) = q_D d_\alpha(t)$, where $d_\alpha(t)$ are the Cartesial components of the displacement vector. If the force on the particle is achieved from the electric field $\mathbf{E}(t)$ acting on the charge q_D, then $\mathbf{F}(t) = q_D \mathbf{E}(t)$. From this relation, one obtains the polarizability of the particle α linking $p_\alpha(t)$ to the field $E_\alpha(t)$ (Eq. (4.152))

$$\alpha = \frac{\alpha_0}{1 - (\omega/\omega_0)^2}, \quad \alpha_0 = \frac{q_D^2}{\omega_0^2 m_D}. \tag{4.168}$$

Polarizability carries the units of length cube.

The Drude polarizability can be used in the formula for dispersion forces to express them in terms of zero-frequency polarizabilities α_{0i} and characteristic frequencies ω_{0i} of two atoms. From Eqs. (4.164) and (4.168), one obtains the London formula for the dispersion interactions

$$E^{(2)} = -\frac{3}{2} \frac{\hbar \omega_{01} \omega_{02}}{\omega_{01} + \omega_{02}} \frac{\alpha_{01} \alpha_{02}}{r^6}. \tag{4.169}$$

The appearance of Planck's constant in this equation points to the quantum origin of dispersion forces.

4.17 *Path integrals

x-, p-representation:
$$\langle x|p\rangle = \frac{1}{\sqrt{2\pi\hbar}}e^{ipx/\hbar}, \quad \sum_x |x\rangle\langle x| = \sum_p |p\rangle\langle p| = 1$$

In Sec. 3.1, the classical equations of motion where derived by choosing from all possible paths connecting $x_0 = x(0)$ and $x = x(t)$ the only path that minimizes the mechanical action. There is an intrinsic uncertainty in measuring trajectories of quantum particles incorporated in the uncertainty relations and in the formalism of quantum mechanical operators. Operators represent classical measurements performed on a quantum state collapsing it to a number of outcomes given by the of set eigenstates of that specific operator. Quantum mechanics postulates that this set is complete and an expansion of a given quantum state in eigenstates of a physical operator is possible. The expansion yields a set of expansion amplitudes assigned to each measurement outcome. The squared amplitudes yield fractions of specific measurement outcomes.

An alternative formulation of quantum mechanics was proposed by Feynman who suggested to use all possible trajectories in the quantum-mechanical action replacing the classical mechanical action (Eq. (3.1)) to produce the time evolution amplitude. It is given as the matrix element of the quantum time evolution operator [31]

$$(x,t|x_0,0) = \langle x|U(t)|x_0\rangle, \quad U(t) = e^{-iHt}. \tag{4.170}$$

As in Sec 4.7 above, we will put $\hbar = 1$ here and suppress curets over the operators for brevity. We will still put $\hbar$ in the final equations to restore correct units. Such a final expression was obtained by Feynman [32] for the time evolution amplitude also called the propagator

$$(x,t|x_0,0) = \int_{\{x_0,x\}} \mathcal{D}x\, e^{iS_t[x]/\hbar}, \tag{4.171}$$

where $\mathcal{D}x$ denotes the path integral over all trajectories starting at x_0 at $t = 0$ and ending at x at $t = t$. The path integral is the notation spcifying a sum over all possible trajectories $x(t)$ connecting these end points. Each such trajectory is weighted with a phase factor $\exp[iS_t[x]/\hbar]$ such that the total transition amplitude is the interference of all amplitudes carrying individual phase factors. The quantum-mechanical action, like the action in the classical mechanics, is a functional of a specific trajectory. It is defined in Feynman's formulation by the Lagrangian (cf. to Eq. (3.1))

$$S_t[x] = \int_0^t d\tau L(x, \dot{x}) = \int_0^t d\tau \left[\tfrac{1}{2} m \dot{x}^2 - U(x) \right] . \qquad (4.172)$$

Similarly to the classical formulation of mechanical equations of motion, one can formulate the action functional based on the Hamiltonian and canonical variables $x(t)$ and $p(t)$. Here this route to derive the propagator is followed, with Feynman's representation appearing as a specific limit of this more general formulation.

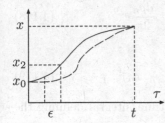

Fig. 4.7 Path starting off at $x_0 = x(0)$ and ending at $x = x(t)$. It is sliced into time intervals of duration ϵ with x_n corresponding to $t_n = n\epsilon$. The dashed line shows an alternative path satisfying the same boundary conditions.

The total path $x(t)$ is sliced into N intervals of duration ϵ such that the time becomes $t = \epsilon N$ (Fig. 4.7). The evolution operator can be viewed as a product of $N + 1$ operators moving the particle forward with the timestep ϵ

$$(x, t | x_0, 0) = \langle x | U(t, t_N) \dots U(t_1) | x_0 \rangle = \sum_{x_1, \dots x_N} \prod_{n=1}^{N+1} (x_n, t_n | x_{n-1}, t_{n-1}). \qquad (4.173)$$

The second equality involves using the completeness relation $\sum_{x_n} |x_n\rangle\langle x_n| = 1$ with the summation over all x_n such that $x_{N+1} = x$, $t_0 = 0$, and $t_{N+1} = t$. Each amplitude here is the matrix element of the evolution operator over the small time interval ϵ

$$(x_n, t_n | x_{n-1}, t_{n-1}) = \langle x_n | e^{-i\epsilon H} | x_{n-1} \rangle \simeq \langle x_n | e^{-i\epsilon U} e^{-i\epsilon T} | x_{n-1} \rangle. \qquad (4.174)$$

In the second step, the evolution operator with the Hamiltonian $H = T + U$ is factored into the operator with the potential energy U and the operator with the kinetic energy T. Corrections to this factorization involve commutators of ϵU and ϵT, which are quadratic in ϵ are neglected at $\epsilon \to 0$. One can again apply the completeness relation for x- and p-representations (Sec. 4.6) to write

$$\langle x_n | e^{-i\epsilon H} | x_{n-1} \rangle = \sum_{x, p_n} \langle x_n | e^{-i\epsilon U} | x' \rangle \langle x' | e^{-i\epsilon T} | p_n \rangle \langle p_n | x_{n-1} \rangle. \qquad (4.175)$$

The state $|p_n\rangle$ is an eigenstate for the kinetic energy operator, $\exp[-i\epsilon T]|p_n\rangle = |p_n\rangle \exp[-i\epsilon T(p_n)]$. We also obtain

$$\langle x_n|e^{-i\epsilon U}|x'\rangle = e^{-i\epsilon U(x_n)}\langle x_n|x'\rangle. \tag{4.176}$$

By applying the completeness relation $\sum_{x'}|x'\rangle\langle x'| = 1$ and substituting the expressions for $\langle x_n|p_n\rangle$ and $\langle p_n|x_{n-1}\rangle$ to Eq. (4.175), one obtains

$$\langle x_n|e^{-i\epsilon H}|x_{n-1}\rangle = \int_{-\infty}^{\infty}\frac{dp_n}{2\pi}\exp\left(i\epsilon\left[p_n\frac{x_n - x_{n-1}}{\epsilon} - T(p_n) - U(x_n)\right]\right). \tag{4.177}$$

With all amplitudes combined in Eq. (4.173), the propagator becomes

$$(x,t|x_0,0) = \prod_{n=1}^{N}\int_{-\infty}^{\infty}dx_n\prod_{n=1}^{N+1}\int_{-\infty}^{\infty}\frac{dp_n}{2\pi}e^{iS_N}, \tag{4.178}$$

where

$$S_N = \sum_{n=1}^{N+1}[p_n(x_n - x_{n-1}) - \epsilon H(p_n, x_n)]. \tag{4.179}$$

In the limit $\epsilon \to 0$, the above sum becomes the time integral defining the quantum-mechanical action

$$S_t[p,x] = \int_0^t d\tau\,[p(\tau)\dot{x}(\tau) - H(p,x)]. \tag{4.180}$$

The product of the infinite number of integrals over the intermediate points along the trajectories $x(t)$ and $p(t)$ in the phase space is called the path integral

$$\prod_{n=1}^{N}\int_{-\infty}^{\infty}dx_n\prod_{n=1}^{N+1}\int_{-\infty}^{\infty}\frac{dp_n}{2\pi} = \int_{\{x_0,x\}}\mathcal{D}'x\int\frac{\mathcal{D}p}{2\pi}. \tag{4.181}$$

For N x-integrals, there are always $N + 1$ p-integrals. This is represented by the prime in the notation used for the $x(t)$ path. We finally obtain

$$\boxed{(x,t|x_0,0) = \int_{\{x_0,x\}}\mathcal{D}'x\int\frac{\mathcal{D}p}{2\pi\hbar}e^{iS_t[p,x]/\hbar},} \tag{4.182}$$

where now $\hbar$ is inserted in the final expression to ensure correct units in the exponent and the integration variables (the action has the units of energy $\times$ time).

Feynman's representation of the path integral is obtained by noticing that the integral over p_n in Eq. (4.177) is formally a Gaussian integral

$$\int\frac{dp_n}{2\pi}\exp\left(ip_n\frac{x_n - x_{n-1}}{\epsilon} - i\epsilon p_n^2/(2m)\right)$$

$$= \frac{1}{\sqrt{2\pi i\epsilon/m}}\exp\left(i\epsilon m\frac{(x_n - x_{n-1})^2}{2\epsilon^2}\right). \tag{4.183}$$

The propagator in the time sliced representation reads

$$(x, t|x_0, 0) = \frac{1}{\sqrt{2\pi i \epsilon/m}} \prod_{n=1}^{N} \int_{-\infty}^{\infty} \frac{dx_n}{\sqrt{2\pi i \epsilon/m}} e^{iS_N}, \qquad (4.184)$$

where S_N in the Lagrangian representation is

$$S_N = \epsilon \sum_{n=1}^{N+1} \left[\frac{m}{2} \frac{(x_n - x_{n-1})^2}{\epsilon^2} - U(x_n) \right]. \qquad (4.185)$$

In the limit $\epsilon \to 0$, $N \to \infty$, $\epsilon N = t$, one recovers Feynman's expression in Eqs. (4.171) and (4.172).

4.18 Propagator of a free particle

The most important specific results for applying path integrals are the propagators for the free particle and for the harmonic oscillator. The more simple case of a free particle is considered here. The harmonic oscillator, involving more algebra, is used to derive the propagator for the stochastic Ornstein-Uhlenbeck process in Sec. 7.8 for which the path-integral representation also applies.

For the free particle, one needs to evaluate the functional integral from Eqs. (4.171) and (4.172) with $U = 0$

$$(x, t|x_0, 0) = \int_{\{x_0, x\}} \mathcal{D}x \, \exp\left[i\frac{m}{2\hbar} \int_0^t d\tau \, \dot{x}^2 \right]. \qquad (4.186)$$

The trajectory $x(t) = \tilde{x}(t) + \delta x(t)$ can be separated in the classical trajectory $\tilde{x}(t)$ that minimizes the action and quantum fluctuations $\delta x(t)$ along the classical path $\tilde{x}(\tau) = x_0 + \Delta x(\tau/t)$, $\Delta x = x - x_0$. The action becomes

$$S_t[x] = \tilde{S}_t + \frac{1}{2} \int_0^t d\tau \, m(\delta \dot{x})^2. \qquad (4.187)$$

Here, the action taken on the classical trajectory is

$$\tilde{S}_t = \frac{m \Delta x^2}{2t}. \qquad (4.188)$$

One can prove [31] that the path integral involving $\delta x(t)$ contributes only the normalization constant to the propagator. One therefore has

$$(x, t|x_0, 0) = C \exp\left[i\frac{m \Delta x^2}{2\hbar t} \right]. \qquad (4.189)$$

The normalization constant C is calculated from the condition $\int dx(x, t|x_0, 0) = 1$, with the result

$$(x, t|x_0, 0) = \sqrt{\frac{m}{2\pi i t \hbar}} \exp\left[i\frac{m\Delta x^2}{2\hbar t}\right]. \tag{4.190}$$

The evolution amplitude from this equation satisfies, at $x \neq x_0$ and $t > 0$, the following differential equation

$$i\hbar\partial_t(x, t|x_0, 0) = -\frac{\hbar^2}{2m}\, \partial_x^2(x, t|x_0, 0). \tag{4.191}$$

This is obviously the Schrödinger equation for a free particle. The evolution amplitude therefore provides the same information as the wave function calculated from the wave equation. The path-integral formulation maintains the structure of the mechanical action of classical mechanics. It greatly helps in gaining intuitive appreciation of the quantum mechanical laws and naturally transitions to classical mechanics in the limit $\hbar \to 0$.

For large t, Eq. (4.189) describes a single quantum of translational energy. It is seen by looking at the exponential part of the amplitude as the phase of a quantum wave packet [32]. The phase change should be 2π on the period of oscillations and one can write

$$2\pi = \frac{m\Delta x^2}{2\hbar t} - \frac{m\Delta x^2}{2\hbar(t + T)} = \frac{m\Delta x^2}{2\hbar t^2}\frac{T}{1 + T/t}. \tag{4.192}$$

The frequency $\omega = 2\pi/T$ becomes at $t \gg T$

$$\omega \simeq \frac{m}{2\hbar}\left(\frac{\Delta x}{t}\right)^2 = \frac{E}{\hbar}, \tag{4.193}$$

where $E = mv^2/2$ is the energy of translational motion. The above equation thus says $E = \hbar\omega$, which is the Einstein equation for the energy of a single quantum (Eq. (4.1)).

Chapter 5

Statistical Mechanics

Suppose one has a system of a very large number of particles, $N \simeq N_A = 6.02 \times 10^{23}$ (Avogadro's number). One can in principle describe the motion of each of these particles by solving Newton's (or Hamilton's) mechanical equations of motion. In old textbooks, the authors would state the obvious: one cannot possibly do that. It is still not possible to do for a macroscopic number of particles, but modern computers allow one to follow the trajectory of each individual particle for hundreds of thousands of them. However, the argument of old statistical mechanics texts remains valid. Even though we can calculate trajectories of a large number of particles, our measurements are still limited to a small number of observables. The rest of the variables are either not available or are just not relevant and only affect those few observables. One therefore has to neglect most of the information pertinent to the system. Keeping a small number of variables reduces the problem to effective potential energies or probabilities of these variables. If only thermodynamic macroscopic variables are retained, the probabilities are converted to thermodynamic free energies. If some of the microscopic variables are kept, the reduced description is cast in terms of potentials of mean force. All these formalisms can be alternatively formulated in the language of random (fluctuating) variables and corresponding probabilities. Dealing with large systems and reduced manifolds of variables in terms of probabilities is the essence of *statistical mechanics* [33–37]. The laws of statistical mechanics emerge in the course of reduction from the full description and complete information content to a small number of variables and reduced information. They cannot be explained in mechanical terms by considering motions of individual particles. The probabilistic character of statistical laws is a result of the reduction of a very large number of degrees of freedom to only a few variables monitored by experiment.

5.1 Probabilities and distributions

Mean and variance:

$$\langle x \rangle = \int dx\, x\, p(x)$$

$$\sigma^2 = \int dx\, (x - \langle x \rangle)^2 p(x)$$

The probability for the purpose of statistical mechanics can be understood as the frequency of occurrence. If the value x_i is measured n_i times out of n_t of total attempts, the probability is $p_i = n_i/n_t$. When the variable x is continuous, one specifies the probability density $p(x)$. It implies that the frequency of x falling in the interval $x \leq x \leq x + \Delta x$ is

$$\Delta p = p(x)\Delta x. \tag{5.1}$$

The probability density satisfies the normalization condition

$$\int dx\, p(x) = 1. \tag{5.2}$$

It can be specified either by its functional form or by its moments, of which the mean $\langle x \rangle$ and variance σ^2 are most commonly considered. The general definition of the moment of nth order is

$$\langle x^n \rangle = \int dx\, x^n p(x). \tag{5.3}$$

In addition, one defines cumulants of nth order

$$\langle (\delta x)^n \rangle = \int dx\, (x - \langle x \rangle)^n p(x). \tag{5.4}$$

Variance, $\sigma^2 = \langle (\delta x)^2 \rangle$, is the second-order cumulant. The square root of the variance is known as the standard deviation. It also comes under the name of root-mean-squared deviation

$$\Delta x_{\text{rms}} = \sqrt{\langle (x - \langle x \rangle)^2 \rangle}. \tag{5.5}$$

When several stochastic (random) variables are involved, one can specify the joint probability density $p(x_1, x_2, \ldots, x_m)$, which gives the probability of measuring $x_1, \ldots, x_m$ simultaneously. When the measurements are independent, one has

$$p(x_1, \ldots, x_m) = p_1(x_1) \times p_2(x_2) \times \cdots \times p_m(x_m). \tag{5.6}$$

When this approximation does not apply, the two variables are said to be correlated. The most common correlations are binary. The binary correlation function $g(x_1, x_2)$ shows how much the joint probability deviates from the assumption of independent stochastic variables

$$p(x_1, x_2) = p_1(x_1)p_2(x_2)g(x_1, x_2). \qquad (5.7)$$

One has $g(x_1, x_2) = 1$ for statistically independent variables. The following notation is used to describe correlations between two variables x_1 and x_2

$$\langle x_1 x_2 \rangle = \int dx_1 dx_2 x_1 x_2 p(x_1, x_2). \qquad (5.8)$$

A number of general rules can be formulated for composite variables or running averages

$$X_N = N^{-1} \sum_{i=1}^{N} x_i. \qquad (5.9)$$

Such types of variables are most common in science. They apply either to results of repeated measurements, when N is the number of data points, or, more commonly, to measurements done on macroscopic objects. In the latter case, N is the number of particles and x_i is a property assigned to each single particle, for instance the particle energy.

One can start by looking at the variance of X_N assuming that variables $\delta x_i = x_i - \langle x_i \rangle$ are statistically independent: $\langle \delta x_i \delta x_j \rangle = 0$ when $i \neq j$. What follows from this assumption is a connection between the variance σ_X^2 of the composite variable X and the variance σ_x^2 of the individual variable x describing each separate particle

$$\sigma_X^2 = \sigma_x^2/N. \qquad (5.10)$$

This is an important result known as the law of large numbers. It states that the result of measurements is improved when more measurements are done, and the standard deviation of those repeated measurements scales down as $1/\sqrt{N}$. Similarly, the reason a well-defined energy can be assigned to a macroscopic sample made of many small parts is because fluctuations of the sample energy scale down as $1/\sqrt{N}$, where N is the number of atoms or molecules making up the sample. Of course, the closest neighbors in that sample will still correlate, but there will always be sufficient number of far away molecules to render fluctuations of any macroscopic property negligible.

Another iconic result of the probability theory used in science across disciplines is the central limit theorem. It states that as the number N

defining the composite variable X_N increases, the distribution of X_N tends to a Gaussian distribution. We have

$$p(X_N) \underset{N \gg 1}{\rightarrow} \frac{1}{\sqrt{2\pi\sigma_X^2}} e^{-(X_N - \langle X_N \rangle)^2/(2\sigma_X^2)}. \tag{5.11}$$

It is obvious that as N increases, $\langle X_N \rangle$ becomes $\langle x \rangle$ and the distribution width shrinks as $1/\sqrt{N}$. The result is an infinitely narrow distribution known as the delta-function.

$$p(X) \underset{N \to \infty}{\rightarrow} \delta(X - \langle x \rangle). \tag{5.12}$$

This equation implies that one reports a single value $\langle x \rangle$ for a property per particle (e.g., energy per particle) when it is measured from a macroscopic sample.

5.2   Elements of thermodynamics

The central target of thermodynamics is the calculation of work done by exchanging heat between the working body (the system) and the surroundings. The mechanical work is the scalar (dot) product of the external force $\mathbf{f}_{\text{ext}}$ applied to the system with the displacement $d\boldsymbol{\ell}$ integrated over the mechanical path ℓ

$$W_{\text{ext}} = \int_\ell \mathbf{f}_{\text{ext}} \cdot d\boldsymbol{\ell}. \tag{5.13}$$

Assuming mechanical equilibrium between the system and the surroundings, $P = P_{\text{ext}}$, one can connect the external force applied to the system with the internal pressure in the system P: $f_{\text{ext}} = PA$, A is the area of the "piston" used to do work. One obtains

$$\delta W_{\text{ext}} = -PdV, \tag{5.14}$$

where $dV = -dV_{\text{ext}}$ is the change of the volume of the system. The first law of thermodynamics states that the change in the internal (potential plus kinetic) energy of the system is the result of combining the heat transferred δQ, external work done on the system, and particles added, each of them carrying its own energy given by its chemical potential μ

$$\boxed{dE = \delta Q + \delta W_{\text{ext}} + \mu dN.} \tag{5.15}$$

The last term is the product of the intensive variable μ and the extensive variable of the number of particles N.

Thermodynamics replaces heat and work with the corresponding products of intensive and extensive variables: $\delta Q = TdS$ and $\delta W_{\text{ext}} = -PdV$, where S is the system entropy (an extensive variable). One obtains the differential form of the first law of thermodynamics

$$dE = TdS - PdV + \mu dN. \tag{5.16}$$

More generally, this equation can be written as

$$dE = TdS - \sum_i f_i d\lambda_i, \tag{5.17}$$

where f_i are generalized forces and λ_i are extensive macroscopic parameters. This equation states that the energy of the system is a unique function of its entropy and of a set of parameters λ_i specifying various ways to either perform work on the system or change its number of particles.

The integral form of the first law uses the fact that there is no change of the internal energy (a state function) on any closed path. One finds at $N = \text{Const}$

$$\delta Q = - \oint \delta W_{\text{ext}}. \tag{5.18}$$

In $P - V$ coordinates, one gets

$$- \oint \delta W_{\text{ext}} = \oint PdV \tag{5.19}$$

and the integral over the closed path is equal to the enclosed area, which determines the heat delivered to the system (Fig. 5.1).

Fig. 5.1 Closed path in $P - V$ coordinates.

Equation (5.16) states that the internal energy $E(S, V, N)$ is a function of three extensive variable: S, V, N. The second-order derivatives of $E(S, V, N)$ do not depend on the order in which derivatives are taken, e.g., $\partial^2 E/\partial S \partial V = \partial^2 E/\partial V \partial S$. This mathematical identity leads to powerful connections between measurements known as Maxwell relations. For instance, by interchanging the derivatives over volume and entropy, one obtains

$$(\partial T/\partial V)_{S,N} = -(\partial P/\partial S)_{V,N}. \tag{5.20}$$

Similarly, for the V, N-pair

$$-(\partial P/\partial N)_{S,V} = (\partial \mu/\partial V)_{S,N}. \tag{5.21}$$

From the differential form of the first law, the energy $E = E(S, V, N)$ is a function of extensive variables only. An extensive variable implies that it doubles when the number of particles in the system doubles. In the mathematical form, this means that if all extensive variables S, V, N are multiplied by a scaling factor λ, the energy should multiply by λ as well: $E(\lambda S, \lambda V, \lambda N) = \lambda E(S, V, N)$. Functions carrying this property are known as homogeneous functions of first order [33]. They satisfy Euler's theorem

$$f(x_1, \ldots, x_n) = \sum_{j=1}^{n} (\partial f / \partial x_j) x_j. \tag{5.22}$$

From this theorem,

$$\boxed{E = TS - PV + \mu N.} \tag{5.23}$$

The intensive parameters are determined by derivatives involving pairs of extensive properties

$$T = (\partial E / \partial S)_{N,V}, \quad P = -(\partial E / \partial V)_{N,S}, \quad \mu = (\partial E / \partial N)_{S,V}. \tag{5.24}$$

Equation (5.23) leads to the definition of the extensive Gibbs free energy

$$G = \mu N = E - TS + PV. \tag{5.25}$$

The first two terms in this equation define the Helmholtz free energy

$$F = E - TS. \tag{5.26}$$

The differential of the Helmholtz free energy follows from Eq. (5.16) as

$$dF = -SdT - PdV + \mu dN. \tag{5.27}$$

The first derivatives of the free energy then define the entropy and pressure

$$S = -(\partial F / \partial T)_{V,N}, \quad P = -(\partial F / \partial V)_{T,N}. \tag{5.28}$$

The second derivatives over the same variables provide connections to the constant-volume heat capacity C_V and the isothermal compressibility $\beta_T = -V^{-1}(\partial V / \partial P)_T$

$$\left(\partial^2 F / \partial T^2\right)_{V,N} = -C_V / T, \quad \left(\partial^2 F / \partial V^2\right)_{T,N} = (\beta_T V)^{-1}. \tag{5.29}$$

Because both of them are positive, $C_V > 0$ and $\beta_T > 0$, $F(T, V)$ is a decreasing function of both T and V, but it is curved up (concave) against T and is curved down (convex) against V (Fig. 5.2). Similarly, the Gibbs energy is concave in both T and P. This result follows from the differential form

$$dG = -SdT + VdP \tag{5.30}$$

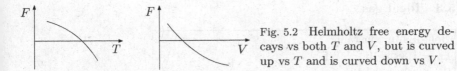

Fig. 5.2 Helmholtz free energy decays vs both T and V, but is curved up vs T and is curved down vs V.

and the second derivative $(\partial^2 G/\partial P^2)_T = -V\beta_T$.

The free energies F and G provide the measure of work done at constant temperature (F) and at constant temperature and pressure (G). These functions tell how much "free energy" is available to do work, that is the energy on the top of the energy dispersed to thermal motion. By differentiating F at constant temperature $(dT = 0)$, one gets at equilibrium conditions

$$dF = \delta W_{ext}. \tag{5.31}$$

The external work δW_{ext} can include both the mechanical expansion work and other kinds of work, such as the chemical work produced by a chemical reaction. By differentiating G at both constant T and constant P yields

$$dG = \delta W_{ext} + PdV. \tag{5.32}$$

Therefore, dG measures the non-expansion work: if $\delta W_{ext} = -PdV$ includes only the work done to expand the system, one gets $dG = 0$.

Differentiating Eq. (5.25) and subtracting the differential form of the first law in Eq. (5.16) leads to the Gibbs-Duhem equation

$$SdT - VdP + Nd\mu = 0. \tag{5.33}$$

It can be conveniently written in terms of the differential of the chemical potential

$$\boxed{d\mu = -sdT + vdP.} \tag{5.34}$$

Here, $s = S/N$ and $v = V/N$ are the molecular entropy and volume, respectively. From this equation, the chemical potential depends only on the intensive variables T, P: $\mu = \mu(T, P)$.

5.3   Ideal gas

Boltzmann constant:
$k_B = 1.38 \times 10^{-23}$ J/K
Stirling's formula:
$N! \simeq (N/e)^N$
Heat capacity:
$C_V = T(\partial S/\partial T)_{V,N} = (\partial E/\partial T)_{V,N}$
Pressure from free energy:
$P = -(\partial F/\partial V)_{T,N}$
Equipartition theorem:
$C_V = (f/2)k_B$, for f degrees of freedom
Indistinguishable particles:
number of states is divided by $N!$

The rules of statistics are applied to specific physical problems where energy and a few other observable properties are measured. These sets of measured quantities define statistical ensembles, which are constructed below. It is instructive to start with a simple case to gain some initial intuition of how such calculations are done. Our first choice is the gas of non-interacting particles, the ideal gas.

The ideal gas can be viewed as N freely moving particles colliding elastically with each other and with the walls of the container with the volume V. Collisions with the walls produce macroscopic pressure, which we derive from the rules of statistics. One can assign the volume v to each particle, which will be specified below. If the volume per particle is known, the number of states that freely moving particles can occupy is given as

$$\Omega_N = \frac{1}{N!}\left(\frac{V}{v}\right)^N. \tag{5.35}$$

In this equation, the number of configurations available through displacing the particles is $(V/v)^N$. This number is divided by the total number of permutations of N particles at each instantaneous configurations, which is $N!$. This factor accounts for the fact that particles are indistinguishable and one cannot experimentally distinguish configurations in which particles have been swapped. The reason for that requirement was first unclear (Gibbs paradox) and was later clarified by quantum mechanics.

Once the number of states is known, one can specify the Boltzmann entropy

$$\boxed{S = k_B \ln \Omega_N.} \tag{5.36}$$

Given that the number of particles N is always large in problems of statistical mechanics, one can apply Stirling's formula, $N! \simeq (N/e)^N$ (e is Euler's number), to come up with the entropy of the ideal gas

$$S = k_B N \ln \left(\frac{Ve}{Nv} \right) + \text{Const.} \tag{5.37}$$

This result is valid up to a constant, and we show below that this constant is equal to $(3/2)k_B N$.

The particles of the ideal gas have zero volume since we have to neglect their potential energy of repulsion (no interactions). Nevertheless, the smallest volume for a moving particle is dictated by quantum mechanics, which stipulates that particle's size cannot decrease below its de Broglie wavelength $\lambda = h/\langle p \rangle$. For particles with mass m moving with the thermal velocity one obtains

$$\boxed{\lambda = h/ \left(2\pi m k_B T \right)^{1/2}.} \tag{5.38}$$

Substituting $v = \lambda^3$ to Eq. (5.37), one arrives at the final expression

$$S = k_B N \ln \left[\left(2\pi m k_B T \right)^{3/2} \frac{Ve}{Nh^3} \right] + \text{Const.} \tag{5.39}$$

From Eq. (5.39), one can define the heat capacity by taking the derivative of the entropy over temperature at constant volume

$$C_V = (\partial E/\partial T)_V = T (\partial S/\partial T)_V. \tag{5.40}$$

By taking the temperature derivative of the ideal-gas entropy one obtains

$$C_V = \tfrac{3}{2} N k_B. \tag{5.41}$$

This result is a specific case of the equipartition theorem, which maintains that each out of f degrees of freedom of a molecule contributes $\tfrac{1}{2} k_B$ to C_V: translations in 3D bring $f = 3$ degrees of freedom. One can next integrate Eq. (5.40) to obtain the internal energy E from the heat capacity

$$E = \tfrac{3}{2} N k_B T. \tag{5.42}$$

Finally, the thermodynamic Helmholtz free energy $F = E - TS$ follows from subtracting Eq. (5.39) from Eq. (5.42). The derivative of the free energy over volume yields the system pressure. The energy E does not depend on volume V and a non-zero derivative comes only from the entropy part of the free energy as $(\partial S/\partial V)_T = k_B N/V$. We therefore obtain for the pressure of the ideal gas

$$\boxed{P = \rho k_B T, \quad \rho = N/V.} \tag{5.43}$$

This relation is the ideal gas law, i.e., the equation of state $P = P(T, V)$ for the gas of non-interacting particles. It describes an arbitrary ensemble on non-interacting particles, e.g., an ideal solution.

We have learned from this derivation that pressure of the ideal gas comes from entropy. This is a specific example of a broad class of phenomena known as entropic forces. Such forces do not exist in mechanics and provide us with an example of the new physics emerging from statistical laws.

These illustrative calculations are designed to show that Boltzmann's definition of entropy provides a valid route to thermodynamic functions and the equation of state if one can calculate the total number of microscopic states consistent with a given macroscopic state. The problem with this formalism is that it is hard to accomplish for most practical problems. The method of statistical ensembles avoids many of the difficulties encountered in direct calculations of the entropy in terms of the number of microscopic states. The derivations of the canonical (Gibbs) and grand canonical ensembles are provided below. This is followed by the generalized canonical ensemble formulated for an arbitrary number of conjugate intensive and extensive thermodynamic variables. One still has to keep in mind that ensembles are idealizations of real systems and deviations from canonical distributions are quite possible when some of the approximations involved are not realized.

5.4   Gibbs ensemble

$$\boxed{\begin{aligned} (\partial S/\partial E)_{N,V} &= 1/T \\ (\partial S/\partial N)_{E,V} &= -\mu/T \end{aligned}}$$

The Gibbs ensemble describes the probability of a small subsystem being in contact with a large thermal bath held at temperature $T_b = T$ to gain the energy E. The probability density $p(E)$ is either a continuous or discrete function of the energy variable E. This problem is an example of the reduction of the entire manifold of a large number of degrees of freedom to a single variable E. This reduction leads to a new statistical law known as the Gibbs distribution.

The Gibbs distribution is a statistical distribution in the phase space of coordinates and momenta $\{\mathbf{x}, \mathbf{p}\}$ (Sec. 3.2) of all particles in the system evolving according to the Hamilton dynamics under the constraint that the Hamiltonian function $H(\mathbf{x}, \mathbf{p})$ is equal to the total energy E

$$H(\mathbf{x}, \mathbf{p}) = E. \tag{5.44}$$

The Gibbs distribution then gives the statistical probability density for finding the specific values $\{\mathbf{x}, \mathbf{p}\}$ in the phase space

$$p(\mathbf{x}, \mathbf{p}) \propto e^{-\beta H(\mathbf{x}, \mathbf{p})}, \tag{5.45}$$

where $\beta = (k_B T)^{-1}$. Equal statistical weights are assigned to all trajectories characterized by the energy E. While this requirement can be viewed as a separate postulate of statistical mechanics, it is also consistent with the idea that laws of statistical mechanics should be invariant in respect to the canonical transformations of the variables. The choice of equal statistical weights assigned to phase-space trajectories does not discriminate between possible choices of canonical variables propagating the classical trajectories according to the Hamilton equations of motion.

Fig. 5.3 Subsystem and the thermal bath in the canonical ensemble.

Once a subspace from the entire phase space is chosen according to its energy E, one wants to determine the probability distribution of different values of E for a subsystem in contact with the thermal bath kept at the temperature $T_b = T$ (Fig. 5.3). Even though the phase-space trajectories are assigned equal weights, different values of E will accumulate different numbers of states $\Gamma(E)$. To proceed, one needs to choose a closed (isolated) system for which the postulate of equal weights in the phase space applies. The probability of achieving the energy E by a small subsystem is obtained by counting the number of states of the entire closed system consistent with this constraint.

The total energy of the closed system $E_t = E + E_b$ is a sum of the energy E of a small subsystem and the energy E_b of a much larger thermal bath. The subsystem is much smaller than the bath in the sense that the number of microstates that the bath can attain at a given total energy E_t is much greater than the number of microstates of the subsystem. The subsystem is not closed: it exchanges energy with the bath, and its energy E is a fluctuating stochastic variable (Fig. 5.3). The goal is to determine the probability density of this stochastic variable given that the number of particles N and the volume of the subsystem V are held constant.

The probability density to find the system in the state with energy E is

$$p(E) = \frac{\Omega(E)}{\Omega_t},$$ (5.46)

where Ω_t is the fixed number of microstates of the total system and $\Omega(E)$ is the number of states of the entire closed system consistent with the energy E maintained in the subsystem. Note that equal statistical weights are assigned to all configurations of the entire system consistent with the energy E of the subsystem.

The number of states of the bath is overwhelmingly larger than the number of states of the subsystem. Therefore, one can put $\Omega(E) \simeq \Omega_b(E_t - E)$ and count only the states of the bath consistent with the constraint that the energy E is left to the subsystem

$$p(E) \propto \Omega_b(E_t - E) = e^{S_b(E_t - E)/k_B}.$$ (5.47)

Here, in the second step, the Boltzmann definition of the entropy (Eq. (5.36)) is applied. Since $E \ll E_t$, one can perform a series expansion of the entropy S_b as a function of energy

$$S_b(E_t - E) \simeq S_b(E_t) - (\partial S_b/\partial E_b)\, E = \text{Const} - E/T_b.$$ (5.48)

In this equation, the definition of the bath temperature was used

$$\frac{1}{T_b} = \left(\frac{\partial S_b}{\partial E_b}\right)_{N_b, V_b}.$$ (5.49)

The intensive variable of temperature is given as the derivative involving two extensive thermodynamic variable. The dependence on the number of particles cancels out in the ratio.

If the temperature is kept equal between the bath and the subsystem, $T_b = T$, one obtains

$$\boxed{p(E) \propto \Gamma(E)e^{-\beta E}.}$$ (5.50)

This is the Gibbs distribution of states of the subsystem characterized by a single collective variable of energy. The reduction of the full mechanical problem to the statistical law is achieved by reducing the information content about the states of the subsystem to only one variable.

The probability density needs to be normalized, and the normalization condition defines the partition function

$$Z = \sum_E \Gamma(E)e^{-\beta E}.$$ (5.51)

Introducing the density of states $\rho(E) = d\Gamma(E)/dE$, one can write the partition function as a continuous integral

$$Z = \int dE \rho(E) e^{-\beta E}. \tag{5.52}$$

The probability density of the Gibbs ensemble becomes

$$p(E) = Z^{-1} \Gamma(E) e^{-\beta E}. \tag{5.53}$$

From this probability, the average (thermodynamic) energy becomes

$$E = \langle E \rangle = \sum_E E p(E) = -\frac{\partial}{\partial \beta} \ln Z. \tag{5.54}$$

Comparing this result to the standard thermodynamic relation $E = (\partial \beta F / \partial \beta)_{N,V}$, one obtains the connection between the partition function and the Helmholtz free energy

$$\boxed{F = -\beta^{-1} \ln Z.} \tag{5.55}$$

For the ideal gas of N particles considered as a subsystem in contact with a large thermal bath, the energy of the gas is the sum of energies ϵ_j of individual particles: $E = \sum_j \epsilon_j$. Therefor the partition function becomes

$$Z_N = \frac{q^N}{N!} \simeq \left(\frac{qe}{N}\right)^N, \tag{5.56}$$

where $N!$ in the denominator accounts for permutations of the particles producing physically indistinguishable configurations and q is the partition function of a single particle in the container with the volume V (Stirling's formula is used in the last term).

To define $\rho(E)$ for a single particle of the ideal gas, one has to recognize that its mechanical state is defined by the position $\mathbf{r}$ and momentum $\mathbf{p}$. According to the uncertainty principle of quantum mechanics, there cannot be more than one state in the element of phase space with $\delta x \delta p_x \simeq h$. Therefore, in 3D space, $d\Gamma(\mathbf{r}, \mathbf{p}) = d\mathbf{r} d\mathbf{p}/h^3$. The kinetic energy of a free particle is $\epsilon = p^2/(2m)$ and $d\epsilon = p dp/m$. One gets

$$\rho(\epsilon) = \frac{V}{h^3} \frac{4\pi p^2 dp}{(p/m) dp} = \frac{4\pi V}{h^3} (2m^3 \epsilon)^{1/2}, \tag{5.57}$$

where the integral over the space occupied by the particle has been taken, $\int d\mathbf{r} = V$.

The partition function of a single particle of the ideal gas becomes

$$q = \frac{4\pi V}{h^3} (2m^3)^{1/2} \int_0^\infty d\epsilon \sqrt{\epsilon} e^{-\beta \epsilon}. \tag{5.58}$$

By using the integral

$$\int_0^\infty d\epsilon \sqrt{\epsilon} e^{-\beta \epsilon} = \frac{\sqrt{\pi}}{2\beta^{3/2}},$$ (5.59)

one obtains

$$q = V h^{-3} (2\pi m k_B T)^{3/2} = \frac{V}{\lambda^3},$$ (5.60)

where λ is the de Broglie thermal wavelength in Eq. (5.38). One obtains for the free energy of the ideal gas

$$F = -k_B T N \ln\left(\frac{Ve}{N\lambda^3}\right).$$ (5.61)

The temperature derivative of this expression gives the entropy

$$S = k_B N \ln\left(\frac{V}{N\lambda^3}\right) + \tfrac{5}{2} k_B N.$$ (5.62)

The missing constant term in Eq. (5.37) is equal to $(3/2)k_B N$.

5.5   Grand canonical ensemble

$$d(PV) = d(\mu N - F) = SdT + Nd\mu,$$
$$\langle N \rangle = (\partial \beta PV / \partial(\beta\mu))_T$$

Another way to form a subsystem is to allow the exchange of both the energy and particles with the bath (Fig. 5.4). The probability of the state with the energy E and the number of particles N is

$$p(E, N) = \frac{\Omega(E, N)}{\Omega_t} \propto e^{S_b(E_t - E, N_t - N)/k_B}.$$ (5.63)

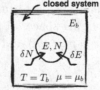

Fig. 5.4   Subsystem and the thermal bath in the grand canonical ensemble.

The entropy of the bath depends on the energy $E_t - E$ and the number of particles $N_t - N$. Since the bath is much larger than the system, $E \ll E_t$ and $N \ll N_t$, the Taylor expansion is applied

$$S_b(E_t - E, N_t - N) \simeq S_b(E_t, N_t) + \partial S_b/\partial E_b(-E) + \partial S_b/\partial N_b(-N). \quad (5.64)$$

With the account for the entropy derivatives over the energy and the number of particles, one gets

$$p(E, N) = \Xi^{-1}\Gamma(E)e^{-\beta(E-\mu N)}. \tag{5.65}$$

The normalization constant Ξ is the grand canonical partition function

$$\Xi = \sum_{E,N} \Gamma(E)e^{-\beta(E-\mu N)} = \sum_N \zeta^N Z_N, \tag{5.66}$$

where $\zeta = \exp[\beta\mu]$ is the fugacity and Z_N is the canonical partition function.

From this definition of the probability, the thermodynamic average number of particles in the system is

$$\langle N \rangle = \sum_{E,N} N p(E, N) = \left(\frac{\partial \ln \Xi}{\partial(\beta\mu)}\right)_T. \tag{5.67}$$

One can apply the thermodynamic relations to obtain

$$d(F - \mu N) = -SdT - Nd\mu = d(-PV) \tag{5.68}$$

From this equation, the thermodynamic number of particles in the subsystem is

$$N = \langle N \rangle = (\partial \beta PV/\partial(\beta\mu))_T. \tag{5.69}$$

One, therefore, obtains the connection between the grand partition function and the thermodynamic parameters

$$\boxed{\beta PV = \ln \Xi.} \tag{5.70}$$

An example of applying the rules of grand canonical ensemble is the calculation for the ideal gas.

$$\Xi = \sum_{N=0}^{\infty} \left(\frac{\zeta V}{\lambda^3}\right)^N \frac{1}{N!} = \exp\left[\zeta V/\lambda^3\right]. \tag{5.71}$$

Therefore, one obtains

$$\beta PV = \ln \Xi = e^{\beta\mu} \frac{V}{\lambda^3} \tag{5.72}$$

and

$$\boxed{\beta\mu = \ln[\rho\lambda^3], \quad \rho = N/V.} \tag{5.73}$$

Further, by taking the derivative over $\beta\mu$ in Eqs. (5.69) and (5.72), one obtains

$$\langle N \rangle = \ln \Xi = \beta PV, \tag{5.74}$$

which is the equation of state for the ideal gas. The free energy of the ideal gas follows from Eqs. (5.73) and (5.74), $\beta F = \beta\mu N - \beta PV = \beta\mu N - N$

$$\beta F = N \ln[\rho\lambda^3] - N. \tag{5.75}$$

The energy of the ideal gas is $\beta E = (3/2)N$ and the entropy $S/k_B = \beta E - \beta F$ becomes

$$S/(Nk_B) = 5/2 - \ln[\rho\lambda^3]. \tag{5.76}$$

This is just another form of Eq. (5.62).

5.6   Chemical reactions

$$\mu = (\partial F/\partial N)_{T,V} = -T\,(\partial S/\partial N)_{E,V}$$
$$P = -\,(\partial F/\partial V)_{T,N}$$
$$S = -\,(\partial F/\partial T)_{V,N}$$

We consider a simple chemical reaction, such as isomerization, in which a molecule changes its state from the energy level ϵ_1 to the energy level ϵ_2. Before the transition, there are N_1 particles in state 1 and N_2 particles in state 2. A single transition will change the energy of the reacting system by $\Delta\epsilon = \epsilon_2 - \epsilon_1$, which is assumed to be positive (Fig. 5.5). The question addressed here is how such a transition can occur spontaneously, by bringing the reacting system in contact with the thermal bath with the temperature T.

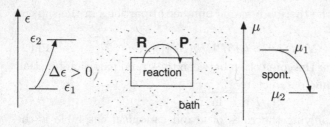

Fig. 5.5   Chemical reaction from reactants (R) to products (P). The system and the bath are isolated ($E_t = $ Const).

The conceptual difficulty of this example is that the energy of the reacting system increases by $\Delta\epsilon$ in this transition. If spontaneous, it should be accompanied by the transfer of heat from the bath to compensate for the energy increase. The question is what are the thermodynamic conditions that would allow a flow of entropy from the bath to the system.

One can apply the first law of thermodynamics (Eq. (5.16)) to calculate the change of the entropy of the reacting system and the bath caused by one isomerization event keeping the total energy and volume constant

$$T\Delta S_t = -\mu_1\Delta N_1 - \mu_2\Delta N_2. \tag{5.77}$$

Since $\Delta N_1 = -1$ and $\Delta N_2 = +1$, one gets

$$\Delta S_t = \mu_1/T - \mu_2/T > 0. \tag{5.78}$$

The spontaneous transition thus requires the reaction to proceed downhill in the chemical potential, from a higher μ_1 to a lower μ_2 (Fig. 5.5). Since one requires $\mu_2 - \mu_1 = \Delta\epsilon - T\Delta s < 0$, a positive $\Delta\epsilon$ must be compensated by a positive change in the entropy per particle Δs.

Where does this positive entropy change come from? Practice suggests that creating a sufficiently large concentration of the reactants in excess to the products tends to drive the reaction in the forward direction. We therefore need to understand how the chemical potential is affected by concentration.

Equation (5.73) suggests that the chemical potential of an ideal gas depends on the logarithm of concentration. One usually determines this dependence by specifying the chemical potential at some reference concentration c_0 (typically 1 M) and leaving the chemical potential μ^0 at that concentration to be measured experimentally. From Eq. (5.73), one arrives at the standard form of the chemical potential

$$\mu_i = \mu_i^0 + k_B T \ln[c_i/c_0], \quad i = 1, 2. \tag{5.79}$$

This equation tells us that chemical intuition is correct and our toy problem of the isomerization reaction can be assigned the chemical potential difference

$$\Delta\mu = \Delta\epsilon + k_B T \ln[c_2/c_1]. \tag{5.80}$$

If $c_1 \gg c_2$, the second term is sufficiently negative ($\Delta s > 0$) to produce $\Delta\mu < 0$ required for a spontaneous transition. The excess of the reactant concentration thus provides sufficient entropic drive to push the reaction into a higher energy state.

5.7 *Virial equation

The connection between the canonical partition function and the system pressure is given by the equation

$$\beta P = \frac{\partial}{\partial V} \ln Z_N. \tag{5.81}$$

We assume that our sample is a cube with the size L and represent $Z_N = Q_N \times Z_N^{\mathrm{id}}$ as a product of the ideal partition function $Z_N^{\mathrm{id}} = (L/\lambda)^{3N}/N!$ and the configuration integral Q_N given in terms of the N-particle interaction energy $V_N(\mathbf{r}_1, \ldots, \mathbf{r}_N)$

$$Q_N = \frac{1}{V^N} \int d\mathbf{r}_1 \ldots d\mathbf{r}_N e^{-\beta V_N}, \tag{5.82}$$

where $V = L^3$ is the volume of the sample. The Cartesian variables in this equation can be scaled with the box size L to introduce the dimensionless variable $x_\alpha = r_\alpha/L$ for all Cartesian components α

$$Q_N = \int d\mathbf{x}_1 \dots d\mathbf{x}_N e^{-\beta V_N(L\mathbf{x}_1,\dots,L\mathbf{x}_N)}. \tag{5.83}$$

This form of the configuration integral can be used to calculate the pressure

$$\beta P = \rho + \frac{L}{3V} \frac{\partial}{\partial L} \ln Q_N, \tag{5.84}$$

where the first summand is the ideal-gas pressure. From Eq. (5.83),

$$\frac{\partial}{\partial L} Q_N = -\frac{\beta}{L} \int d\mathbf{x}_1 \dots d\mathbf{x}_N \sum_{\alpha=1}^{3N} \frac{\partial V_N}{\partial r_\alpha} r_\alpha e^{-\beta V_N}. \tag{5.85}$$

The equation for the pressure reduces to

$$\beta P = \rho - \frac{\beta}{3V} \sum_\alpha \left\langle r_\alpha \frac{\partial V_N}{\partial r_\alpha} \right\rangle. \tag{5.86}$$

This result is known as the virial equation. It expresses the pressure as the sum of the collisional ideal-gas pressure and the average virial $\langle \mathbf{r} \cdot \mathbf{f} \rangle$

$$\beta P = \rho + \frac{\beta}{3V} \sum_{i=1}^{N} \langle \mathbf{r}_i \cdot \mathbf{f}_i \rangle, \tag{5.87}$$

where $\mathbf{f}_i = -\partial V_N/\partial \mathbf{r}_i$.

The second term in Eq. (5.87) defines the virial expansion, that is the expansion of the virial in powers of the liquid density

$$\boxed{\frac{\beta P}{\rho} = 1 + \frac{\beta}{3}\langle \mathbf{r} \cdot \mathbf{f} \rangle,} \tag{5.88}$$

where

$$\frac{\beta}{3}\langle \mathbf{r} \cdot \mathbf{f} \rangle = B(T)\rho + C(T)\rho^2 + \dots. \tag{5.89}$$

The most important term here is $B(T)$ known as the second virial coefficient. The temperature T_B at which $B(T_B) = 0$ is called the Boyle temperature because the gas at T_B obeys Boyle's ideal-gas law to the cubic order in the gas density (see Sec. 6.3 for a detailed discussion).

One can gain some intuitive insight into the molecular origin of the macroscopic pressure by considering the interaction between two particles at $\mathbf{r}_1$ and $\mathbf{r}_2$ interacting with the energy $U(r_{12})$ (Fig. 5.6). The force acting on particle 1 is $\mathbf{f}_1 = -(\partial U/\partial r_{12})(\partial r_{12}/\partial \mathbf{r}_1) = -\mathbf{f}$, where $\mathbf{r}_{12} = \mathbf{r}_2 - \mathbf{r}_1$. The

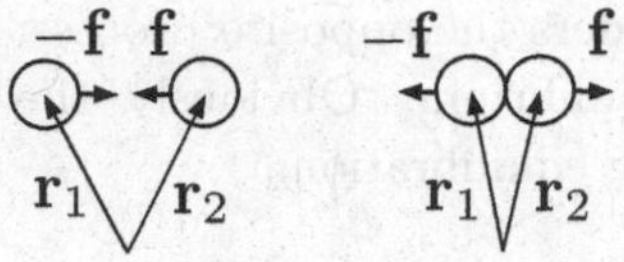

Fig. 5.6 Attraction (left) and repulsion (right) forces acting between two particles.

force acting on particle 2 is $\mathbf{f}_2 = -(\partial U/\partial \mathbf{r}_{12})(\partial \mathbf{r}_{12}/\partial \mathbf{r}_2) = \mathbf{f}$. One obtains for the virial

$$\sum_{i=1}^{2} \langle \mathbf{r}_i \cdot \mathbf{f}_i \rangle = \langle \mathbf{r}_{12} \cdot \mathbf{f} \rangle. \tag{5.90}$$

It is clear from the figure that attractions at larger distances will contribute negative values to the virial, while repulsions at shorter distances will contribute positive values. Correspondingly, attractions are responsible for a negative pressure and repulsions produce a positive pressure. The magnitudes of repulsive forces are greater than attractive forces since repulsion potentials are steeper and the overall pressure is positive. The separation between attractive and repulsive components of the total pressure is the basis of the van der Waals equation of state discussed in Sec. 6.4.

5.8 Generalized canonical distribution

The first law of thermodynamics (differential form in Eq. (5.16)) shows that the entropy is a function of three extensive variables $X = \{X_1, X_2, X_3\} = \{E, V, N\}$. These variables are split between the subsystem and the bath such that total value $X_t = X + X_b$ of the entire closed system is fixed and $dX_i = -dX_{bi}$ (Fig. 5.7). The change of the total entropy is a sum of changes of the entropy of the subsystem and the bath

$$\Delta S_t = \Delta S(X) + \Delta S_b(X_b). \tag{5.91}$$

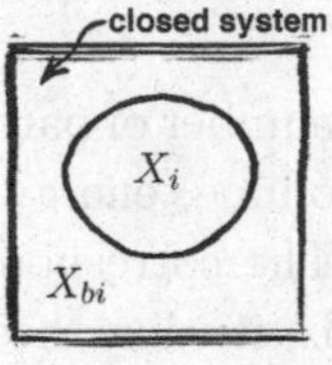

Fig. 5.7 A subsystem separated from a closed (isolated) system.

The total entropy is at the maximum at equilibrium (Fig. 5.8). This statement is the second law of thermodynamics, which states that any deviation from equilibrium lowers the total entropy of a closed system.

The more common formulation of the law considers the opposite process: from a nonequilibrium state to the state of equilibrium. Obviously, the entropy of a closed system has to increase during equilibration.

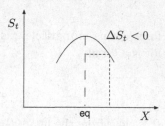

Fig. 5.8   Entropy maximum at equilibrium.

Given that the total entropy is at maximum at equilibrium, all first order expansion terms disappear from ΔS_t when variables X deviate from their equilibrium values X_e. This condition can be formalized as

$$y_{ei} = \left.\frac{\partial S}{\partial X_i}\right|_{X_e} = -\left.\frac{\partial S_b}{\partial X_i}\right|_{X_e} = \left.\frac{\partial S_b}{\partial X_{bi}}\right|_{X_e}. \tag{5.92}$$

One finds that the intensive variables $y_i = (\partial S/\partial X_i)$ are equal between the subsystem and the bath at equilibrium. This set of y_i is equal to $\{1/T, P/T, -\mu/T\}$.

One can now generalize the derivation of ensemble probabilities described above by noting that the change in the entropy of the bath is much greater than that of the subsystem and one can write

$$p \propto e^{\Delta S_t/k_B} \propto e^{\Delta S_b/k_B}. \tag{5.93}$$

By expanding the entropy change of the bath as

$$\Delta S_b(X_t - X) = -\sum_i \frac{\partial S_b}{\partial X_{bi}} X_i \tag{5.94}$$

one arrives at the generalized canonical distribution

$$\boxed{p \propto e^{-k_B^{-1} \sum_i y_{ei} X_i}.} \tag{5.95}$$

The canonical and grand canonical ensembles considered above are just special cases of this equation.

To define a subsystem one needs to specify either the number of particles $N = \text{Const}$ or the volume $V = \text{Const}$. At constant volume, energy and the number of particles become fluctuating variables. The corresponding intensive variables are $y_{e1} = 1/T$ and $y_{e3} = -\mu/T$. Substituting those to Eq. (5.93), one arrives at the grand canonical ensemble (Eq. (5.65))

$$p \propto e^{\beta(\mu N - E)}. \tag{5.96}$$

One similarly can derive the canonical constant-pressure ensemble by allowing energy and volume fluctuate while keeping $N = \text{Const}$.

5.9   Fluctuations of thermodynamic variables

Isothermal compressibility:
$\beta_T = -V^{-1}(\partial V/\partial P)_{T,N}$
Isobaric expansivity:
$\alpha_P = V^{-1}(\partial V/\partial T)_{P,N}$

Here we address the problem of fluctuations of thermodynamic variables driven by thermal agitation in a small subsystem in contact with the thermal bath. Combined, they form a closed system. Any fluctuation in the subsystem is reflected by the alteration of the total entropy ΔS_t of the closed system. The probability of this fluctuation is given by Eq. (5.93).

We define the subsystem by requiring $N = $ Const and ask what is the probability of a given fluctuation? Since the volume defines the number of available states, it is clear that fluctuations of the volume should also produce fluctuations of the entropy. Therefore, any finite system of N particles must show fluctuations of both the volume and the entropy (Fig. 5.9).

Fig. 5.9   Volume and entropy fluctuations of the subsystem.

We will now keep both the entropy change of the subsystem and that of the bath and use the equilibrium condition from Eq. (5.92) to write the total entropy change as

$$\Delta S_t = \Delta S + \frac{1}{T}\left(-\Delta E - P\Delta V + \mu\Delta N\right), \qquad (5.97)$$

where all three intensive variables y_{ei} are used to express the change of the entropy of the bath. According to the first law of thermodynamics, $\Delta E = T\Delta S - P\Delta V + \mu\Delta N$ and $\Delta S_t = 0$ in the first-order expansion, in agreement with the condition of equilibrium. To find the probability of a fluctuation, one needs a second-order expansion in thermodynamic variables. To obtain fluctuations of the entropy and volume, we perform a series expansion of the energy function $E = E(S, V, N)$, which, similarly to the entropy, is defined by three extensive variables.

The second-order expansion term for the energy function is

$$\Delta E^{(2)} = \tfrac{1}{2} \sum_{i,j} \frac{\partial^2 E}{\partial X_i \partial X_j} \Delta X_i \Delta X_j, \tag{5.98}$$

where now one has $X = \{S, V, N\}$ and the set of intensive parameters $y_{ei} = (\partial E/\partial X_i)_e$ is equal to $\{T, -P, \mu\}$ (differential form of the first law of thermodynamics, Eq. (5.16)). One can also observe that the change in the intensive parameter is expressed by the identity

$$\Delta y_i = \sum_j \frac{\partial^2 E}{\partial X_i \partial X_j} \Delta X_j \tag{5.99}$$

and one can alternatively write

$$\Delta E^{(2)} = \tfrac{1}{2} \sum_i \Delta y_i \Delta X_i. \tag{5.100}$$

For the thermodynamic parameters in the set X, one obtains

$$\Delta E^{(2)} = \tfrac{1}{2} \left(\Delta T \Delta S - \Delta P \Delta V + \Delta \mu \Delta N \right). \tag{5.101}$$

To proceed, one needs to define the subsystem, which requires fixing either N or V. If the subsystem is defined by the number of particles, $N = \text{Const}$, one gets ΔS_t by substituting the expansion of $E(S, V)$ into Eq. (5.97)

$$\boxed{p \propto \exp \left[\frac{\beta}{2} (\Delta P \Delta V - \Delta S \Delta T) \right].} \tag{5.102}$$

Because the probability should be small, each product in the exponent needs to be negative. This also follows from the concave shape of the entropy function (Fig. 5.8). It is easy to convince oneself that this is true: for instance, an increase in the volume, $\Delta V > 0$, should lead to a decrease in the pressure, $\Delta P < 0$.

We have four variables, S, T, P, V, of which only two are independent. By taking V and T, one can expand the entropy and pressure

$$\Delta S = \frac{C_V}{T} \Delta T + (\partial P/\partial T)_V \, \Delta V,$$

$$\Delta P = -\frac{1}{\beta_T V} \Delta V + (\partial P/\partial T)_V \, \Delta T, \tag{5.103}$$

where the Maxwell relation $(\partial S/\partial V)_T = (\partial P/\partial T)_V$ was used in the first equation. Substituting this result into the probability equation, one arrives at the Gaussian distribution of the fluctuations of ΔV and ΔT

$$p \propto \exp \left[-\frac{\beta C_V}{2T} \Delta T^2 - \frac{\beta}{2 \beta_T V} \Delta V^2 \right]. \tag{5.104}$$

The variances of two variables are given by

$$\langle \Delta T^2 \rangle = k_B T^2 / C_V,$$
$$\langle \Delta V^2 \rangle = k_B T V \beta_T, \quad (5.105)$$
$$\langle \Delta V \Delta T \rangle = 0.$$

The last equation tells us that fluctuations of temperature and volume are uncorrelated.

For the ideal gas, $\beta_T = P^{-1}$ and one obtains $\langle \Delta V^2 \rangle / V = \rho^{-1}$. One can further assume that the density is constant to transform ΔV to ΔN as

$$\Delta V = \rho^{-1} \Delta N. \quad (5.106)$$

The variance of the number of particles in a given volume then becomes

$$\langle \Delta N^2 \rangle = N. \quad (5.107)$$

This is the statistics of a Poisson process describing multiple independent events occurring with a constant average frequency. The fluctuation of the number of particles of an ideal gas in a given volume is given by a Poisson process of independent arrivals of noninteracting particles.

The general rule following from this derivation is that the variance of an extensive variable is proportional to N, while the variance of an intensive variable is proportional to N^{-1}. Therefore, properties such as temperature are well defined in the thermodynamic limit of $N \to \infty$ since $\langle \Delta T^2 \rangle \to 0$. On the other hand, reduced variances, such as $\langle \Delta V^2 \rangle / V$, are finite and depend on intensive variables only.

The last result reconciles the statistical and thermodynamic definitions of intensive variables, such as temperature. In thermodynamics, temperature is defined by establishing equilibrium with an infinitely large thermal bath, and it never fluctuates. On the contrary, the microcanonical definition of temperature, $T^{-1} = (\partial S / \partial E)_{N,V}$ (Eq. (5.49)), allows fluctuations of temperature. Nevertheless, these fluctuations disappear in the thermodynamic limit $N \to \infty$ when the standard thermodynamic definition is attained.

One can next adopt P and S as two fluctuating thermodynamic variables in Eq. (5.102). This choice leads to the fluctuation relation for the heat capacity at constant pressure in terms of the variance of the system entropy

$$k_B C_P = \langle \Delta S^2 \rangle. \quad (5.108)$$

One can further use Eq. (5.103) to obtain the cross-correlation between ΔV and ΔS. Since $\langle \Delta V \Delta T \rangle = 0$, one has

$$\langle \Delta S \Delta V \rangle = (\partial P / \partial T)_V \langle \Delta V^2 \rangle = k_B T V \beta_T (\partial P / \partial T)_V. \quad (5.109)$$

This equation can be further simplified by applying the following result [34]

$$\left(\frac{\partial P}{\partial T}\right)_V = -\frac{\partial(P,V)}{\partial(P,T)}\frac{\partial(P,T)}{\partial(V,T)} = \frac{\alpha_p}{\beta_T}. \tag{5.110}$$

The cross-correlation between fluctuations of the volume and entropy becomes

$$\langle \Delta S \Delta V \rangle = k_B T V \alpha_p. \tag{5.111}$$

As an example of a cross-correlation between extensive and intensive variables, one calculates from Eq. (5.103)

$$\langle \Delta S \Delta T \rangle = k_B T. \tag{5.112}$$

One generally finds that cross-correlations between extensive and intensive variables do not carry a dependence on the number of particles in the subsystem.

5.10 Heat capacity

Quantization of the harmonic oscillator:
$E_n = \hbar\omega(n + 1/2)$
Constant-volume heat capacity:
$C_V = (\partial E/\partial T)_V$

Heat capacity at constant pressure C_P or at constant volume C_V are commonly tabulated thermodynamic quantities. Being a higher-order thermodynamic derivative, heat capacity often brings deep insights into the physics of thermal excitations in a macroscopic material. The Einstein model of heat capacity for an ensemble of quantum harmonic oscillators is a good starting point.

The energy of a quantum harmonic oscillator E_n can be used to define the Gibbs partition function (see the Box)

$$Z = \sum_{n=0}^{\infty} e^{-\hbar\omega(n+1/2)} = e^{-\hbar\omega/2}\left(1 - e^{-\hbar\omega}\right)^{-1}$$
$$= [2\sinh\beta\hbar\omega/2]^{-1}. \tag{5.113}$$

The average energy of the harmonic oscillator is

$$E = -\frac{\partial \ln Z}{\partial \beta} = \frac{\hbar\omega}{2}\coth\frac{\beta\hbar\omega}{2}. \tag{5.114}$$

In the classical limit $\beta\hbar\omega \ll 1$, $\coth \beta\hbar\omega/2 \approx 2/(\beta\hbar\omega)$ and one gets the standard result of the equipartition theorem applied to the harmonic oscillator, $E = k_B T$. In the opposite quantum limit, $\beta\hbar\omega \gg 1$, the harmonic oscillator is in its ground state (zero point energy) with $E = \hbar\omega/2$.

The high-temperature limit for the entropy is also revealing. One anticipates that many vibrational states are occupied when $\beta\hbar\omega \ll 1$. The number of states occupied is T/T_v, where $T_v = \hbar\omega/k_B$ is the characteristic vibrational temperature. According to the Boltzmann definition of entropy (Eq. (5.36)), one can estimate the high-temperature entropy

$$S = k_B \ln [T/T_v]. \tag{5.115}$$

The logarithmic scaling of the entropy with temperature is a signature of the increasing number of states involved.

Heat capacity follows by differentiating $E(\beta)$ over β

$$C_V/k_B = -\beta^2 \frac{\partial E}{\partial \beta} = \left(\frac{T_v}{2T}\right)^2 [\sinh(T_v/(2T))]^{-2}. \tag{5.116}$$

This is a rising function of temperature, leveling off to the classical limit $C_V = k_B$ at high temperatures when $T_v/T \ll 1$ (Fig. 5.10).

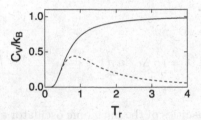

Fig. 5.10 Heat capacity of the harmonic oscillator (solid line) and of the two-state system (dashed line) vs the reduced temperature T_r: $T_r = 2k_B T/(\hbar\omega)$ for the harmonic oscillator and $T_r = 2k_B T/\Delta\epsilon$ for the two-state system.

Another classical model allowing a closed-form expression for the heat capacity is the two-state system. If the ground and excited states of the system are separated by the energy difference $\Delta\epsilon$ (Fig. 5.5), one can put the energy of the ground state equal to zero. As a matter of practice, the system Hamiltonian is written in Dirac notations

$$\hat{H} = \Delta\epsilon|2\rangle\langle2|, \tag{5.117}$$

where $|2\rangle$ is the ket for the "excited" state separated from the "ground" state $|1\rangle$ by the energy gap $\Delta\epsilon$ (note that $\Delta\epsilon$ can be negative). The partition function of the system is then

$$Z = \sum_{n=0,1} \langle n|e^{-\beta\hat{H}}|n\rangle = 1 + e^{-\beta\Delta\epsilon}. \tag{5.118}$$

Here, $\exp[-\beta\hat{H}]$ is understood as a corresponding series of operators

$$e^{-\beta\hat{H}} = 1 - \beta\hat{H} + \tfrac{1}{2}\beta^2\hat{H}^2 + \dots . \tag{5.119}$$

We define the energy from the partition function by taking the derivative over β

$$E = -\frac{\partial \ln Z}{\partial \beta} = \Delta\epsilon n_2, \tag{5.120}$$

where

$$n_2 = \left[e^{\beta\Delta\epsilon} + 1\right]^{-1} \tag{5.121}$$

is the population of the excited state.

The reason why the average energy takes this simple form is our setting of the ground-state energy equal to zero. In a more general definition when the energy of the ground state is set equal to ϵ_1 and the energy of the excited state is ϵ_2, one obtains the average energy as the sum of two states weighted with their corresponding populations

$$E = \epsilon_1(1 - n_2) + \epsilon_2 n_2. \tag{5.122}$$

In either definition, the heat capacity is obtained by taking another derivative over β

$$\begin{aligned}
C_V/k_{\mathrm{B}} &= -\beta^2 \frac{\partial E}{\partial \beta} \\
&= \left(\frac{\beta\Delta\epsilon}{2}\right)^2 [\cosh\beta\Delta\epsilon/2]^{-2} = (\beta\Delta\epsilon)^2 n_2(1 - n_2).
\end{aligned} \tag{5.123}$$

Figure 5.10 illustrates the heat capacities of the harmonic oscillator and of the two-state system vs the reduced temperature T_r ($T_r = 2k_BT/(\hbar\omega) = 2T/T_v$ for the harmonic oscillator and $T_r = 2k_BT/\Delta\epsilon$ for the two-state system). The heat capacity of the two-state system goes through a maximum as a function of temperature, and then decays to zero at high temperatures. The high-temperature drop is explained by the saturation of the number of states available to the two-state system at high temperatures. Since the probabilities of occupying both states are equal at high temperatures, the high-temperature entropy tends to a constant value

$$S = k_{\mathrm{B}} \ln 2, \tag{5.124}$$

thus producing zero heat capacity $C_V = T(\partial S/\partial T)_V$. This is in contrast to the entropy of the harmonic oscillator, which keeps increasing logarithmically (Eq. (5.115)). The result is a non-zero limit $C_V \to k_{\mathrm{B}}$ at $T \to \infty$.

The maximum of the heat capacity for the two-state system is reached at about $T_r \simeq 0.83$. For the physical understanding of this result one has to turn to the fluctuation formula for the heat capacity: the maximum of C_V corresponds to the maximum of energy fluctuations.

This result follows directly from the statistical definition of the thermodynamic energy as the average Hamiltonian of the system

$$E = \langle H \rangle = Z^{-1} \int H e^{-\beta H} d\Gamma. \tag{5.125}$$

Taking the derivative over β in this equation, we get

$$C_V / k_B = -\beta^2 \frac{\partial E}{\partial \beta} = \beta^2 \langle (\delta H)^2 \rangle, \tag{5.126}$$

where $\delta H = H - \langle H \rangle$. The maximum of the heat capacity of the two-state system is therefore a trade-off between the quadratic $(\beta \Delta \epsilon)^2$ and $n_2(1-n_2)$ exponentially decaying with increasing $\beta \Delta \epsilon$. The heat capacity approximately following the two-state behavior is often observed for unfolding of proteins, from which one concludes that many proteins behave thermodynamically as two-state systems, continuously changing their populations between folded and unfolded states with increasing temperature.

The energy of the system is the average of the system Hamiltonian. Can one write the Helmholtz free energy $F = \langle f \rangle$ as an average of some fluctuating variable f? This definition is allowed if f is defined as

$$f = H + \beta^{-1} \ln p, \quad p = Z^{-1} e^{-\beta H}. \tag{5.127}$$

Here, p is the probability density of a given configuration of the system. Equation $F = \langle f \rangle$ leads to Shannon's definition of the entropy

$$\boxed{S = -k_B \langle \ln p \rangle.} \tag{5.128}$$

From this equation, $\delta f = 0$ and $\langle \delta f \delta H \rangle = 0$. One therefore obtains

$$k_B^{-1} \langle \delta H \delta S \rangle = \beta \langle \delta H^2 \rangle, \tag{5.129}$$

which leads to yet another definition of the heat capacity as the cross-correlation between the fluctuations of energy (Hamiltonian) and entropy

$$C_V = \beta \langle \delta H \delta S \rangle. \tag{5.130}$$

5.11 *Legendre transformation

The generalized canonical distribution derived in Eq. (5.93) gives the probability density for a given realization of macroscopic variables X_i in terms of intensive variables $y_i = \partial S / \partial X_i$

$$p = e^{\Theta - k_{\mathrm{B}}^{-1} \sum_i y_{ei} X_i}. \tag{5.131}$$

Here, y_{ei} is the intensive parameter taken at equilibrium and Θ is the generalized thermodynamic potential required to normalize the probability density

$$\mathrm{Tr}[p] = 1, \tag{5.132}$$

where Tr implies integration over the entire phase space of the canonical ensemble. This normalization condition gives for Θ

$$\Theta = -\ln \left(\mathrm{Tr} \left[e^{-k_{\mathrm{B}}^{-1} \sum_i y_{ei} X_i} \right] \right). \tag{5.133}$$

The thermodynamic potential $\Theta = \Theta(y_{ei})$ is a function of intensive variables y_{ei} with the derivatives defined in terms of averages of the extensive variables

$$\langle X_i \rangle = k_{\mathrm{B}} \partial \Theta / \partial y_{ei}. \tag{5.134}$$

One can also produce the entropy of the system by using Shanon's definition of the entropy (Eq. (5.128))

$$S = -k_{\mathrm{B}} \langle \ln p \rangle = -k_{\mathrm{B}} \Theta + \sum_i y_{ei} \langle X_i \rangle. \tag{5.135}$$

The entropy $S = S(\langle X_i \rangle)$ is a function of average values of extensive variables X_i. Taking the derivative of the entropy yields the result consistent with the definition of the intensive variables y_{ei}

$$dS = \sum_i y_{ei} d\langle X_i \rangle. \tag{5.136}$$

The thermodynamic potential as a function of intensive variables and the entropy as a function of extensive variables are related by the Legendre transformation which follows directly from Eq. (5.135) by rearrangement

$$\boxed{\Theta(y_{ei}) + k_{\mathrm{B}}^{-1} S(\langle X_i \rangle) = k_{\mathrm{B}}^{-1} \sum_i y_{ei} \langle X_i \rangle.} \tag{5.137}$$

One can consider special cases. If $y_{e1} = 1/T$ and $\langle X_1 \rangle = E$, one associates the thermodynamic potential with the Helmholtz free energy

$$\Theta = \beta E - k_{\mathrm{B}}^{-1} S = \beta F. \tag{5.138}$$

If the number of variables is extended by adding $y_{e3} = -\mu/T$ and $\langle X_3 \rangle = N$, one obtains

$$\Theta = \beta(E - G) - k_B^{-1} S = -\beta PV, \tag{5.139}$$

which is the grand canonical thermodynamic potential. The use of $y_{e1} = 1/T$, $y_{e2} = P/T$ and $\langle X_1 \rangle = E$, $\langle X_2 \rangle = V$ yields the $\Theta = \beta G$, where G is the Gibbs free energy. The general rules applies here: increasing the number of extensive parameters X_i in the entropy function leads to a longer list of intensive variables y_{ei} specifying the generalized thermodynamic potential.

The Legendre transformation is used to derive thermodynamic potentials by extending the list of pairs of intensive variables and extensive parameters. As an example, one can consider the transformation from the Helmholtz free energy to the Gibbs free energy in terms of the elastic stress $\sigma_{\alpha\beta}$ applied to the body (Sec. 9.1). The application of the stress results in the variation of the strain tensor $\epsilon_{\alpha\beta}$ such that the change in the Helmholtz free energy is (N is kept constant)

$$\delta F = -S\delta T + \int d\mathbf{r}\, \sigma_{\alpha\beta} \delta\epsilon_{\alpha\beta}. \tag{5.140}$$

The Gibbs energy is obtained from the Helmholtz free energy by the Legendre transformation

$$G = F - \int d\mathbf{r}\, \sigma_{\alpha\beta} \epsilon_{\alpha\beta}. \tag{5.141}$$

One obtains for the variation

$$\delta G = -S\delta T - \int d\mathbf{r}\, \epsilon_{\alpha\beta} \delta\sigma_{\alpha\beta}. \tag{5.142}$$

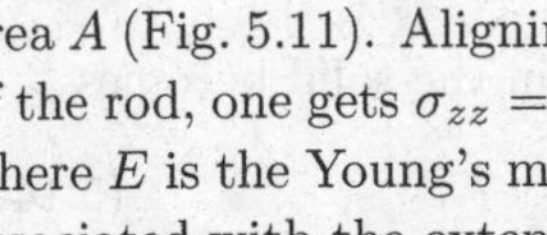

Fig. 5.11 Stretching of a rod by the force $f = pA$ applied to two ends.

As a specific example one can consider stretching of a rod (Sec. 9.3) at constant temperature by the force $f = pA$ applied to two ends with the area A (Fig. 5.11). Aligning the z-axis of the laboratory frame with the axis of the rod, one gets $\sigma_{zz} = p$ and, according to the Hooke's law, $\sigma_{zz} = E\epsilon_{zz}$, where E is the Young's modulus. One finds the change of the Gibbs energy associated with the extension ΔL of the rod ($\int d\mathbf{r} = AL$)

$$\Delta G = -\tfrac{1}{2}ALE\epsilon_{zz}^2 = -\tfrac{1}{2}f\Delta L, \tag{5.143}$$

where $\epsilon_{zz} = \Delta L/L$. The negative sign of ΔG tells us that under the constant external force f the rod will spontaneously stretch by $\Delta L = pL/E$ (E carries the units of pressure).

5.12   Debye model

The Debye model of heat capacity of an atomic crystal sums up the heat capacities of acoustic phonons $C_V(\omega)$ as given by Eq. (5.116). One needs to account for the density of states of the phonons. It is the same as the density of states of photons (light quanta) confined in the box with the volume $V = L^3$, except that there are 3 polarizations of acoustic phonons (one longitudinal and two transverse), in contrast to 2 polarizations of electromagnetic photons. One can therefore adopt Eq. (4.3) multiplied with $3/2$

$$N(\lambda) = 4\pi \frac{L^3}{\lambda^3} = \frac{V\omega^3}{2\pi^2 c_s^3}, \tag{5.144}$$

where c_s is the speed of sound averaged between transverse, c_t, and longitudinal, c_ℓ, polarizations: $3c_s^{-3} = c_\ell^{-3} + 2c_t^{-3}$. The density of states $g(\omega) = dN(\omega)/d\omega$ thus becomes

$$g(\omega) = \frac{3V}{2\pi^2 c_s^3}\omega^2 \propto \omega^2, \tag{5.145}$$

which is the Rayleigh-Jeans theory. The density of states should integrate to the maximum frequency, which is defined by the number of oscillators in the crystal $3N - 6 \simeq 3N$

$$\int_0^{\omega_m} d\omega g(\omega) = 3N. \tag{5.146}$$

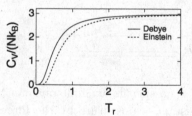

Fig. 5.12 $3C_V/(Nk_{\mathrm{B}})$ for the harmonic oscillator (dashed line, Einstein) compared to the Debye model (solid line) vs the reduced temperature T_r: $T_r = 2k_BT/(\hbar\omega)$ for the harmonic oscillator and $T_r = 2T/\theta_D$ for the Debye model.

The total heat capacity of phonon vibrations in the solid becomes

$$C_V = \int_0^{\omega_m} d\omega g(\omega) C_V(\omega), \tag{5.147}$$

where the heat capacity of a single lattice vibration (phonon) is from Eq. (5.116)

$$C_V(\omega)/k_{\mathrm{B}} = \left(\frac{\beta\hbar\omega}{2}\right)^2 \frac{1}{\sinh^2(\beta\hbar\omega/2)}. \tag{5.148}$$

The integral over frequencies can be written in terms of the Debye temperature $k_B \theta_D = \hbar \omega_m$ as follows

$$C_V/k_B = 9N \left(\frac{T}{\theta_D}\right)^3 \int_0^{\theta_D/T} dz \frac{z^4}{4 \sinh^2(z/2)}. \tag{5.149}$$

At $T \ll \theta_D$, $\theta_D/T \to \infty$ and the integral becomes a constant. One obtains in this limit $C_V \propto (T/\theta_D)^3$. Alternatively, at $T \gg \theta_D$, one can use the expansion $\sinh^2(z/2) \simeq (z/2)^2$ producing the following integral

$$C_V/k_B = 9N \left(\frac{T}{\theta_D}\right)^3 \int_0^{\theta_D/T} dz z^2. \tag{5.150}$$

In this high-temperature limit $T \gg \theta_D$ the result is $C_V \simeq 3Nk_B$, which is the Dulong-Petit law. Full integration of the Debye model is compared to the Einstein model based on three harmonic oscillators (for each lattice vibration) in Fig. 5.12.

5.13 Phase equilibria and Clausius-Clapeyron equation

The equilibrium between two phases held at equal temperature and pressure is established by the equality of their corresponding chemical potentials

$$\mu_1(T, P) = \mu_2(T, P). \tag{5.151}$$

This equation establishes the line $P_c(T)$ in the pressure-temperature plane of co-existence between two phases (Fig. 5.13). One can re-write the above equation as the equality between full temperature derivatives (Eq. (5.34)) $d\mu_i/dT = -s_i + v_i dp_c/dT$ $(i = 1, 2)$

$$-s_1 + v_1(dP_c/dT) = -s_2 + v_2(dP_c/dT), \tag{5.152}$$

where s_i and v_i are the entropy and volume of each phase per particle (or per mole). Solving this equation for the temperature slope of the transition line, one gets the Clausius-Clapeyron equation

$$\boxed{dP_c/dT = \Delta s/\Delta v = \Delta h/(T\Delta v),} \tag{5.153}$$

where $\Delta h = T\Delta s$ $(\Delta \mu = 0)$ is the change of the enthalpy (per particle or per mole) between two phases (latent heat).

For the liquid-vapor coexistence line, the volume of the vapor v_g is much greater than the volume of the liquid v_l. One can adopt the ideal-gas law for the vapor volume and neglect the liquid volume in the difference $\Delta v = v_g - v_l \simeq v_g = k_B T/P$. By substituting this approximation to Eq. (5.153) one arrives at the Clapeyron formula

$$dP_c/dT = \beta \Delta h P_c/T. \tag{5.154}$$

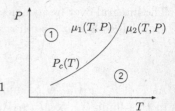

Fig. 5.13 Transition line $P_c(T)$ between phases 1 and 2.

If additionally Δh is viewed as a constant, this equation predicts an exponential dependence of the vapor pressure on $\beta \Delta h$

$$\boxed{P_c \propto \exp[-\beta \Delta h].}$$ (5.155)

The vapor volume along the coexistence line scales with temperature as

$$v_g \propto T \exp[\beta \Delta h].$$ (5.156)

At sufficiently low temperatures the volume decreases with increasing temperature. This somewhat counterintuitive result comes from the fact that heating along the coexistence line brings more molecules to the vapor phase, thus decreasing the volume available to each molecule.

When applied to melting phase transition, Clausius-Clapeyron equation gives the pressure dependence of the melting temperature

$$dT_m/dP = (v_l - v_s)/s_m,$$ (5.157)

where $s_m = \Delta h_m/T_m > 0$ is the melting entropy and v_s is the volume per particle in the solid phase. For instance, the negative slope $dT_m/dP < 0$ of the melting line for water is caused by a higher molecular volume of ice, $v_l < v_s$.

The Clausius-Clapeyron equation can be extended to other first-order phase transitions, for instance to spontaneous magnetization and ferroelectricity. The chemical potential of a polar substance placed in the external electric field E_0 follows from the Gibbs-Duhhem equation (Eq. (5.34))

$$d\mu = -sdT + vdP - \langle m \rangle dE_0,$$ (5.158)

where $\langle m \rangle$ is the average molecular dipole in the material along the direction of the field. The average dipole moment is zero in the paraelectric phase, but becomes nonzero when spontaneous polarization appears in the ferroelectric phase. Following the arguments leading to the Clausius-Clapeyron equation, one obtains for the change of ferroelectric transition temperature T_f caused by the external field

$$(\partial T_f/\partial E_0)_P = \langle m \rangle/\Delta s,$$ (5.159)

where $\Delta s = s_p - s_f$ is the change of the entropy between the unordered (paraelectric) and dipole-ordered (ferroelectric) phase. Given that $\Delta s > 0$, applying an external field raises the temperature of transition to the dipole-ordered phase.

Both entropy (enthalpy) and volume change discontinuously when crossing the transition line of the first-order phase transition. In contrast, for the second-order phase transition, there is no change in the volume or enthalpy at the transition point. This condition can be used to derive two Ehrenfest relations.

One can first take the temperature derivative of the condition $\Delta v = 0$ for the second-order phase transition

$$\Delta(\partial v/\partial T)_P + (dP_c/dT)\Delta(\partial v/\partial P)_T = 0. \tag{5.160}$$

By using the definition of the isothermal expansivity and isobaric compressibility, one arrives at the first Keesom-Ehrenfest relation

$$\boxed{dP_c/dT = \Delta\alpha_p/\Delta\beta_T.} \tag{5.161}$$

The second Ehrenfest relation is derived by taking the temperature derivative of the condition $\Delta h = 0$

$$\Delta(\partial h/\partial T)_P + (dP_c/dT)\Delta(\partial h/\partial P)_T = 0. \tag{5.162}$$

The first term in this equation gives the difference in heat capacities at constant pressure Δc_p. For the second derivative one can apply the Maxwell relation

$$(\partial h/\partial P)_T = v + T(\partial s/\partial P)_T = v - Tv\alpha_p. \tag{5.163}$$

Since $\Delta v = 0$, on obtains the second Keesom-Ehrenfest relation

$$\boxed{dP_c/dT = \Delta c_p/(Tv\Delta\alpha_p).} \tag{5.164}$$

The combination of two Keesom-Ehrenfest relations is the Prigogine-Defay ratio

$$\Pi = \frac{\Delta c_p \Delta\beta_T}{Tv(\Delta\alpha_p)^2} = 1. \tag{5.165}$$

Fluctuation relations for the thermodynamic variables derived in Sec. 5.9 can be used to re-write Π in terms of the variances

$$\Pi = \frac{\Delta\langle\delta s^2\rangle\Delta\langle\delta v^2\rangle}{(\Delta\langle\delta v\delta s\rangle)^2}, \tag{5.166}$$

where δx are used to specify fluctuations of thermodynamic variables in a given phase to distinguish from "Δ" specifying the difference in the corresponding properties between two phases.

The two forms of Prigogine-Defay ratio establish the relation between the caloric (heat capacity) and mechanical (expansivity and compressibility) properties across the equilibrium second-order phase transition.

5.14 Gaussian distribution

$$(2\pi)^3 \delta(\mathbf{k}_1 - \mathbf{k}_2) = V \delta_{\mathbf{k}_1, \mathbf{k}_2}$$
$$S(0) = \langle(\delta N)^2\rangle/N = k_{\mathrm{B}} T \rho \beta_T$$

Consider the problem of finding the distribution of a variable composed of many individual contributions

$$A = \sum_{i=1}^{N} a_i. \tag{5.167}$$

As an example, one can consider the interaction energy of a target particle with the surrounding medium. Each a_i then becomes the interaction energy with molecule i and the sum would run over all N molecules in the medium. To be specific, we assume that $A = A(\mathbf{r}_1, \ldots, \mathbf{r}_N)$ is a function of coordinates $\mathbf{r}_i$ of N participating particles.

As the number of contributions N to the stochastic variable A grows, one expects, according to the central limit theorem, to find the distribution of A approaching the Gaussian distribution. The distribution function $P(A)$ is found by averaging the delta-function

$$p(A) = \langle \delta(A - A(\mathbf{r}_1, \ldots, \mathbf{r}_N)) \rangle = Z^{-1} \int d\Gamma \delta(A - A(\mathbf{r}_1, \ldots, \mathbf{r}_N)) e^{-\beta H}, \tag{5.168}$$

where $d\Gamma = d\mathbf{r}_1 \ldots d\mathbf{r}_N$. One can next use the Fourier integral for the delta-function to write

$$p(A) = \int_{-\infty}^{\infty} \frac{dx}{2\pi} e^{ixA} \left\langle e^{-ixA(\mathbf{r}_1, \ldots, \mathbf{r}_N)} \right\rangle. \tag{5.169}$$

The function

$$G(x) = \left\langle e^{-ixA(\mathbf{r}_1, \ldots, \mathbf{r}_N)} \right\rangle \tag{5.170}$$

is known as the cumulant generating function. The reason for this name is that the average in Eq. (5.170) can be represented by an infinite series of cumulants $\langle(\delta A)^n\rangle$, $\delta A = A - \langle A\rangle$ ($\langle(\delta A)^n\rangle = \langle A\rangle$ for $n = 1$)

$$G(x) = \exp\left[\sum_{n=0}^{\infty} \frac{(-ix)^n}{n!} \langle(\delta A)^n\rangle\right]. \tag{5.171}$$

Each cumulant is then given as the nth-order derivative of $G(x)$

$$\langle(\delta A)^n\rangle = \frac{1}{(-i)^n} \frac{d^n}{dx^n} \ln[G(x)]\Big|_{x=0}. \tag{5.172}$$

If A is assumed to follow the Gaussian statistics, only first two cumulants are non-vanishing and, by integrating over x in Eq. (5.169), one gets the Gaussian distribution

$$p(A) = \left[2\pi\sigma_G^2\right]^{-1/2} \exp\left[-\frac{(A - \langle A \rangle)^2}{2\sigma_G^2}\right], \tag{5.173}$$

where

$$\sigma_G^2 = \langle (\delta A)^2 \rangle = \langle A^2 \rangle - \langle A \rangle^2. \tag{5.174}$$

In many applications, the probability of achieving a given value A is associated with the free energy $F(A)$, also known as the potential of mean force

$$p(A) \propto e^{-\beta F(A)}, \tag{5.175}$$

where only the exponential term is considered and the normalization constant is dropped. From the Gaussian distribution of A one obtains a parabolic form for the potential of mean force

$$\boxed{F(A) = \frac{(A - \langle A \rangle)^2}{2\beta\sigma_G^2}.} \tag{5.176}$$

The dependence on temperature through β in the denominator in this equation deserves special comment. In a number of cases, the Gaussian variance is proportional to temperature, $\sigma_G \propto T$. This result follows from the fluctuation-dissipation theorem discussed in Sec. 8.3 and is given by Eq. (8.37). When this result applies to the variable A, the free energy function $F(A)$ is the potential energy of a harmonic oscillator with the average displacement $\langle A \rangle$ (assumed depending weakly on temperature) and the temperature-independent force constant $\kappa = (\beta\sigma_G^2)^{-1}$. The entropy component of the free energy is zero in that case. The example of a freely jointed chain discussed in the next section shows that this prediction of the fluctuation-dissipation theorem does not always hold. The variance of the end-to-end distance for this model of a polymer is given by a Brownian walk and is independent of temperature. The resulting free energy is proportional to temperature, $F(A) \propto T$, and becomes fully entropic, with zero energy component. The general situation with the temperature dependence of the force constant κ depends on specifics of medium fluctuations projected on the variable A.

The variance of A can be related to statistics of density fluctuations in the medium. To establish this connection, we write A as an integral with the fluctuating scalar density field $\rho(\mathbf{r})$

$$A = \int d\mathbf{r} A(\mathbf{r})\rho(\mathbf{r}), \quad \rho(\mathbf{r}) = \sum_{i=1}^{N} \delta(\mathbf{r} - \mathbf{r}_i). \tag{5.177}$$

The variance becomes

$$\sigma_G^2 = \int dr dr' A(\mathbf{r}) A(\mathbf{r}') K(\mathbf{r} - \mathbf{r}), \tag{5.178}$$

where

$$K(\mathbf{r} - \mathbf{r}') = \langle \delta\rho(\mathbf{r})\delta\rho(\mathbf{r}')\rangle. \tag{5.179}$$

We have assumed here that the correlation function of the density fluctuations at space points $\mathbf{r}$ and $\mathbf{r}'$ depends only on the difference $\mathbf{r} - \mathbf{r}'$. By using the convolution theorem (Sec. 1.7), one can write

$$\sigma_G^2 = \int \frac{d\mathbf{k}}{(2\pi)^3} \tilde{A}(\mathbf{k})\tilde{A}(-\mathbf{k})\tilde{K}(\mathbf{k}). \tag{5.180}$$

Here, the spatial Fourier transform of the correlation function is

$$\tilde{K}(\mathbf{k}) = \int d\mathbf{r} K(\mathbf{r})e^{i\mathbf{k}\cdot\mathbf{r}} = \frac{1}{V}\int d\mathbf{r} d\mathbf{r}' \langle\delta\rho(\mathbf{r}+\mathbf{r}')\delta\rho(\mathbf{r}')\rangle e^{i\mathbf{k}\cdot(\mathbf{r}+\mathbf{r}')-i\mathbf{k}\cdot\mathbf{r}'}. \tag{5.181}$$

This equation can be written as

$$\tilde{K}(\mathbf{k}) = \rho S(k), \tag{5.182}$$

where $\rho = N/V$ and

$$S(k) = N^{-1}\langle\rho_{\mathbf{k}}\rho_{-\mathbf{k}}\rangle \tag{5.183}$$

is the density structure factor of the material, which is discussed in more detail in Sec. 6.6 in application to structural properties of liquids. One finally obtains for the variance of A

$$\boxed{\sigma_G^2 = \rho \int \frac{d\mathbf{k}}{(2\pi)^3}\tilde{A}(\mathbf{k})\tilde{A}(-\mathbf{k})S(k).} \tag{5.184}$$

In the extreme limit of an infinitely-ranged scalar field $A(\mathbf{r}) = 1$, the variable A becomes the number of particles in the volume, $A = N$ (Eq. (5.177)). The Fourier transform becomes $\tilde{A} = (2\pi)^3\delta(\mathbf{k}) = V\delta_{\mathbf{k},0}$. One gets for the variance (see Eq. (1.55))

$$\langle(\delta N)^2\rangle = \frac{\rho}{V}\sum_{\mathbf{k}} V^2\delta_{\mathbf{k},0}S(k) = NS(0). \tag{5.185}$$

This result establishes the connection between the $k = 0$ structure factor and the relative variance of the number of particles in a fixed volume

$$\boxed{S(0) = \langle(\delta N)^2\rangle/N.} \tag{5.186}$$

The ratio on the right-hand side is unity for the ideal gas (Eq. (5.107)), thus establishing $S(0) = 1$ in that limit. In fact, this result extends to all values

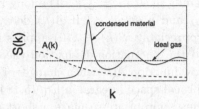

Fig. 5.14 Screened potential $\tilde{A}(k)$ (Eq. (5.187)) (dashed lines) compared to the structure factor (solid line). The dotted line indicates $S(k)$ for the ideal gas.

of k for an ideal gas and one has $S(k) = 1$ (Fig. 5.14). This result can be directly used to derive the ideal gas law. Given that $S(0) = k_{\mathrm{B}} T \rho \beta_T = 1$ (β_T is the isothermal compressibility, Eq. (6.43)), one obtains $\beta P = \rho$ by integrating over pressure.

We next note that the equation for the variance of the number of particles in a given volume (Eq. (5.107)) is derived in Sec. 5.9 from purely thermodynamic arguments. It is therefore expected that a long-range function $A(\mathbf{r})$ should produce the Gaussian variance consistent with thermodynamics and the fluctuation-dissipation theorem. This expectation is met by requiring $S(0) \propto T$. When this condition is imposed, the explicit dependence on temperature cancels out in the force constant of the parabola describing the potential of mean force (Eq. (5.176)) and one arrives at an effective harmonic-oscillator form for $\dot{F}(A)$. This line of thought also implies that a short-range $A(\mathbf{r})$ should produce a more complex temperature dependence of σ_G^2. The free energy $F(A)$ gains both the energy and entropy components in that case. In the limiting case when $A(\mathbf{r})$ involves only the nearest neighbors, such as in the problem of a freely jointed chain considered in the next section, σ_G^2 becomes temperature independent and $F(A)$ is fully entropic.

A general rule that one can derive from these considerations is that a long-range direct-space function $A(\mathbf{r})$ will produce a short-range reciprocal-space $\tilde{A}(\mathbf{k})$ for which $S(k) \simeq S(0)$ can be adopted. In contrast, a short-range $A(\mathbf{r})$ will result in a long-range $\tilde{A}(\mathbf{k})$ for which the entire dependence $S(k)$ needs to be considered, with a more complex temperature dependence of the potential of mean force (Fig. 5.14).

For a specific example, one can consider the screened Coulomb potential $A(r) = r^{-1} \exp[-r/\lambda_D]$ discussed in more detail in Secs. 6.13 and 10.5. Its range is defined by the screening length λ_D. The Fourier transform of this potential

$$\tilde{A}(k) = \frac{4\pi}{k^2 + \kappa^2} \tag{5.187}$$

is given in terms of the inverse screening length $\kappa = \lambda_D^{-1}$. The long-range potential corresponds to a large λ_D and a small κ. If $\tilde{A}(k)$ decays to zero before $S(k)$ starts to change appreciably, $S(k) \simeq S(0)$ applies (Fig. 5.14). This is the result relevant to the fluctuation-dissipation theorem. Otherwise, a small λ_D creates a shallow function $\tilde{A}(k)$ when changes of $S(k)$ have to be accounted for in the reciprocal-space integral in Eq. (5.184).

In the end of this section, it seems appropriate to make a short comment on the physical significance of reciprocal-space calculations adopted here. This approach is widely used in theories of condensed materials and will be applied multiple times in the sections below. From the mathematical perspective, the transition to reciprocal space is nothing more than a convenient use of the Fourier transform reducing the real-space convolution to a single integral in reciprocal space (convolution theorem, Sec. 1.7). This is clearly true, but there are additional, more physically motivated, benefits of this approach. The $k = 0$ values of correlation and susceptibility functions describe collective effects of materials not obvious at the single-particle level. Therefore, adopting $k = 0$ values for these functions as the zero-order approximation allows one to account for collective effects, which otherwise could have been missed if approximate schemes were used in real space more appropriate to study fluctuations and dynamics at the single-particle level. For instance, adopting $S(k) \simeq S(0)$ in Eq. (5.184) incorporates the information about collective density fluctuations into an approximate solution, with its potentially nontrivial dependence on thermodynamic parameters through the isothermal compressibility.

The situation here is somewhat similar to the adoption of phonons in place of atomic vibrations in solid-state physics. What mathematically appears as merely a Fourier transformation leads to the emergence of a new physical reality of long-lived collective lattice vibrations. This approach is less fruitful for disordered materials (liquids, etc) since elementary excitations are often strongly damped, but in this case as well much can be accomplished by switching to correlation functions defined in reciprocal space.

5.15 Freely jointed chain

The simplest model of a linear polymer is a chain of segments with the length b freely rotating around the joints [38] (Fig. 5.15). This model implies that the probability to find each joint is a uniform function of the rotation angles and is a delta-function of the radius-vector (rigid segments).

The probability density for the 3D radius vector **r** along a single segment becomes

$$p(\mathbf{r}) = (4\pi b^2)^{-1}\delta(r - b).$$ (5.188)

The total probability to find the chain in a specific conformation is the product of probabilities for all independently rotating segments

$$p(\mathbf{r}_1,\ldots,\mathbf{r}_N) = \prod_{i=1}^{N} p(\mathbf{r}_i).$$ (5.189)

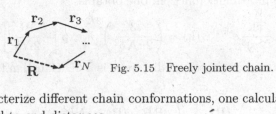

Fig. 5.15   Freely jointed chain.

To characterize different chain conformations, one calculates the distribution of end-to-end distances

$$\mathbf{R} = \sum_{i=1}^{N} \mathbf{r}_i.$$ (5.190)

The collective variable **R** is a specific example of the variable A considered in the previous section. The distribution is given by the expression in which the positions of N joints are integrated under the constraint than the sum of the segment vectors forms the vector **R** (compare to Eq. (5.168))

$$p(\mathbf{R}) = \int \delta\left(\mathbf{R} - \sum_{i=1}^{N} \mathbf{r}_i\right) \prod_{i=1}^{N} p(\mathbf{r}_i) d\mathbf{r}_i.$$ (5.191)

Instead of calculating the probability itself, one can calculate its cumulant generating function

$$G(\mathbf{k}) = \left\langle e^{-i\mathbf{k}\cdot\sum_i \mathbf{r}_i} \right\rangle.$$ (5.192)

The connection between the two functions is provided by the inverse Fourier transform

$$p(\mathbf{R}) = \int \frac{d\mathbf{k}}{(2\pi)^3} G(\mathbf{k}) e^{i\mathbf{k}\cdot\mathbf{R}}.$$ (5.193)

One obtains

$$G(\mathbf{k}) = [g(\mathbf{k})]^N,$$ (5.194)

where

$$g(\mathbf{k}) = \left\langle e^{-i\mathbf{k}\cdot\mathbf{r}_i} \right\rangle.$$ (5.195)

By using the one-segment probability function from Eq. (5.188) one obtains

$$g(\mathbf{k}) = \frac{\sin kb}{kb} \simeq 1 - \tfrac{1}{6}(kb)^2. \tag{5.196}$$

Therefore, the cumulant generating function of the chain is approximated by the exponential function

$$G(\mathbf{k}) = \left[1 - \tfrac{1}{6}(kb)^2\right]^N \simeq \exp\left[-(N/6)(kb)^2\right]. \tag{5.197}$$

By performing the inverse Fourier transform from the cumulant generating function to the probability function, one obtains

$$p(\mathbf{R}) = \int \frac{d\mathbf{k}}{(2\pi)^3} e^{i\mathbf{k}\cdot\mathbf{R}} G(\mathbf{k}) = \left(\frac{3}{2\pi Nb^2}\right)^{3/2} \exp\left[-\frac{3\mathbf{R}^2}{2Nb^2}\right]. \tag{5.198}$$

The distribution of end-to end distances is a 3D Gaussian function with the variance

$$\langle \mathbf{R}^2 \rangle = \int d\mathbf{R}\, \mathbf{R}^2 p(\mathbf{R}) = Nb^2. \tag{5.199}$$

This is the result for a 3D random walk, which can be obtained directly from the condition of statistical independence of $\mathbf{r}_i$ and $\mathbf{r}_j$ when $i \neq j$

$$\boxed{\langle \mathbf{R}^2 \rangle = \sum_{i,j=1}^{N} \langle \mathbf{r}_i \cdot \mathbf{r}_j \rangle = Nb^2.} \tag{5.200}$$

The probability $p(\mathbf{R})$ can be identified with the potential of mean force, i.e., the free energy (reversible work) required to change the distance between the ends of the chain

$$F(\mathbf{R}) = -\beta^{-1}\ln p(\mathbf{R}) = k_{\mathrm{B}}T\frac{3\mathbf{R}^2}{2Nb^2} + \text{Const.} \tag{5.201}$$

One sees that stretching the linear chain of non-interacting segments requires changing its entropy. The corresponding elastic force of the chain becomes

$$\mathbf{f} = -\frac{\partial F}{\partial \mathbf{R}} = -\frac{3k_{\mathrm{B}}T}{Nb^2}\mathbf{R}. \tag{5.202}$$

Entropy is the origin of the force and it is called the entropic force. Similarly to the pressure of the ideal gas (Sec. 5.3) and the entropic osmotic pressure (Sec. 10.2), the force required to stretch the chain is proportional to temperature. Increasing temperature increases the restoring force acting from the chain on an external load. The result is the well-known contraction of rubber at higher temperatures [21].

5.16 Coil-globule transition

The freely jointed chain does not account for interactions between the segments. If these interactions are repulsive, the chain should expand. It will contract if the interactions are attractive. The condensation of the coil into a more compact polymer globule is called the coil-globule transition (Fig. 5.16).

Consider beads interacting with the potential $u(r)$ combining a repulsive and attractive branches such that the potential well, that is the most negative value of the interaction energy, is $-\epsilon$. The free energy of the coil will contain the entropic component described by Eq. (5.201) and the free energy arising from interactions between the segments. The second component is difficult to calculate in a general case, but in cases when the density of segments ρ is low one can apply the virial expansion (Eq. (5.89)). The free energy of interaction between the beads becomes

$$F_{\text{int}}/N = \rho k_{\text{B}} T B(T). \tag{5.203}$$

Here, $B(T)$ is the second virial coefficient.

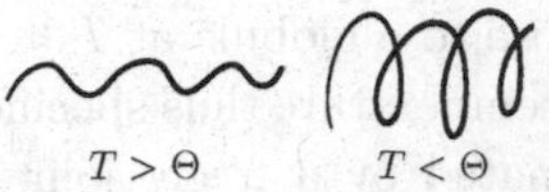

Fig. 5.16   Coil-globule transition at the transition temperature Θ.

$T > \Theta$ $T < \Theta$

The expression for $B(T)$ is derived in a somewhat different way in application to real gases below (Sec. 6.3). Here, the derivation in terms of Mayer functions is outlined for a pair-wise decomposable interaction potential

$$V_N = \sum_{i<j=1}^{N} u(r_{ij}). \tag{5.204}$$

The configuration integral can be written as

$$Q_N = V^{-N} \int d\mathbf{r}_1 \ldots d\mathbf{r}_N \prod_{i<j} [1 + f(r_{ij})], \tag{5.205}$$

where $f(r)$ is the Mayer function

$$f(r) = e^{-\beta u(r)} - 1. \tag{5.206}$$

Expanding the product of the Mayers functions in the configuration integral, one obtains

$$Q_N = 1 + \tfrac{1}{2} N \rho \int d\mathbf{r} f(r) + \ldots. \tag{5.207}$$

If the remaining terms in the expansion are truncated, the free energy arising from interactions becomes

$$F_{\text{int}} = -\beta^{-1} \ln Q_N \simeq N \rho k_{\text{B}} T B(T), \qquad (5.208)$$

where we have for the second virial coefficient

$$B(T) = -\tfrac{1}{2} \int d\mathbf{r} f(r). \qquad (5.209)$$

Similarly to the Boyle temperature for gases, one can define the temperature $T = \Theta$ at which $B(\Theta) = 0$ and write

$$B(T) \simeq v \frac{T - \Theta}{T}. \qquad (5.210)$$

Here, v carries the meaning of an effective volume of interaction. At $T > \Theta$, $B(T) > 0$ and repulsions dominate in the interactions between the segments. For instance if the interaction potential between the segments is approximated by repulsion of hard spheres with the diameter σ_s, one gets

$$B = \frac{2\pi}{3} \sigma_s^3 = 4v_0 > 0, \qquad (5.211)$$

where $v_0 = (\pi/6)\sigma_s^3$ is the volume of the segment (see Eq. (6.20)). In contrast, at $T < \Theta$, $B(T) < 0$ and interactions between the segments are dominated by attractions. The coil can collapse to a globule at $T < \Theta$ resulting in the coil-to-globule transition. The Θ-temperature thus specifies the condition at which the coil is well approximated by a freely jointed chain.

One can use the combination of the virial expansion with the entropic part of the free energy to estimate how interactions alter the scaling of the root-mean squared radius of the polymer with the number of segments. Far above the Θ-temperature, one can put $B \simeq v$ and write the total free energy as a function of the coil radius R

$$F(R) = k_{\text{B}} T \frac{3R^2}{2Nb^2} + k_{\text{B}} T \frac{vN^2}{(4\pi/3)R^3}, \qquad (5.212)$$

where, in the second term, the density of the segments ρ is replaced with $N/[(4\pi/3)R^3]$.

Minimizing $F(R)$ in respect to R, one obtains $(F'(R_0) = 0)$

$$\frac{R_0}{Nb^2} = \frac{3vN^2}{4\pi R_0^4}. \qquad (5.213)$$

This equation leads to the following scaling of the effective radius of the coil with the number of segments

$$\boxed{R_0 \propto N^{3/5}.} \qquad (5.214)$$

This result is different from the scaling suggested by the freely jointed chain (Eq. (5.200))

$$R_0 \simeq \sqrt{\langle \mathbf{R}^2 \rangle} \propto N^{1/2}. \tag{5.215}$$

The difference arises from the excluded-volume effects neglected when the uniform probability for the angles of each segment is assumed in Eq. (5.188).

Chapter 6

Liquids

Liquids are much denser than real gases. Each molecule in the liquid phase is within the interaction length with a sufficiently large number of neighbors, of the order of ~ 12 for a close packing. This high density leads to a sufficiently high cohesive energy that holds the liquid particles together. At the same time, the potential and kinetic energies of binary interactions are close in magnitude, leading to continuous mutual exchange of the neighbors and less frequent rearrangements of the local structure. Liquids flow as a result and do not sustain static stress, demonstrating viscosity instead. High density and close magnitudes of the kinetic and potential energies make two established references of physics, the ideal gas (for real gasses) and the harmonic oscillator (for solids), poor stating points for developing practical theories of liquids. This difficulty historically presented the main obstacle to the development of quantitative *theories of the liquid state*. The situation changed dramatically with the development of the model of the fluid of hard spheres, which has become the reference system for liquid state theories in the last decades [15, 39–42]. The goal of such theories is to calculate thermodynamic and dynamic properties of liquids assuming that the intermolecular interaction potentials, either binary or multiparticle, have been determined separately. This Chapter does not cover modern formalisms for thermodynamics of the liquid state. The goal here is to explain the mean-field theory of the liquid state going back to van der Waals and to show how this theory is modified when more accurate models of packing of molecular repulsive cores are introduced. This Chapter also serves as an introduction to the structure of liquids, which builds the necessary background for molecular hydrodynamics considered in Chap. 8 and for solvation theories covered in Chap. 12.

6.1   Properties of liquids

Compressibility:
$\beta_T \sim 0.5 - 1$ GPa^{-1} (liquids)
$\beta_T \sim 10^4$ GPa^{-1} (gas at 1 atm)
Kinematic viscosity ($\nu = \eta/\rho$):
$\nu \sim 10^{-6}$ m^2/s (liquids)
$\nu \simeq D \sim 10^{-5}$ m^2/s (gas at 1 atm)

The ideal gas considered in Sec. 6.3 is an idealized limit of the low-density phase in which the collision mean free path, i.e., the average distance between binary collisions far exceeds the molecular diameter σ. The equation of state for the ideal gas can be given in terms of the compression factor Z

$$Z = (\beta P/\rho) = 1. \tag{6.1}$$

The real gas on the $P - T$ phase diagram is shown as the vapor (V), which condenses to the liquid (L) with lowering temperature at constant pressure below the critical pressure P_c (Fig. 6.1). The point of equilibrium between the vapor and the liquid is reached at T^* of condensation for which P^* is the pressure of saturated vapor. If $P^* = 1$ atm, the temperature T^* is equal to the normal boiling temperature T_b. The phase coexistence line is described by the Clausius-Clapeyron equation covered in Sec. 5.13.

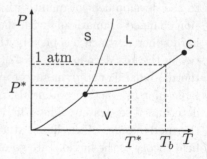

Fig. 6.1 Phase diagram showing the solid (S), liquid (L), and vapor (V) phases. The liquid-vapor equilibrium, shown by P^*, T^* points on the phase diagram, terminates at the critical point (C).

The liquid density is much higher than the density of the vapor in equilibrium with the liquid. The reduced density of the liquid state

$$\rho^* = \rho\sigma^3 \tag{6.2}$$

is typically around unity. The parameter used alongside with ρ^* is the packing fraction equal to the ratio of the molecular volume $v_0 = (\pi/6)\sigma^3$

to the volume per molecule in the liquid $v = 1/\rho$

$$\eta = v_0/v = (\pi/6)\rho^*. \tag{6.3}$$

Accordingly, many liquids have the packing density of $0.4 - 0.5$. This value needs to be compared with the highest packing density for spherical molecules $\eta_{FCC} = \pi\sqrt{2}/6 \simeq 0.74$ reached for the face-centered cubic (FCC) lattice. Since this is the highest packing fraction that can be achieved, $\eta_{FCC} - \eta$ provides the measure of the free volume available to molecular thermal motion. It is clear that molecules in the liquid are sufficiently close to each other to experience continuous many-body collisions and intermolecular forces. The kinetic energy of liquid molecules is nearly equal to the potential energy of binary interaction $K \simeq |U|$. This relation puts liquids as an intermediate phase between dilute gases, for which $K \gg |U|$, and dense solids with $K \ll |U|$. The close values of the kinetic and potential energies imply that no stationary local structure can be maintained, leading to a continuous structural rearrangement of the local order in the liquid and complete absence of the global order.

The binary interaction energy between two molecules in the liquid is usually composed of two branches: a short-range and steep repulsion and a long-range and less steep attraction. The repulsion branch U_{rep} is positive and is falling with increasing distance, while the attraction part U_{att} is negative, becoming less negative with increasing distance and approaching zero at infinite separation. The total potential energy is the sum of two branches (Fig. 6.2)

$$U = U_{rep} + U_{att}. \tag{6.4}$$

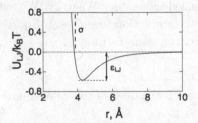

Fig. 6.2 Lennard-Jones binary interaction potential for methane: $\epsilon_{LJ}/k_B = 170$ K and $\sigma_{LJ} = 3.8$ Å. The interaction energy is scaled with $k_B T$, $T = 298$ K. The vertical dashed line shows the hard-sphere diameter σ.

The leading term in the attraction between nonpolar molecules (no charge or dipole moment) is the dispersion interaction discussed in Sec. 4.15. This interaction is caused by spontaneous fluctuations of the electric charge distribution in the molecules caused by virtual transitions to the excited electronic states. The leading term decaying as r^{-6} represents interactions between induced dipoles. Short-range repulsion interactions are caused by Pauli exclusion principle. They generally decay exponentially with the distance, but for the reasons of mathematical convenience they are represented by power laws, such as in the 12-6 Lennard-Jones (LJ) potential

$$U_{\text{LJ}} = 4\epsilon_{\text{LJ}} \left[(\sigma_{\text{LJ}}/r)^{12} - (\sigma_{\text{LJ}}/r)^6 \right]. \tag{6.5}$$

The LJ diameter σ_{LJ} in this equation is the distance at which the interaction energy is equal to zero. The LJ energy ϵ_{LJ} defines the depth of the attraction well at the equilibrium distance $r_0 = 2^{1/6}\sigma_{\text{LJ}}$ (Fig. 6.2). The size of the molecule is associated with the diameter σ defined as the hard-sphere diameter discussed in more detail below. The attraction well is close to the thermal energy for the liquid state and one finds $k_{\text{B}}T/\epsilon_{\text{LJ}} \approx 1$. This relatively weak attraction energy implies that cages in the liquid can be formed only fleetingly and their rearrangement is associated with the structural relaxation time of the liquid.

Fig. 6.3 $Z^* = \beta^* P^*/\rho^*$ vs $\beta^* \Delta H_v$ at P^* equal to the saturated vapor pressure and $T^* = 298$ K for 87 molecular liquids; ΔH_v is the enthalpy of vaporization per molecule.

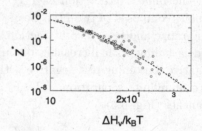

The importance of two separate length scales responsible for repulsions and attractions in the liquid was recognized by van der Waals (vdW), who suggested to project the additivity of two interaction branches in Eq. (6.4) to the corresponding additivity of the compression factors

$$\boxed{Z = Z_{\text{rep}} - Z_{\text{att}}.} \tag{6.6}$$

This representation of the equation of state is the basis of the vdW equation of state considered below.

It is important to note that Z_{rep} and Z_{att} are both positive and they strongly compensate each other. Figure 6.3 illustrates this compensation by showing $Z^* = \beta^* P^*/\rho^*$ for 87 common liquids [14] estimated at the pressure of the saturated vapor with $T^* = 298$ K. The compression factor is plotted agains the scaled vaporization enthalpy $\beta^* \Delta H_v$, $\beta^* = (k_{\text{B}}T^*)^{-1}$.

The vaporization enthalpy roughly represents the strength of attraction interactions in the liquid that are overcome when the molecule is moved to the vapor. The plot shows the scale of compression factors $Z^* \ll 1$. As discussed below, the repulsion compression factor Z_{rep} is about $10 - 30$ for these liquids (Fig. 6.11). The equilibrium compression factor Z^* is then the result of a nearly complete compensation of two big numbers in Eq. (6.6), and it is lowered by increasing the strength of attractions. This strong compensation is an ample evidence of complexity facing theories of the liquid state, which have to deal with cancellations of large numbers and no small parameters to construct the theory by perturbation methods.

The vaporization enthalpy represents the cohesive energy of the liquid, that is the energy of intermolecular interactions holding the molecules together. While the binary potential energy at equilibrium U_0 is close to the thermal energy $k_B T$, the cohesive energy $z U_0$ is estimated as the binary energy multiplied by the number of neighbors z for each molecule in the liquid. The number of neighbors can change significantly depending on how open is the liquid structure: from $z \simeq 4 - 5$ for water connected by a network of hydrogen bonds to $z \simeq 12$ for the FCC and hexagonal close packing. As is seen from Fig. 6.3, the cohesive energy among common molecular liquids is in the range of $(10 - 30) k_B T$ at $T = 298$ K.

6.2 Radial distribution function

Liquids lack global structure, but local preferences in mutual positions of the molecules still exist and can be characterized with the radial distribution function. It specifies the probability density to find a particle within the radial shell with the volume $4\pi r^2 dr$ at the distance r from a given target particle. The target particle can be arbitrarily chosen within a macroscopic sample provided the sample is translationally and rotationally isotropic (no physical properties are affected by translations or rotations of the entire sample). Isotropy can be alternatively described as the invariance of the observables in respect to translations or rotations of the laboratory system of coordinates associated with the sample (translational-rotational invariance).

The probability to find a particle at the distance r from a target particle becomes

$$dp = g(r)dV/V, \quad dV = 4\pi r^2 dr \tag{6.7}$$

with the probability density $g(r)/V$ in which $g(r)$ is the radial distribution function. Since the probability is normalized to unity, $g(r)$ is normalized to the sample volume V

$$\int_0^\infty dr 4\pi r^2 g(r) = V. \tag{6.8}$$

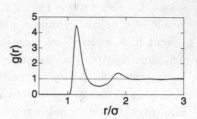

Fig. 6.4 Radial distribution function of a molecular liquid.

Molecules have strongly repulsive molecular cores, which require large
energy to penetrate. Therefore, $g(r)$ is close to zero at $r < \sigma$, where σ is
the effective molecular diameter (Fig. 6.2). In the opposite limit of large
distances r, the probability to find a particle in the volume $d\Omega$ is mostly
determined by the bulk density of the material and $g(r) \to 1$ at $r \gg \sigma$ (Fig.
6.4).

The product of bulk number density $\rho = N/V$ with $g(r)$ defines the
local density profile

$$\rho(r) = \rho g(r). \tag{6.9}$$

At low densities, when binary molecular interactions dominate, $\rho(r)$ can be
given by the Boltzmann distribution in terms of the pairwise interaction
potential energy between the molecules $\phi(r)$

$$\rho(r) = \rho e^{-\beta\phi(r)}. \tag{6.10}$$

This equation does not apply at densities typical for liquids and solids since
many-particle interactions dominate in these materials.

6.3 Second virial coefficient

Corrections to the ideal-gas law are given in the form of the virial expansion
in terms of the gas number density $\rho = N/V$

$$\beta P = \rho + B(T)\rho^2 + \dots. \tag{6.11}$$

The expansion is infinite, but the most important coefficient in the series
is the second virial coefficient $B(T)$.

The expression for $B(T)$ can be derived from the low-density approximation $g(r) \simeq \exp[-\beta\phi(r)]$. Since $V^{-1}g(r)$ gives the probability density

of finding the molecules at a given distance in the real gas (Eq. (6.7)), one can use this physical meaning to calculate the average potential energy of interactions

$$U = \frac{N(N-1)}{2V} \int d\mathbf{r}\phi(r)g(r). \tag{6.12}$$

Here, $N(N-1)/2$ is the number of distinct pairs of molecules in the system of N molecules. The total internal energy of the gas follows from adding the average kinetic energy $(3/2)Nk_BT$. Since $N \gg 1$, $N(N-1) \approx N^2$ and one obtains

$$E/N = \tfrac{3}{2}k_BT + 2\pi\rho \int_0^\infty dr r^2 g(r)\phi(r). \tag{6.13}$$

For the next step, one needs the thermodynamic relation following from the first law of thermodynamics, $dE = TdS - PdV$:

$$(\partial E/\partial V)_T = -P + T\,(\partial S/\partial V)_T. \tag{6.14}$$

From the Maxwell relation, $(\partial S/\partial V)_T = (\partial P/\partial T)_V$, one obtains

$$(\partial E/\partial V)_T = -\,(\partial \beta P/\partial \beta)_V. \tag{6.15}$$

From Eq. (6.13) and by substituting the Boltzmann distribution form for $g(r)$ (Eq. (6.10)) one obtains

$$(\partial \beta P/\partial \beta)_V = 2\pi\rho^2 \int_0^\infty dr r^2 \phi(r)e^{-\beta\phi(r)}. \tag{6.16}$$

The limit of $\beta = 0$ corresponds to $T \to \infty$ when the ideal-gas law applies. Therefore,

$$\beta P = \rho + 2\pi\rho^2 \int_0^\beta d\beta' \int_0^\infty dr r^2 \phi(r)e^{-\beta'\phi(r)}. \tag{6.17}$$

Integration over β' results in the following equation for the second virial coefficient

$$\boxed{B(T) = 2\pi \int_0^\infty dr r^2 \left(1 - e^{-\beta\phi(r)}\right).} \tag{6.18}$$

The interaction potential between the molecules is a combination of a strong repulsion at short distances with a slowly-varying attraction at larger distances (Eq. (6.4)). This property can be used to construct approximations for the second virial coefficients, which can be used as the basis for constructing more general equations of states. One can approximate $\phi(r)$ as a hard-sphere repulsion, $\phi(r) \to \infty$, at $r < \sigma$ and a soft attraction $\phi(r) < 0$

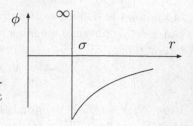

Fig. 6.5 Separation of the intermolecular interaction energy into the hard-sphere repulsion at $r = \sigma$ and a weak attraction at $r > \sigma$.

at $r > \sigma$ (Fig. 6.5). Upon expanding the Boltzmann factor in small $\beta\phi$ at $r > \sigma$, the second virial coefficient can then be written as

$$B(T) = 2\pi \int_0^\sigma drr^2 - 2\pi\beta \int_\sigma^\infty drr^2 |\phi(r)|. \tag{6.19}$$

The first integral is equal to $4v_0$, where $v_0 = (\pi/6)\sigma^3$ is the hard-sphere volume of a single molecule. The second integral is a constant, which does not depend on either temperature or density. Therefore, $B(T)$ can be written in the form

$$B(T) = 4v_0 \left(1 - \frac{T_B}{T} \right). \tag{6.20}$$

In this equation, T_B is the Boyle temperature at which the second virial coefficient becomes equal to zero, $B(T_B) = 0$. The deviation of the pressure of a real gas from the ideal-gas prediction is very small at this temperature since the virial expansion starts from the third virial coefficient.

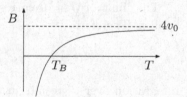

Fig. 6.6 Temperature dependence of the second virial coefficient $B(T)$.

Given Eq. (6.20), the overall temperature dependence of the second virial coefficient can be represented by a hyperbolic function which crosses zero at $T = T_B$ and saturates to the high-temperature limit $4v_0$ at $T \to \infty$ (Fig. 6.6). The range of $B < 0$ is where attractions dominate in the intermolecular interactions, and they make the volume occupied by the real gas smaller than the volume of the ideal gas at the same temperature and pressure. When $B > 0$, the volume of the gas is higher than of the corresponding ideal gas since the molecular repulsions dominate in the intermolecular interactions.

6.4   van der Waals equation

By substituting Eq. (6.20) for the second virial coefficient back to the virial expansion, one obtains

$$\beta P = \rho + 4v_0\rho^2 - \beta a\rho^2, \tag{6.21}$$

where

$$a = 4v_0 k_B T_B \tag{6.22}$$

is known as the van der Waals (vdW) constant. The reason for this assignment is that only one step separates Eq. (6.21) from the vdW equation of state, which is capable to predict the gas-liquid phase transition. One notices that the first two terms in Eq. (6.21) represent the linear and quadratic terms of the infinite series expansion of the function

$$\frac{\rho}{1 - 4v_0\rho} = \rho + 4v_0\rho^2 + 16v_0^2\rho^3 + \ldots . \tag{6.23}$$

This function can be viewed as representing molecular repulsions, which do not allow the gas to be compressed above the density specified by the condition: $4v_0\rho = 1$. If this function is substituted in place of first two terms on right-hand-side of Eq. (6.21), one obtains the vdW equation of state

$$\boxed{\beta P = \frac{\rho}{1 - b\rho} - \beta a\rho^2,} \tag{6.24}$$

where

$$b = 4v_0 \tag{6.25}$$

is the second vdW constant.

The vdW equation predicts the existence of the gas-liquid phase transition, which terminates with the critical point. The ability of the vdW equation to predict the existence of the terminal (critical) point of the liquid-vapor coexistence line allows one to connect the vdW parameters a and b to the critical thermodynamic variables $v_c = V_c/N$, T_c, and P_c

$$3b = v_c = 3k_B T_c/(8P_c), \quad a = (9/8)v_c k_B T_c. \tag{6.26}$$

One also finds the relation for the Boyle temperature

$$T_B = (27/8)T_c. \tag{6.27}$$

The two terms in the vdW equation represent the repulsion and attraction compression factors in the equation of state (Eq. (6.6))

$$Z_{\text{rep}} = \frac{1}{1 - b\rho} = \frac{1}{1 - 4\eta}, \quad Z_{\text{att}} = \beta a\rho. \tag{6.28}$$

The vdW repulsion term was introduced here from the virial expansion. One can alternatively demonstrate that it arises from entropic arguments related to the number of microstates available to the molecule.

By repeating the arguments leading to Eq. (5.37), one can determine the entropy of the system of particles with the total volume V

$$S = k_{\mathrm{B}} N \ln \left(\frac{(V - 4Nv_0)e}{N\lambda^3} \right) + \text{Const}, \tag{6.29}$$

where λ is the Broglie thermal wavelength (Eq. (5.38)). As above, the volume excluded from the access of in the liquid is $4Nv_0$. The pressure due to the intermolecular repulsive interactions is calculated from the entropy by taking the volume derivative

$$P_{\mathrm{rep}} = T \left(\frac{\partial S}{\partial V} \right)_{N,T} = \frac{Nk_{\mathrm{B}}T}{V - 4v_0 N}. \tag{6.30}$$

This equation leads to Z_{rep} in Eq. (6.28).

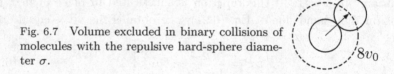

Fig. 6.7 Volume excluded in binary collisions of molecules with the repulsive hard-sphere diameter σ.

The excluded volume $4Nv_0$ in the equation for the entropy comes from the volume $8v_0$ excluded from access of molecular centers in pairwise collisions (Fig. 6.7). Since two molecules are involved in each pairwise collision, the total excluded volume is $(N/2) \times 8v_0$. This picture also makes clear that the entropy equation and the vdW equation of state must miss the entropy effects arising from many-particle collisions which are very important in the liquid phase. Alternatives to the vdW repulsion term correcting for this deficiency are discussed below for the fluid of hard spheres.

6.5 Direct correlation function

The Ornstein-Zernike (OZ) equation allows one to construct long-range correlations in liquids, as represented by the (spatial) correlation function

$$h(r) = g(r) - 1 \tag{6.31}$$

from much shorter correlations represented by the direct correlation function $c(r)$. The equation that produces $h(r)$ from $c(r)$ is the sum of the direct correlation function and the direct-space convolution of $c(|\mathbf{r} - \mathbf{r}'|)$ with the correlation function

$$h(r) = c(r) + \rho \int d\mathbf{r}' c(|\mathbf{r} - \mathbf{r}'|)h(\mathbf{r}'). \qquad (6.32)$$

The physical meaning of this equation can be understood by substituting $h(r)$ back to the integral on the right-hand side of the OZ equation

$$
\begin{aligned}
h(r) =& c(r) + \rho \int d\mathbf{r}' c(|\mathbf{r} - \mathbf{r}'|)c(r') \\
&+ \rho^2 \int d\mathbf{r}' c(|\mathbf{r} - \mathbf{r}'|) \int d\mathbf{r}'' c(|\mathbf{r}' - \mathbf{r}''|)h(\mathbf{r}'') + \dots.
\end{aligned}
\qquad (6.33)
$$

This mathematical result is illustrated in Fig. 6.8. Each short-range correlation between the neighboring liquid molecules contributes to the overall long-range correlation specified by $h(r)$ between target particles 1 and 2. These short-range correlations are combined into all possible paths, involving intermediate molecules, which connect 1 to 2. Each of the short segments of such trajectories is assigned to $c(r)$. Different trajectories propagate short-range correlations between the neighboring molecules into long-range correlations. An advantage of this representation is that $c(r)$ can be a fairly simple and smooth function to produce complex long-range correlations oscillating in space due to the granular microscopic nature of molecular materials.

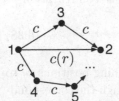

Fig. 6.8   Chains of short-range direct correlation functions $c(r)$ producing the long-range correlation function $h(r)$.

The distinction between the short-range and long-range components of the molecular potential is also reflected by approximations typically used for $c(r)$. By separating the potential into the repulsive core and a soft attraction, one can arrive [39,43] at a popular approximation for $c(r)$ known as the random phase approximation

$$c(r) = c_0(r) - \beta\phi(r). \qquad (6.34)$$

where $\phi(r)$ is the long-range part of the potential (Fig. 6.5). The first term, $c_0(r)$, in the direct correlation function describes short-range correlations produced by the repulsion interaction potential. A simple low-density approximation for this term is discussed below, but an accurate solution for $c_0(r)$ is achieved by solving the Percus-Yevick closure of the QZ equation for the fluid of hard spheres [15, 39].

The OZ equation is converted from the integral convolution equation to an algebraic equation by transforming the direct-space functions $c(r)$ and $h(r)$ to reciprocal-space functions $\tilde{c}(k)$ and $\tilde{h}(k)$ (space Fourier transform). The algebraic equation becomes (convolution theorem, Sec. 1.7)

$$\tilde{h}(k) = \tilde{c}(k) + \rho\tilde{c}(k)\tilde{h}(k). \tag{6.35}$$

One can alternatively write this equation in the form

$$(1 + \rho\tilde{h}(k))(1 - \rho\tilde{c}(k)) = 1. \tag{6.36}$$

This equation defines the density structure factor of the liquid

$$\boxed{S(k) = 1 + \rho\tilde{h}(k) = (1 - \rho\tilde{c}(k))^{-1}.} \tag{6.37}$$

6.6 Structure factor

An alternative definition of the structure factor, which connects it to scattering of radiation and neutrons from liquids, is through the variance of the reciprocal-space microscopic density. The dynamic structure factor is defined in Sec. 8.8. Time is not considered here and one starts with the microscopic density of the liquid at a given instantaneous configuration

$$\rho(\mathbf{r}) = \sum_{j=1}^{N} \delta(\mathbf{r} - \mathbf{r}_j). \tag{6.38}$$

This function specifies the liquid configuration by assigning positions $\mathbf{r}_j$ to N molecules. These positions are precisely defined through the corresponding delta functions $\delta(\mathbf{r} - \mathbf{r}_j)$. Thermal agitation alters the particle positions and one has to consider an ensemble average $\rho(r) = \langle\rho(\mathbf{r})\rangle$. It there are no preferences in an isotropic liquid, one gets

$$\rho(r) = V^{-1} \int d\mathbf{r}\rho(\mathbf{r}) = N/V = \rho. \tag{6.39}$$

The average density of the isotropic liquid is simply the macroscopic number density ρ. One cannot learn about correlations between the molecules in the liquid from the first statistical moment of the microscopic density and higher statistical moments have to be considered. The most important is the second statistical moment, the variance.

One considers the Fourier transform of the microscopic density

$$\rho_\mathbf{k} = \int d\mathbf{r}\rho(\mathbf{r})e^{i\mathbf{k}\cdot\mathbf{r}}. \tag{6.40}$$

From Eq. (6.38), it is given by

$$\rho_\mathbf{k} = \sum_{j=1}^{N} e^{i\mathbf{k}\cdot\mathbf{r}_j}. \tag{6.41}$$

The density structure factor of the liquid is the second statistical moment of $\rho_\mathbf{k}$

$$\boxed{S(k) = N^{-1}\langle|\rho_\mathbf{k}|^2\rangle.} \tag{6.42}$$

The $k = 0$ value of the structure factor is related to the isothermal compressibility β_T (Eq. (5.186))

$$\boxed{S(0) = \rho k_\mathrm{B}T\beta_T,} \tag{6.43}$$

where

$$\beta_T = \rho^{-1}\left(\partial\rho/\partial P\right)_T. \tag{6.44}$$

The constraint imposed on $S(0)$ allows one to construct approximations for the direct correlation function. From Eqs. (6.37) and (6.44), one obtains

$$1 - \rho\tilde{c}(0) = (\partial\beta P/\partial\rho)_T. \tag{6.45}$$

If the density derivative on the right-hand-side is estimated based on the second-order virial expansion, one gets

$$\tilde{c}(0) = -2B(T). \tag{6.46}$$

This is the low-density expansion for the direct correlation function connecting it to the second virial coefficient.

Turning back to Eq. (6.19), one can write

$$\tilde{c}(0) = -8v_0 - \beta\int d\mathbf{r}\phi(r). \tag{6.47}$$

This is the $k = 0$ value for the Fourier transform of the random phase approximation in Eq. (6.34). The repulsive part of the direct correlation function satisfies the condition

$$\int d\mathbf{r}c_0(r) = -8v_0. \tag{6.48}$$

This equation suggests a simple approximation for $c_0(r)$

$$\begin{aligned}
c_0(r) &= -1, \ r \leq \sigma \\
&= 0, \ r > \sigma.
\end{aligned} \tag{6.49}$$

From this equation one obtains for the Fourier transform

$$\rho\tilde{c}_0(k) = 4\pi\rho \int_0^\infty dr\, r^2 c(r) \frac{\sin(kr)}{kr} \tag{6.50}$$

the following result

$$\rho\tilde{c}_0(k) = -24\eta \left[\frac{\sin(k\sigma)}{(k\sigma)^3} - \frac{\cos(k\sigma)}{(k\sigma)^2} \right]. \tag{6.51}$$

Despite the simplistic form of $c_0(r)$ in the direct space, $S(k)$ obtained from $c_0(k)$ according to Eq. (6.37) carries many characteristic features of the structure factors reported for simple atomic fluids from x-ray and neutron scattering (Fig. 6.9). This derivation illustrates the importance of direct correlation functions for the theories of liquids. They carry the information about short-range, "direct" correlations between the molecules, which is propagated through the integral convolution of the OZ equation to many-particle correlations described by the more complex correlation function $h(r)$ and by the corresponding radial distribution function $g(r)$.

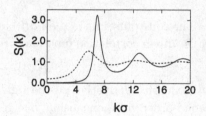

Fig. 6.9 Structure factor calculated according to Eqs. (6.37) and (6.51) (dashed line) and from the Percus-Yevick solution in Eqs. (6.60) and (6.61) (solid line).

6.7   Speed of sound

It is instructive to derive the connection between the structure factor and the isothermal and adiabatic speed of sound. The isothermal speed of sound is given by

$$c_T = (m\rho\beta_T)^{-1/2}, \tag{6.52}$$

while the adiabatic speed of sound c_S is connected to c_T by the ratio of the constant-pressure and constant-volume heat capacities $\gamma = c_p/c_v$ (the adiabatic constant)

$$c_S^2 = \gamma c_T^2. \tag{6.53}$$

For the ideal gas,

$$\beta_T^{\text{id}} = P^{-1}, \tag{6.54}$$

$\rho \beta_T^{\rm id} = \beta$ and one gets

$$mc_S^2 = \gamma k_{\rm B} T, \quad mc_T^2 = k_{\rm B} T. \tag{6.55}$$

This result can be identified with the equipartition theorem.

There is an enhancement of the kinetic energy stored in the sound wave in condensed materials

$$mc_T^2 = k_{\rm B} T / S(0), \quad S(0) \ll 1. \tag{6.56}$$

Sound waves carry more energy in condensed materials compared to gases. One can estimate the speed of sound in the water vapor at $T = 300$ K as $c_T \simeq 371$ m/s. With the liquid water compressibility of $\beta_T = 0.457 \times 10^{-9}$ Pa^{-1}, one obtains for the isothermal speed of sound in room-temperature liquid water $c_T \simeq 1480$ m/s.

6.8 Hard-sphere fluid

Like the ideal gas for the gas phase and the harmonic oscillator for the crystalline phase, the fluid of hard spheres establishes the idealized reference system which can be used to develop advanced theories of the liquid state. This reference system is used to built both perturbation theories and integral-equation theories of liquids [15]. For the hard-sphere fluid, the intermolecular potential $U_{\rm HS}$ does not include the attraction branch and the repulsive part of the potential is replaced by the hard wall, i.e., the interaction energy is infinite at distances below the hard-sphere diameter σ.

Since there is no attraction potential to hold the liquid molecules, the hard sphere fluid needs external pressure to keep it at a given density ρ. This density is kept equal to the density of the real liquid when using hard spheres as the reference system. Thermodynamic properties of the fluid of hard spheres are not affected by temperature because the potential energy is infinite within the repulsive core. The only parameter, besides density, that can be varied is the hard-sphere diameter. There are several formalisms available to map the hard-sphere diameter on a Lennard-Jones liquid to minimize the errors of perturbation theories [15, 39]. Empirically, the hard-sphere diameter for nonpolar liquids can be related [44] to the vdW parameter b (Fig. 6.10), which is directly measured through critical properties of real liquids (Eq. (6.26)).

From the virial equation (Eq. (5.86)), the compression factor of the liquid is related to the average virial, i.e., the average of the scalar product of the force and the position of the particle

Fig. 6.10 $b^{1/3}$ (b is in cm^3/mol) vs the hard-sphere diameter σ for nonpolar gases and n-alkanes [44]. The dashed line is the polynomial fit $b^{1/3} = 0.9337 + 0.5828\sigma + 0.03315\sigma^2$.

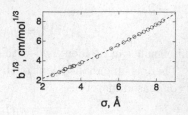

$$Z_{\text{HS}} = 1 - \frac{4\pi\beta}{6} \int_0^\infty dr\, r^3 g(r) dU_{\text{HS}}/dr. \qquad (6.57)$$

One can next use the relation

$$-\beta g(r)\frac{dU_{\text{HS}}}{dr} = g(r)e^{\beta U_{\text{HS}}}\frac{d}{dr}e^{-\beta U_{\text{HS}}} = y(r)\frac{d}{dr}e^{-\beta U_{\text{HS}}}, \qquad (6.58)$$

where $y(r) = g(r)\exp[\beta U_{\text{HS}}]$ is a continuous function of r. One can further notice that $\exp[-\beta U_{\text{HS}}]$ is equal to the Heaviside function $h(r - \sigma)$ and $d\exp[-\beta U_{\text{HS}}]/dr = \delta(r - \sigma)$. This relation establishes the connection between the pressure of the hard-sphere fluid and the contact value $g(\sigma)$ of the radial distribution function

$$Z_{\text{HS}} = 1 + 4\eta g(\sigma). \qquad (6.59)$$

An exact mathematical solution for the radial distribution function can be obtained in the Percus-Yevick (PY) closure. This is an approximation for the direct correlation function $c(r)$, which, like Eq. (6.49), assumes that $c(r) = 0$ at $r > \sigma$, but solves for the functional form of $c(r)$ inside the repulsive hard-sphere core $r < \sigma$. The closure is established by additionally requiring $g(r) = 0$ at $r < \sigma$.

The solution by Baxter of the PY closure [15, 39] finds the structure factor of the hard-sphere fluid $S(k)$ in terms of the function $Q(k)$ such that

$$S(k) = [Q(k)Q(-k)]^{-1}. \qquad (6.60)$$

The function $Q(k)$ is given by the following integral

$$Q(k) = 1 - 12\eta \int_0^1 e^{ik\sigma r} \qquad (6.61)$$
$$\left[\tfrac{1}{2}a(r^2 - 1) + b(r - 1)\right] dr,$$

where the functions of the packing density $a(\eta)$ and $b(\eta)$ are given by the relations

$$a = \frac{1 + 2\eta}{(1 - \eta)^2},$$
$$b = -\frac{3\eta}{2(1 - \eta)^2}. \qquad (6.62)$$

This solution is compared to the low-density expansion of the direct correlation function (Eq. (6.51)) in Fig. 6.9.

The PY solution also provides the contact value of the radial distribution function

$$g^{\mathrm{PY}}(\sigma) = \frac{2+\eta}{2(1-\eta)^2} \tag{6.63}$$

and, from Eq. (6.59), of the compression factor

$$Z_v = \frac{1+2\eta+3\eta^2}{(1-\eta)^2}, \tag{6.64}$$

where the subscript "v" stands for the virial solution.

However, this is not the only route to obtain the compression factor. One can alternatively use the relation between the structure factor and isothermal compressibility (Eq. (6.43))

$$S(0) = k_{\mathrm{B}}T\rho\beta_T = k_{\mathrm{B}}T\left(\frac{\partial\rho}{\partial P}\right)_T = a(\eta)^{-2}. \tag{6.65}$$

From this expression, one can integrate over the packing fraction to obtain an alternative relation for the compression factor

$$Z_c = \frac{1+\eta+\eta^2}{(1-\eta)^3}, \tag{6.66}$$

where the subscript "c" stands for the compressibility route.

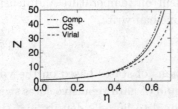

Fig. 6.11
Compression factors for hard-sphere fluids from virial (Eq. (6.64)), compressibility (Eq. (6.66)), and Carnahan-Starling (Eq. (6.67)) solutions.

Two different solutions for the compression factor of the hard-sphere fluid indicate that the PY solution is not thermodynamically consistent. A very successful empirical solution suggested by Carnahan and Starling is to take a linear combination of two routes to Z

$$\boxed{Z^{\mathrm{CS}} = \frac{1}{3}\left[2Z_c + Z_v\right] = \frac{1+\eta+\eta^2-\eta^3}{(1-\eta)^3}.} \tag{6.67}$$

This solution also requires to change the contact value of the radial distribution function according to Eq. (6.59)

$$g^{\mathrm{CS}}(\sigma) = \frac{2-\eta}{2(1-\eta)^3}. \tag{6.68}$$

The Carnahan-Starling solution can be used for the repulsion compression factor to generalize the vdW equations of state (Eq. (6.24)) [45]

$$Z = Z^{\text{CS}} - \beta a \rho. \tag{6.69}$$

The repulsion compression factor is fairly large at the liquid packing fraction $\eta \simeq 0.5$ (Fig. 6.11) and needs to be nearly identically compensated by $Z_{\text{att}} = \beta a \rho$ to arrive at small overall compression factors typical for liquids at normal conditions (Fig. 6.3).

6.9   van der Waals fluid

We now return to the vdW picture of the intermolecular potential in the liquid given as the combination of the hard repulsive wall and a soft attraction (Fig. 6.5). The internal energy of a molecule in such a liquid is a sum of the average kinetic energy ($(3/2)k_{\text{B}}T$ for an atomic liquid) and the average potential energy

$$U/N = u = 2\pi\rho \int_0^\infty dr r^2 g(r)\phi(r). \tag{6.70}$$

If the potential energy is sufficiently long-ranged, the details of the local structure of the liquid represented by wiggles of $g(r)$ will average out in the integral, which becomes little dependent on temperature and density. Since $\phi < 0$, the integral is negative and one can write

$$\boxed{u = -a\rho.} \tag{6.71}$$

This relation is the basis for all equations of states based on the vdW picture. The physical picture is that of a fluid of hard repulsive cores swimming in the uniform potential u proportional to liquid's density and thus vanishing in the gas phase (Fig. 6.12). The equation of state for this physical model is given by the generalized vdW equation (6.69) [45] in which all theory's sophistication is devoted to improving the repulsion compression factor, but the second term arising from the uniform background potential stays the same.

Fig. 6.12   Schematic picture of the vdW liquid of hard-core particles residing in a container with the macroscopic length L. The particles experience a constant mean-field potential $-a\rho$.

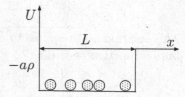

The compression factor can be used to derive the free energy of the generalized vdW liquid by using the thermodynamic relation

$$\left(\frac{\partial \beta F/N}{\partial \rho}\right)_T = \frac{Z(\rho)}{\rho}. \tag{6.72}$$

One can integrate this expression over density between the vapor phase viewed as an ideal gas and the liquid phase to obtain

$$\Delta F/N = f_{\text{vdW}} + \Delta\mu_{\text{id}}. \tag{6.73}$$

Here, $\Delta\mu_{\text{id}} = \mu_{\text{id}}^l - \mu_{\text{id}}^g = k_B T \ln(V_g/V_l)$ is the change of the ideal-gas chemical potential between a less dense vapor and a more dense liquid (Eq. (5.73)). The free energy of the vdW liquid is obtained by integrating the deviation of the compression factor from that of the ideal gas

$$\beta f_{\text{vdW}} = \int_0^\rho \frac{d\rho'}{\rho'}(Z(\rho') - 1). \tag{6.74}$$

The result is

$$f_{\text{vdW}} = f^{\text{CS}} - a\rho, \tag{6.75}$$

where the free energy of the repulsive molecular cores as described by the Carnahan-Starling equation of state is given as

$$f^{\text{CS}} = k_B T \frac{4\eta - 3\eta^2}{(1 - \eta)^2}. \tag{6.76}$$

The free energy of the repulsive cores of the liquid is proportional to temperature and is thus entropic in character. This is consistent with the view that the reversible work done on bringing the hard molecular cores into the restricted space of the liquid volume is spent on altering the number of microscopic states, i.e., the entropy. The attraction term in the free energy of the vdW liquid does not contribute to its entropy and one gets

$$\frac{s_{\text{vdW}}}{k_B} = \frac{S_{\text{vdW}}}{N k_B} = -\frac{4\eta - 3\eta^2}{(1 - \eta)^2}. \tag{6.77}$$

6.10 Trouton's rule

Trouton's rule is an empirical observation that the entropy of vaporization ΔS_v of simple (aprotic) liquids at the normal boiling temperature T_b stays approximately constant at the value

$$\boxed{\Delta s_v/k_B = \Delta S_v/(N k_B) \simeq 9.7.} \tag{6.78}$$

Fig. 6.13 Experimental vaporization entropies (Exp.) for 22 liquids compared to vdW estimates (Eq. (6.80)). Protic liquids show an upward deviation from the vdW result. The horizontal line is drawn through the vdW values at $\Delta s_v/k_B = 9.7$.

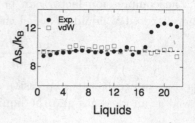

The normal boiling temperature T_b is defined by the liquid-vapor coexistence line at the saturated vapor pressure of 1 atm (Fig. 6.1). The value of the vaporization entropy is the entropy change in the process of moving a single molecule from the liquid at T_b to the vapor at the same temperature.

Trouton's rule is a beautiful experimental proof of the validity of the generalized vdW view of simple liquids. Upward deviations from Trouton's rule have been observed for hydrogen-bonding (protic) liquids (alcohols and water). Molecules in these liquids are connected by the network of hydrogen bonds. Local and highly directional interactions produce lower entropy in the liquid phase and require more entropy to move the molecules to the vapor. Similar observations apply to ionic liquids made of oppositely charged ions. The potential landscape of all these liquids is complex and must deviate from the picture of a flat potential well drawn in Fig. 6.12.

Since vaporization of the liquid is a first-order phase transition, the entropy of vaporization is given it terms of enthalpy (latent heat) of vaporizing the liquid

$$\Delta S_v = \Delta H_v/T_b. \tag{6.79}$$

It is the latent heat and the boiling temperature T_b that are typically reported to produce ΔS_v.

The entropy view of the vaporization process is convenient since the vdW liquid has no entropy of attractions, and the entire entropy change in going from the liquid phase to the vapor is given as

$$\frac{\Delta s_v^{\text{vdW}}}{k_B} = -\frac{s_{\text{vdW}}}{k_B} + \ln(V_g/V_l), \tag{6.80}$$

where the entropy of the vdW liquid is from Eq. (6.77). Deviations from this rule quantify the distinction between the structure of a real liquid at T_b and the idealized vdW view.

The data [46] collected for a number of protic and aprotic liquids (Fig. 6.13) support the accuracy of the vdW picture at $T = T_b$. There is little deviation of the experimental entropy from the entropy of hard-sphere unpacking, which mostly determines the entropy of removing a liquid molecule from the bulk. The upward deviations at the end of the list plotted in Fig. 6.13 mark protic liquids (water is the last liquid on the list).

6.11 Density functional theory

The free energy of the ideal gas in Eq. (5.61) can be written as a function of the uniform number density ρ

$$\beta F^{\text{id}}(\rho)/V = \rho \left(\ln[\rho\lambda^3] - 1\right). \tag{6.81}$$

One can extend this result to an ideal gas placed in an external field with the potential energy ϕ^{ext} by noting that nothing in the statistical derivation should change if instead of ρ one has an inhomogeneous density $\rho(\mathbf{r})$. The condition of non-interacting particles is sufficient for the statistical derivation. Since $\rho(\mathbf{r})$ is a scalar field, the function $F^{\text{id}}(\rho)$ should become a functional of the scalar density field $F^{\text{id}}[\rho]$. One needs only to add the energy of interaction with the external field $\phi^{\text{ext}}(\mathbf{r})$ to obtain

$$\beta F^{\text{id}}[\rho] = \int d\mathbf{r}\rho(\mathbf{r}) \left(\ln[\rho(\mathbf{r})\lambda^3] - 1\right) + \beta \int d\mathbf{r}\rho(\mathbf{r})\phi^{\text{ext}}(\mathbf{r}). \tag{6.82}$$

One next wonders if this expression can be further extended to include interactions between particles in the liquid. The virial expansion helps here again to pave the way to a quantitative theory. Equation (6.74) allows one to add interactions through the compression factor. The total free energy as a function of the density becomes

$$\beta F(\rho) = \beta F^{\text{id}}(\rho) + N \int_0^\rho \frac{d\rho'}{\rho'} \left(Z(\rho') - 1\right), \tag{6.83}$$

where N in front of the integral accounts for the fact that $F(\rho)$ is the extensive free energy, which scales linearly with the number of particles.

From the virial expansion, one finds

$$Z(\rho) - 1 = B(T)\rho, \quad B(T) = \tfrac{1}{2} \int d\mathbf{r} \left(1 - e^{-\beta\phi(r)}\right). \tag{6.84}$$

By adopting this equation in the functional of the density field and writing $N = \int d\mathbf{r}\rho(\mathbf{r})$, one finds

$$\beta F[\rho] = \beta F^{\text{id}}[\rho] + \tfrac{1}{2} \int d\mathbf{r}d\mathbf{r}' \rho(\mathbf{r})\rho(\mathbf{r}') \left(1 - e^{-\beta\phi(\mathbf{r}-\mathbf{r}')}\right)$$
$$\simeq \beta F^{\text{id}}[\rho] + (\beta/2) \int d\mathbf{r}d\mathbf{r}' \rho(\mathbf{r})\rho(\mathbf{r}')\phi(\mathbf{r} - \mathbf{r}'), \tag{6.85}$$

where the first-order Taylor expansion of the Boltzmann factor is used in the second step.

The equilibrium density function can be found by minimizing the free energy in respect to the density $\rho(\mathbf{r})$ with the constraint that the total number of particles is preserved when the density is varied. This condition requires minimizing the grand canonical potential

$$\Omega[\rho] = F[\rho] - \mu \int d\mathbf{r}\rho(\mathbf{r}), \tag{6.86}$$

where μ is the chemical potential of the liquid.

By substituting Eq. (6.85) to Eq. (6.86), one obtains

$$\beta\Omega[\rho] = \int d\mathbf{r}\rho(\mathbf{r}) \left(\ln[\rho(\mathbf{r})\lambda^3] - 1\right) + \beta \int d\mathbf{r}\rho(\mathbf{r}) \left(\phi^{\text{ext}}(\mathbf{r}) - \mu\right)$$
$$+ (\beta/2) \int d\mathbf{r}d\mathbf{r}' \rho(\mathbf{r})\rho(\mathbf{r}')\phi(\mathbf{r} - \mathbf{r}'). \tag{6.87}$$

The equilibrium density of an inhomogeneous liquid is obtained by minimizing $\Omega[\rho]$ in terms of the density function ρ

$$\left.\frac{\delta\Omega}{\delta\rho(\mathbf{r})}\right|_{\rho=\rho^{(1)}} = 0. \tag{6.88}$$

The solution of this equation defines the density profile $\rho^{(1)}(\mathbf{r})$

$$\rho^{(1)}(\mathbf{r})\lambda^3 = \exp\left[\beta(\mu - \phi^{\text{eff}}(\mathbf{r}))\right], \tag{6.89}$$

which is determined by the effective potential

$$\phi^{\text{eff}}(\mathbf{r}) = \phi^{\text{ext}}(\mathbf{r}) + \int d\mathbf{r}'\phi(\mathbf{r} - \mathbf{r}')\rho^{(1)}(\mathbf{r}'). \tag{6.90}$$

Equation (6.90) establishes an effective potential acting on a given particle in the liquid, which includes an average potential energy due to all its neighbors distributed with the density $\rho(\mathbf{r})$, which needs to be determined by solving the self-consistent relations given by Eq. (6.89). A practical solution of this problem requires finding suitable trial functions for the density profile with the parameters calculated from a minimization protocol.

The last term in Eq. (6.87) was obtained from the low-density virial expansion. The next step is to extend it to more realistic situations of dense liquids. The connection can be seen from the appearance of $-\beta\phi$ in the random-phase approximation for the direct correlation function in Eq. (6.34). It suggests that $-\beta\phi$ in the density functional can be replaced with the direct correlation function. This turns out to be correct and a successful approximation for the grand canonical density functional reads

$$\beta\Omega[\rho] = \int d\mathbf{r}\,[\rho(\mathbf{r})\ln[\rho(\mathbf{r})/\rho] - \delta\rho(\mathbf{r})] + \beta \int d\mathbf{r}\delta\rho(\mathbf{r})\left(\phi^{\text{ext}}(\mathbf{r}) - \mu\right)$$
$$- \frac{1}{2}\int d\mathbf{r}d\mathbf{r}'\delta\rho(\mathbf{r})\delta\rho(\mathbf{r}')c(\mathbf{r} - \mathbf{r}'), \tag{6.91}$$

where $\delta\rho(\mathbf{r}) = \rho(\mathbf{r}) - \rho$ is the deviation of the position-dependent density from the uniform density ρ.

The functional $\Omega[\rho]$ can be expanded in terms of the small deviation $\delta\rho$ from the uniform density ρ. If the expansion is truncated after the second-order terms, one can write the resulting quadratic functional

$$\beta\Omega[\rho_{\mathbf{k}}] = \frac{1}{2N}\sum_{\mathbf{k}}\frac{|\rho_{\mathbf{k}}|^2}{S(k)} + \frac{\beta}{V}\sum_{\mathbf{k}}\rho_{\mathbf{k}}\phi^{\text{ext}}_{-\mathbf{k}}. \tag{6.92}$$

Here, we have converted the density variation to reciprocal space $\rho_{\mathbf{k}}$ and dropped the term with the chemical potential contributing only to $\mathbf{k} = 0$. A shorthand notation is also used for the integral in reciprocal space (Sec. 1.7)

$$\int \frac{d\mathbf{k}}{(2\pi)^3} \longrightarrow \frac{1}{V}\sum_{\mathbf{k}}. \tag{6.93}$$

Finally, the density structure factor in the denominator in (6.92) is replacing $1 - \rho\tilde{c}(k)$ according to Eq. (6.37). Equation (6.92) is a Gaussian functional for the density fluctuations, which are shifted from their zero equilibrium values by the external potential. The equilibrium values for the density field are then given by minimizing the functional

$$\rho_{\mathbf{k}} = -\beta\rho S(k)\phi^{\text{ext}}_{\mathbf{k}}. \tag{6.94}$$

The interaction energy of the external field with the liquid becomes

$$u = \frac{1}{V}\sum_{\mathbf{k}}\rho_{\mathbf{k}}\phi^{\text{ext}}_{-\mathbf{k}} = -\frac{\beta\rho}{V}\sum_{\mathbf{k}}S(k)\left|\phi^{\text{ext}}_{\mathbf{k}}\right|^2. \tag{6.95}$$

This result can be applied to problems of solvation when an impurity (solute) in the liquid is the source of the external field. The energy of this interaction is calculated from Eq. (6.95) (Sec. 12.7).

6.12   Ionic liquids

One can apply the density functional theory to derive an approximate expression for the structure factor of an ionic liquid. Section 10 presents the Debye-Hückel theory of electrolytes focused on dilute solutions of positive and negative ions in a polar liquid. The screening of electrostatic interactions by the ionic atmosphere self-consistently constructed around each selected ion is fully described by the Debye-Hückel screening parameter κ. It enters the solution thermodynamics, but also becomes the only parameter required to specify the structure factor of the ions in the electrolyte (Eq. (10.56))

$$S(k) = \frac{k^2}{k^2 + \kappa^2}. \tag{6.96}$$

The Debye-Hückel theory is not applicable to ionic liquids. Those are composed of positive and negative ions, which constantly persist in close contact with each other [40]. The charge fluctuations are, therefore, strongly affected by the repulsive cores of the ions and attraction intermolecular interactions. This change of the physics compared to the standard Debye-Hückel framework projects itself to a major change of the structure factor captured by the density-functional theory.

One-component plasma is considered here as an illustrative example of applying the density-functional theory to systems with long-range Coulomb interactions [40]. This fluid is made of ions of equal charge z and number density ρ placed in a neutralizing uniform background of opposite charge to preserve the system neutrality. The external interaction energy is dropped here and we consider the density functional for the Helmholtz free energy $F[\rho_\mathbf{k}]$. The direct correlation function for the one-component plasma is taken in the random phase approximation (Eq. (6.34)) as

$$\tilde{c}(k) = \tilde{c}_0(k) - 4\pi\beta z^2 e^2 / k^2. \tag{6.97}$$

In this equation, $4\pi/k^2$ is the Fourier transform of the Coulomb potential and $\tilde{c}_0(k)$ denotes the direct correlation function of the reference system represented by the repulsive cores of the ions and excluding the long-range Coulomb interactions. The latter contribute the last term, $-\beta\tilde{\phi}$, to the direct correlation function in Eq. (6.97).

By repeating the steps of the previous section, we can immediately write the density functional of the reciprocal-space density field $\rho_\mathbf{k}$

$$F[\rho_\mathbf{k}] = \frac{1}{2V\rho} \sum_\mathbf{k} \frac{|\rho_\mathbf{k}|^2}{S(k)}, \tag{6.98}$$

where for the density structure factor of the ions one has

$$S(k)^{-1} = 1 - \rho \tilde{c}_0(k) + \kappa^2/k^2. \tag{6.99}$$

Here, $\kappa^2 = 4\pi\beta\rho e^2 z^2$ is the Debye-Hückel screening parameter such that $\lambda_D = \kappa^{-1}$ is the Debye-Hückel screening length.

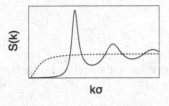

Fig. 6.14 Debye-Hückel structure factor (Eq. (6.96), dashed line) compared to the modified structure factor given by Eq. (6.100) (solid line).

The structure factor $S_0(k) = (1 - \rho \tilde{c}_0(k))^{-1}$ defines density fluctuations in the reference system of repulsive ionic cores. By separating the reference and Coulomb components in the structure factor, we obtain

$$S(k) = S_0(k) \frac{k^2}{k^2 + \kappa^2 S_0(k)}. \tag{6.100}$$

One sees that restrictions imposed by the repulsive cores of the closely packed ions lead to the appearance of the density structure factor modifying the screening length in the ionic liquid (Fig. 6.14). At small wave vectors, one applies the $k \to 0$ limit to the density structure factor (Eq. (6.43)), which can be written as

$$S_0(0) = \beta_T/\beta_T^{\mathrm{id}}, \tag{6.101}$$

where $\beta_T^{\mathrm{id}} = (\rho k_B T)^{-1}$ is the compressibility of the ideal gas (Eq. (6.54)). Liquids are much less compressible than gases and one anticipates $S_0(0) \ll 1$ (see Fig. 6.9).

In the range of low wave vectors, one can take $k \to 0$ in $S_0(k)$ to obtain

$$S(k) = \frac{\beta_T}{\beta_T^{\mathrm{id}}} \frac{k^2}{k^2 + \Lambda^{-2}}, \tag{6.102}$$

where

$$\Lambda = \lambda_D \sqrt{\beta_T^{\mathrm{id}}/\beta_T} \gg \lambda_D. \tag{6.103}$$

The screening of ions in an ionic liquid occurs on a much longer length scale than in the Debye-Hückel electrolyte. The reason is that densely packed ions are restricted, by the repulsive cores, in their ability to rearrange to achieve optimal electrostatic screening.

6.13 Screened Coulomb potential

It is instructive to use Eq. (6.94) to calculate the screened electrostatic potential. Assume that one of the ions in the ionic liquid is viewed as the source of external electric field polarizing the surrounding liquid. If the charge $z_0 e$ is assigned to this ion, the interaction energy between the target ion and an ion in the ionic liquid $\phi_{\mathbf{k}}^{\text{ext}}$ becomes

$$\phi_{\mathbf{k}}^{\text{ext}} = \frac{4\pi z_0 z e^2}{k^2}. \qquad (6.104)$$

This external source of electrostatic energy induces a change in the surrounding density, which is given by Eq. (6.94)

$$\delta\rho_{\mathbf{k}} = -\frac{z_0}{z}\frac{\kappa^2}{k^2}S(k). \qquad (6.105)$$

This change of density in turn leads to an additional electrostatic potential

$$\delta\varphi_{\mathbf{k}} = \frac{4\pi e z}{k^2}\delta\rho_{\mathbf{k}}, \qquad (6.106)$$

which adds to the external electrostatic potential $\varphi_{\mathbf{k}}^{\text{ext}} = 4\pi e z_0/k^2$. The overall screened electrostatic potential in the liquid is

$$\varphi_{\mathbf{k}} = \delta\varphi_{\mathbf{k}} + \varphi_{\mathbf{k}}^{\text{ext}} = \frac{4\pi z_0 e}{k^2 + S_0(k)\kappa^2}. \qquad (6.107)$$

By taking the inverse Fourier transform and putting $S_0(k) \simeq S_0(0)$, one obtains

$$\boxed{\varphi(r) = \frac{z_0 e}{r}e^{-r/\Lambda},} \qquad (6.108)$$

where Λ is the modified screening length in Eq. (6.103). This is the screened Coulomb potential with the screening length Λ.

Chapter 7

Diffusion

The statistical nature of matter considered in previous Chapters leads to fluctuations, i.e., to deviations of observable properties from ensemble-averaged values. The rise and decay of fluctuations occur in time, with characteristic time scales specific to thermal energy flows between different degrees of freedom. *Diffusion* is the simplest among these phenomena. It arises from fluctuations of forces acting on a target particle from the surrounding medium, which lead to the observable random motion. The reduction of the entire statistical ensemble to a single particle offers an opportunity for simple models operating in terms of just a few observables, the coordinates and velocities of the target particle for diffusion. This reduction is more consistently achieved in molecular hydrodynamics considered in the next Chapter. For target diffusion, the Langevin dynamics and the Fokker-Planck equation become productive tools to develop models of diffusional dynamics [47–52]. Fundamentally, these theories introduce the idea of energy dissipation, which is absent in the classical Hamilton mechanics. Diffusional dynamics is reached when friction dominates over the ballistic motion studied by classical mechanics (Chap. 3). The dynamics of condensed materials is ruled either by short-time (sub-picosecond to picoseconds) local excitations or by collective hydrodynamic modes. We are concerned here with the former, leaving the dynamics of collective modes to the next Chapter. In both cases, the use of continuum equations of motion (Langevin or hydrodynamic equations) is allowed by time and length scales explored, which involve a large number of individual molecular collisions.

7.1 Diffusion

Diffusion constant:
$D \sim 10^{-9}$ m^2/s (liquids)
$D \sim 10^{-5}$ m^2/s (gas at 1 atm)
Flux:
$\mathbf{j} = \rho \mathbf{v}$
Mobility:
$v_x = \mu f_x$

Diffusion occurs spontaneously as spreading of particles caused by the gradient of the number density $\rho(x)$, which is the local number of particles per unit volume. The flux of particles is the number of them crossing a unit area perpendicular to the direction of the flux per unit time. Number flux is formally defined as $\mathbf{j} = \rho \mathbf{v}$, its projection along the x-axis is $j_x = \rho v_x$ (Fig. 7.1).

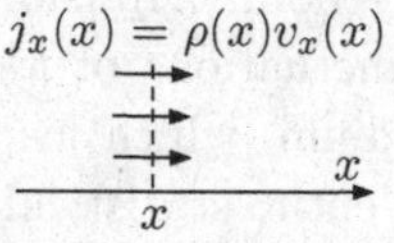

$$j_x(x) = \rho(x) v_x(x)$$

Fig. 7.1   Number flux.

According to Fick's law, the number flux can be related to the gradient of the number density: j_x is the negative of the product of the gradient of ρ (often expressed as the gradient of concentration) and the diffusion constant D

$$j_x = -D d\rho/dx, \tag{7.1}$$

where diffusion is along the x-axis (Fig. 7.2).

It is instructive to derive the Einstein relation connecting the diffusion constant with particle's mobility before we proceed to a general problem of fluctuation-dissipation relations and the linear response theory. One needs to counterbalance the diffusive flow with some external force acting on the particles. One can assume, for instance, that the particles carry charge q and an external electric field is applied to prevent the particles from moving. Such a stationary condition implies that the diffusive flux and the driven flux j_{qx} add up to zero (Fig. 7.2)

$$j_x + j_{qx} = 0. \tag{7.2}$$

The driven flux is created by the external force f_x such that j_{qx} is related to the external force through the particle mobility μ

$$j_{qx} = \rho \mu f_x. \tag{7.3}$$

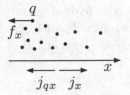

Fig. 7.2 Diffusion and driven fluxes.

We now assume that a stationary concentration profile has been established, and it is given by the Boltzmann distribution consistent with the external potential $U(x)$ such that $f_x = -dU/dx$

$$\rho(x) = \rho_0 e^{-\beta U(x)}. \tag{7.4}$$

Substituting this equation to Fick's law yields

$$j_x = -\beta D\rho f_x. \tag{7.5}$$

Combining with Eq. (7.3) in Eq. (7.2), one arrives at the Einstein equation [53]

$$\boxed{D = k_{\mathrm{B}} T \mu.} \tag{7.6}$$

This relation is a special case of the general fluctuation-dissipation theorem derived in Sec. 8.3. It connects the spontaneous process (diffusion) with the driven process (mobility). The proportionality coefficient is temperature. This relation can be used as an operative definition of a "dynamic temperature" specified by transport coefficients, in contrast to the thermodynamic temperature connecting entropy S to energy E

$$\frac{1}{T} = \left(\frac{\partial S}{\partial E}\right)_V. \tag{7.7}$$

Equation (7.6) establishes the equality between the "dynamic" and thermodynamic temperatures at the condition of detailed balance specified by zero total flux in the system (Eq. (7.2)).

7.2 Langevin equation

Assume that the external drag is equal to the friction force $|f_{H,x}| = \zeta v_x$ (Sec. 8.7). Equality of forces implies zero net force and constant velocity of the target particle. Since $f_x = |f_{H,x}| = \mu^{-1} v_x$, the friction coefficient ζ is equal to the inverse mobility μ and Einstein's equation becomes

$$D = \frac{k_{\mathrm{B}} T}{\zeta}, \quad \mu = \zeta^{-1}. \tag{7.8}$$

The Stokes-Einstein equation follows from applying the hydrodynamic Stokes friction $\zeta_H = 6\pi\eta R$ experienced by a spherical particle with the radius R (Sec. 8.7) in a liquid with the shear viscosity η. By using this expression in Einstein's equation one obtains

$$D = \frac{k_B T}{6\pi\eta R}. \tag{7.9}$$

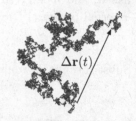

Fig. 7.3 Stochastic Brownian motion.

We now write the Langevin equation for a Brownian particle moving with velocity $\mathbf{v}(t)$ under the action of the random force $\mathbf{R}(t)$ (Fig. 7.3)

$$\dot{\mathbf{v}} + \gamma\mathbf{v} = m^{-1}\mathbf{R}, \tag{7.10}$$

where $\gamma = \zeta/m$. We assume that the velocity and the random force are uncorrelated, $\langle \mathbf{v}(0)\cdot\mathbf{R}(t)\rangle = 0$, and that the correlation time for the random force is infinitely short. This is specified by the delta-function in the time correlation function of the random force: $\langle \mathbf{R}\rangle = 0$, $\langle \mathbf{R}(t)\cdot\mathbf{R}\rangle = C\delta(t)$. The assumption of no correlation between the velocity and the force is consistent with the statistics of the Gibbs ensemble if one assumes that the random force $\mathbf{R} = -\partial U_{0s}/\partial\mathbf{r}$ is produced by fluctuating microscopic interaction energy U_{0s} between the particle and the surrounding medium. In the Gibbs ensemble, velocities are uncorrelated from coordinates and this assumption is fully justified.

The solution of the Langevin equation of motion reads

$$\mathbf{v}(t) = \mathbf{v}(0)e^{-\gamma t} + m^{-1}\int_0^t dt' e^{-\gamma(t-t')}\mathbf{R}(t'). \tag{7.11}$$

One obtains two relations. The first one is

$$\langle \mathbf{v}(t)\cdot\mathbf{v}\rangle = \langle \mathbf{v}^2\rangle e^{-\gamma t}, \tag{7.12}$$

where $\mathbf{v} = \mathbf{v}(0)$ is used here and below. The second relation is obtained by taking $t \to \infty$ and squaring both sides of Eq. (7.11)

$$\langle \mathbf{v}(\infty)^2\rangle = m^{-2}\lim_{t\to\infty}\int_0^t dt'\int_0^t dt'' e^{-2\gamma t+\gamma t'+\gamma t''}\langle \mathbf{R}(t')\cdot\mathbf{R}(t'')\rangle. \tag{7.13}$$

Substituting the delta-function for the correlation of the random force, one obtains

$$\langle \mathbf{v}(\infty)^2 \rangle = \langle \mathbf{v}^2 \rangle = \frac{C}{2\gamma m^2}. \tag{7.14}$$

Since $\langle \mathbf{v}^2 \rangle = 3k_\mathrm{B}T/m$ for particle speeds distributed according to the Maxwell distribution, one gets $C = 6\zeta k_\mathrm{B}T$.

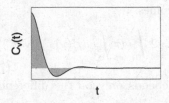

Fig. 7.4 Velocity autocorrelation function $C_v(t) = \langle \mathbf{v}(t) \cdot \mathbf{v} \rangle$. The area under the curve is equal to $3D$, where D is the diffusion constant.

One can next integrate Eq. (7.12) over time to obtain

$$\int_0^\infty dt \langle \mathbf{v}(t) \cdot \mathbf{v} \rangle = 3k_\mathrm{B}T/(m\gamma). \tag{7.15}$$

From Eq. (7.8), this result can be converted to the connection between the diffusion constant and the velocity autocorrelation function $C_v(t) = \langle \mathbf{v}(t) \cdot \mathbf{v} \rangle$

$$\boxed{D = \tfrac{1}{3} \int_0^\infty dt \langle \mathbf{v}(t) \cdot \mathbf{v} \rangle.} \tag{7.16}$$

This equation establishes the connection between the diffusion constant and the area under the curve of the velocity autocorrelation function (Fig. 7.4). The velocity autocorrelation function is typically oscillatory in dense liquids, leading to the mutual cancellation between positive and negative parts in the overall area under the curve.

Equation (7.14) for the constant C can be converted to the integral involving the time correlation function of the random force

$$\zeta = (3k_\mathrm{B}T)^{-1} \int_0^\infty dt \langle \mathbf{R}(t) \cdot \mathbf{R} \rangle. \tag{7.17}$$

This result, known as the second fluctuation-dissipation theorem [54], is discussed in more detail in Sec. 7.10. It connects the friction coefficient to the time integral of the autocorrelation function of the random force.

The white noise of the random force is described by the correlation function

$$\langle \mathbf{R}(t) \cdot \mathbf{R} \rangle = 6\zeta k_\mathrm{B}T\delta(t). \tag{7.18}$$

In order to obtain Eq. (7.17) from (7.18), one needs to postulate

$$\int_0^\infty dt\delta(t) = \tfrac{1}{2}. \tag{7.19}$$

7.3 Mean-squared displacement

The displacement of a particle can be represented by integrating its velocity (Fig. 7.3)

$$\Delta \mathbf{r}(t) = \mathbf{r}(t) - \mathbf{r}(0) = \int_0^t \mathbf{v}(\tau)d\tau. \tag{7.20}$$

The mean-squared displacement then becomes

$$\langle \Delta r(t)^2 \rangle = \int_0^t d\tau \int_0^t d\tau' \langle \mathbf{v}(\tau) \cdot \mathbf{v}(\tau') \rangle = 2 \int_0^t d\tau \int_0^\tau d\tau' \langle \mathbf{v}(\tau) \cdot \mathbf{v}(\tau') \rangle. \tag{7.21}$$

Since the velocity correlation function depends only on the difference of times $s = \tau - \tau'$, we can write

$$\langle \Delta r(t)^2 \rangle = 2 \int_0^t d\tau \int_0^\tau ds \langle \mathbf{v}(s) \cdot \mathbf{v} \rangle = 2 \int_0^t ds(t - s)\langle \mathbf{v}(s) \cdot \mathbf{v} \rangle. \tag{7.22}$$

By taking the limit $t \to \infty$, one obtains

$$\lim_{t \to \infty} (2t)^{-1} \langle \Delta r(t)^2 \rangle = \int_0^\infty ds \langle \mathbf{v}(s) \cdot \mathbf{v} \rangle. \tag{7.23}$$

By combining this result with Eq. (7.16), one obtains the diffusion constant in terms of the mean-squared displacement

$$\boxed{D = \lim_{t \to \infty} \frac{\langle \Delta r(t)^2 \rangle}{6t}.} \tag{7.24}$$

7.4 Rotational Brownian motion

Consider Brownian rotations of a long rod around the axis perpendicular to the rod's long axis specified by the unit vector $\hat{\mathbf{u}}$ (Fig. 7.5). The angular velocity of this rotation is the vector $\mathbf{\Omega}$ which is formally defined by the equation (Eq. (3.50))

$$\frac{d\hat{\mathbf{u}}}{dt} = \mathbf{\Omega} \times \hat{\mathbf{u}}. \tag{7.25}$$

Since $\mathbf{\Omega}$ is perpendicular to $\hat{\mathbf{u}}$, one can alternatively write

$$\mathbf{\Omega} = \hat{\mathbf{u}} \times \frac{d\hat{\mathbf{u}}}{dt}. \tag{7.26}$$

To obtain this result one uses the identity $\mathbf{a} \times (\mathbf{b} \times \mathbf{c}) = \mathbf{b}(\mathbf{a} \cdot \mathbf{c}) - \mathbf{c}(\mathbf{a} \cdot \mathbf{b})$: $\hat{\mathbf{u}} \times (\mathbf{\Omega} \times \hat{\mathbf{u}}) = \mathbf{\Omega}(\hat{\mathbf{u}} \cdot \hat{\mathbf{u}})$.

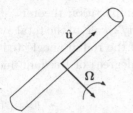

Fig. 7.5   Rotational motion of a rod.

The angular momentum of rotation (Secs. 3.5 and 4.10) is the product of the moment of inertia I_0 (Eq. (3.55)) and the angular velocity

$$\mathbf{L} = I_0\boldsymbol{\Omega}. \tag{7.27}$$

For a long, uniform rod with the mass m and the length l, $I_0 = ml^2/12$.

The Langevin equation is written by adding a source of random noise to the mechanical equation of motion. In the case of the linear momentum, the equation of motion is Newton's equation and one obtains the Langevin equation by adding a random force. For rotations, the equation of motion is the change of the angular momentum due to torques applied to the body. Correspondingly, writing the Langevin equation required adding a random torque $\mathbf{T}(t)$ (Eq. (3.58)). The Langevin equation for rotations becomes

$$\dot{\mathbf{L}} = -\zeta_r\boldsymbol{\Omega} + \mathbf{T}, \tag{7.28}$$

where ζ_r is the rotation friction coefficient. This equation can be rewritten in terms of the angular velocity to make it similar to Eq. (7.10) for the translational Brownian motion

$$\dot{\boldsymbol{\Omega}} + \gamma_r\boldsymbol{\Omega} = I_0^{-1}\mathbf{T}, \tag{7.29}$$

where $\gamma_r = \zeta_r/I_0$. As in the case of random forces producing displacements of particles, we assume that random torques are zero on average, $\langle\mathbf{T}\rangle = 0$, and are delta-function correlated in time

$$\langle\mathbf{T}(t)\cdot\mathbf{T}\rangle = C_r\delta(t). \tag{7.30}$$

From Eq. (7.29), the angle at time t is

$$\boldsymbol{\Omega}(t) = \boldsymbol{\Omega}(0)e^{-\gamma_r t} + I_0^{-1}\int_0^t dt' e^{-\gamma_r(t-t')}\mathbf{T}(t'). \tag{7.31}$$

The amplitude of the random torque C_r in Eq. (7.30) can be determined by taking the $t \to \infty$ limit in $\langle\boldsymbol{\Omega}(t)^2\rangle$, which becomes

$$\langle\boldsymbol{\Omega}(t)^2\rangle\big|_{t\to\infty} = I_0^{-2}\lim_{t\to\infty}\int_0^t dt'\int_0^t dt'' e^{-\gamma_r(2t-t'-t'')}\langle\mathbf{T}(t')\cdot\mathbf{T}(t'')\rangle$$
$$= \frac{C_r}{2\gamma_r I_0^2}. \tag{7.32}$$

The same property can be found from the equipartition theorem stating that each degree of freedom should contribute $(1/2)k_B T$ to the total average energy. Since rotations around the long axis of the rod are neglected here, only two angles contribute to the rotational degrees of freedom and one finds

$$\tfrac{1}{2}I_0\langle \mathbf{\Omega}^2\rangle = k_B T. \tag{7.33}$$

Equating Eqs. (7.32) and (7.33), one obtains

$$\langle \mathbf{T}(t) \cdot \mathbf{T}\rangle = 4\zeta_r k_B T \delta(t). \tag{7.34}$$

With this relation, the Langevin equation for random rotations of a rod [50] is complete. The factor of four in this equation, replacing the factor of six in Eq. (7.18), is due to two rotational degrees of freedom, in contrast to three translational degrees of freedom for a Brownian particle.

7.5   *Fokker-Planck equation

One can apply the Langevin equation to derive the probability density $p(v, t)$ to find velocity v at time t. To simplify the notations, we consider the one-dimensional case

$$p(v, t) = \langle \delta\left(v(t) - v\right)\rangle. \tag{7.35}$$

The average here is taken over the stochastic fluctuations of the random force $R(t)$. The stochastic velocity is connected to the random force through the one-dimensional Langevin equation

$$\dot{v} + \gamma v = m^{-1}R, \tag{7.36}$$

where $v(t) = v_x(t)$.

The time derivative of the probability function can be converted to the partial derivative over the velocity magnitude

$$\partial_t p(v, t) = \left\langle \frac{\partial}{\partial v(t)}\delta\left(v(t) - v\right)\dot{v}\right\rangle = -\frac{\partial}{\partial v}\langle \delta\left(v(t) - v\right)\dot{v}\rangle. \tag{7.37}$$

Substituting the Langevin equation, one obtains

$$\partial_t p = \gamma\frac{\partial}{\partial v}(vp) - m^{-1}\frac{\partial}{\partial v}\langle \delta\left(v(t) - v\right)R(t)\rangle. \tag{7.38}$$

Since fluctuations of velocity $v(t)$ are cased by fluctuations of the random force, we can write the correlation between the delta-function and the random force in terms of a functional derivative

$$\langle \delta\left(v(t) - v\right)R(t)\rangle = \int_0^t dt'\left\langle \frac{\delta}{\delta v(t)}\delta\left(v(t) - v\right)\frac{\delta v(t)}{\delta R(t')}\right\rangle\langle R(t')R(t)\rangle. \tag{7.39}$$

The functional derivative over $v(t)$ is equal to the negative of the ordinary derivative over v. In addition, for the functional derivative of the stochastic velocity over the random force one obtains from Eq. (7.11) $\delta v(t)/\delta R(t') = m^{-1} \exp[-\gamma(t - t')]h(t - t')$, where $h(t - t')$ is the Heaviside function (Sec. 2.2) making the derivative equal to zero at $t' > t$.

The correlation function becomes

$$\langle \delta\left(v(t) - v\right) R(t)\rangle = -m^{-1}\frac{\partial}{\partial v}\int_0^t dt'\, p(v, t)e^{-\gamma(t-t')}\langle R(t)R(t')\rangle. \quad (7.40)$$

The one-dimensional white noise is described by the correlation function

$$\langle R(t)R(t')\rangle = 2\zeta k_{\mathrm{B}}T\delta(t - t'). \quad (7.41)$$

Substituting this relation to Eq. (7.40), one obtains

$$\langle \delta\left(v(t) - v\right) R(t)\rangle = -\gamma k_{\mathrm{B}}T\frac{\partial}{\partial v}p(v, t), \quad (7.42)$$

where we again used Eq. (7.19). The final equation for the probability density is

$$\frac{\partial p}{\partial(\gamma t)} = \frac{\partial}{\partial v}\left(vp\right) + \frac{k_{\mathrm{B}}T}{m}\frac{\partial^2 p}{\partial v^2}. \quad (7.43)$$

This is the Fokker-Planck equation describing time evolution of the probability density as diffusion in the momentum phase space. This derivation shows that the Langevin equation written in terms of stochastic fluctuating forces can be alternatively transformed to an ordinary differential equation for the probability density. The white noise of random forces is transformed to the diffusion (last) term in the Fokker-Planck equation. This is clearly seen when $T = 0$ is adopted in Eq. (7.43) and the last term vanishes. The resulting equation is the continuity relation for the probability density, which can be written as

$$\partial_t p + \partial_v(\dot{v}p) = 0. \quad (7.44)$$

The $T = 0$ limit of Eq. (7.43) follows from this equation upon substituting $\dot{v} = -\gamma v$.

7.6 Overdamped stochastic dynamics

If the Brownian particle in placed in the external potential $U(x)$, Eq. (7.36) is modified to

$$m\dot{v} + \zeta v = f + R, \quad (7.45)$$

where $f = f_x = -\partial_x U$. In the overdamped limit, one neglects particle's acceleration and assumes that the position of the particle $x(t)$ is fully determined by the potential force f and the random force R

$$\dot{x} = \mu f + \mu R, \tag{7.46}$$

where $\mu = \zeta^{-1}$ according to Eq. (7.8). The random force $R(t)$ is a Gaussian noise with zero average and the time correlation function proportional to the delta-function (Eq. (7.41)).

The probability $p(x, t)$ to find the particle at the position x can be derived following the steps similar to those presented above for $p(v, t)$. Alternatively, one notes that every conservative property must satisfy the differential balance equation. The condition $\int dx p(x, t) = 1$ requires that changing the fraction of particles $p(x, t)dx$ in a small interval dx must arise from the flux of particles from that region. This is given by the equation balancing the change of the number of particles with the change in the flux (continuity relation)

$$\partial_t p + \partial_x j = 0. \tag{7.47}$$

Here, $j(x, t)$ is the sum of the particle flux with the average velocity $\langle \dot{x} \rangle$ and the diffusional flux $-D\partial_x p$

$$j = \langle \dot{x} \rangle p - D\partial_x p. \tag{7.48}$$

Given that $\langle \dot{x} \rangle = \mu f = -\mu \partial_x U$ and $D = \mu k_B T$, one obtains

$$\boxed{D^{-1}\partial_t p = \partial_x^2 p + \partial_x \left(p\beta\partial_x U \right).} \tag{7.49}$$

This is the Fokker-Planck equation for the probability density $p(x, t)$. It can be alternatively written in a more familiar form consistent with the Schrödinger equation of quantum mechanics by adopting the transformation $p = \bar{p} \exp[-\beta U/2]$. The equation for $\bar{p}(x, t)$ then becomes

$$D^{-1}\partial_t \bar{p} = -\bar{H}\bar{p}, \tag{7.50}$$

where the effective Hamiltonian operator reads

$$\bar{H} = -\frac{\partial^2}{\partial x^2} - \frac{1}{2}\frac{\partial^2 \beta U}{\partial x^2} + \frac{1}{4}\left(\frac{\partial \beta U}{\partial x} \right)^2. \tag{7.51}$$

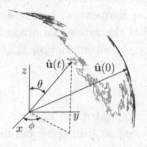

Fig. 7.6 Brownian motion on the unit sphere.

7.7 Rotational diffusion

An example of overdamped stochastic motion described by the Fokker-Planck equation is the diffusional rotational dynamics of a rigid body with axial symmetry. It can be viewed as rotational diffusion of the unit vector $\hat{u}(t)$ on a three-dimensional sphere [51] (Fig. 7.6). The Langevin equation describing this motion was derived in Sec. 7.4.

The basis for this solution is the Fokker-Planck equation for overdamped dynamics in Eqs. (7.50) and (7.51). This problem is distinct from the one-dimensional case considered above in two regards: (i) there is no potential energy affecting the dynamics of the unit vector $\hat{u}$ and (ii) diffusion on the unit sphere happens in three dimensions but involves only two degrees of freedom specified by the azimuthal angle ϕ and polar angle θ (Fig. 7.6). One therefore needs to replace the second derivative over x in Eq. (7.51) with the Laplace operator in spherical coordinates where only the derivatives over ϕ and θ are taken (Eq. (1.38)). Accounting for the fact that diffusion occurs on the surface of unit radius $r = |\hat{u}| \doteq 1$, the equation for the probability density $p(\hat{u}, t) = p(\theta, \phi, t)$ transforms from Eq. (7.50) to the following diffusion equation

$$\partial_t p = D_r \nabla^2_{\theta\phi} p, \tag{7.52}$$

where D_r is the rotational diffusion constant. The Stokes law for rotational friction in Eq. (7.28) requires $\zeta_r = 8\pi\eta R^3$ and according to Einstein's equation (7.8) one gets

$$D_r = \frac{k_B T}{8\pi\eta R^3}. \tag{7.53}$$

Here, R is the effective (Stokes) radius of the particle with the axial symmetry specified by some internal coordinates, e.g., the dipole moment of the particle.

The eigenfunctions of the angular component of the Laplace operator are considered in solving the problem of angular momentum in quantum mechanics (Sec. 4.10), from which one knows that the azimuthal quantum number ℓ establishes the eigenvalue $-\ell(\ell+1)$ for the eigenfunction $Y_{\ell m}(\hat{\mathbf{u}})$ (spherical harmonic)

$$\nabla^2_{\theta\phi}Y_{\ell m}(\hat{\mathbf{u}}) = -\ell(\ell+1)Y_{\ell m}(\hat{\mathbf{u}}). \tag{7.54}$$

Given that the eigenfunctions of the Laplace operator are known, any solution of Eq. (7.52) can be given as a linear expansion in spherical harmonics with time-dependent expansion coefficients. It is, however, more convenient to find the solution of the problem for the rotational diffusion starting with a specific orientation of the unit vector $\hat{\mathbf{u}}_0 = \hat{\mathbf{u}}(0)$. The probability $P(\hat{\mathbf{u}}, t|\hat{\mathbf{u}}_0, 0)$ to find the unit vector with orientation $\hat{\mathbf{u}}$ at time t given that it started with orientation $\hat{\mathbf{u}}_0$ at $t = 0$ is known as the transition probability or propagator. The formal solution for the transition probability follows from rewriting Eq. (7.52) as an operator equation

$$p(\hat{\mathbf{u}}, t|\hat{\mathbf{u}}_0, 0) = \exp[tD_r\nabla^2_{\theta\phi}]\delta\left(\hat{\mathbf{u}} - \hat{\mathbf{u}}_0\right), \tag{7.55}$$

where $\delta\left(\hat{\mathbf{u}} - \hat{\mathbf{u}}_0\right)$ specifies the initial probability density. The delta-function here is the mathematical description of the assumed certainty about the particle orientation at $t = 0$.

The delta-function of a unit vector can be expanded in spherical harmonics (see Box summarizing properties of the delta-function in Sec. 2.2)

$$\delta\left(\hat{\mathbf{u}} - \hat{\mathbf{u}}_0\right) = \sum_{\ell=0}^{\infty} \sum_{m=-\ell}^{\ell} Y_{\ell m}(\hat{\mathbf{u}}_0)Y_{\ell m}^*(\hat{\mathbf{u}}). \tag{7.56}$$

The result of action of the exponential operator on the spherical harmonic follows from expanding it in Taylor's series

$$\exp[tD_r\nabla^2_{\theta\phi}]Y_{\ell m}^*(\hat{\mathbf{u}}) = \sum_{n=0}^{\infty} \frac{(tD_r)^n}{n!}(\nabla^2_{\theta\phi})^n Y_{\ell m}^*(\hat{\mathbf{u}})$$

$$= \sum_{n=0}^{\infty} \frac{(-\ell(\ell+1)tD_r)^n}{n!}Y_{\ell m}^*(\hat{\mathbf{u}}) = \exp[-\ell(\ell+1)tD_r]Y_{\ell m}^*(\hat{\mathbf{u}}). \tag{7.57}$$

One can next combine Eqs. (7.55), (7.56), and (7.57) to obtain the transition probability

$$p(\hat{\mathbf{u}}, t|\hat{\mathbf{u}}_0, 0) = \sum_{\ell m} Y_{\ell m}(\hat{\mathbf{u}}_0)Y_{\ell m}^*(\hat{\mathbf{u}}) \exp[-\ell(\ell+1)tD_r]. \tag{7.58}$$

The importance of the transition probability is that it allows one to calculate the time correlation functions depending on orientational dynamics. The most important is the autocorrelation function of the unit vector $\hat{\mathbf{u}}(t)$. We write it in the form of the second-rank tensor in which the correlation between Cartesian projections $\hat{u}_\alpha(t)$ and $\hat{u}_\beta = u_\beta(0)$ is considered

$$C_{\alpha\beta}(t) = \langle \hat{u}_\alpha(t)\hat{u}_\beta \rangle. \tag{7.59}$$

With the use of the transition probability, one needs to calculate the integral

$$C_{\alpha\beta}(t) = \int d\hat{\mathbf{u}}d\hat{\mathbf{u}}_0 \hat{u}_\alpha \hat{u}_{0\beta} p(\hat{\mathbf{u}}, t | \hat{\mathbf{u}}_0, 0) p(\hat{\mathbf{u}}_0), \tag{7.60}$$

where integration is performed over the orientations of the unit vectors at time t ($\hat{\mathbf{u}}$) and $t = 0$ ($\hat{\mathbf{u}}_0$). The probability density $p(\hat{\mathbf{u}}_0)$ specifies the distribution of orientations $\hat{\mathbf{u}}_0$ at $t = 0$. If one assumes a uniform distribution of orientations, then $p(\hat{\mathbf{u}}_0)$ is $(4\pi)^{-1}$ (Fig. 7.7), which is the inverse of the area of the unit sphere sampled by the unit vector (Fig. 7.6).

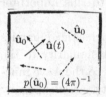

Fig. 7.7 Rotational dynamics producing an ensemble of unit vectors $\hat{\mathbf{u}}(t)$ from random orientations $\hat{\mathbf{u}}_0$ at $t = 0$.

The integral in Eq. (7.60) is calculated by using the addition theorem for spherical harmonics

$$(\hat{\mathbf{u}} \cdot \hat{\mathbf{r}}_\alpha) = \frac{4\pi}{3} \sum_{m=-1,0,1} Y_{1m}(\hat{\mathbf{u}}) Y_{1m}^*(\hat{\mathbf{r}}_\alpha) \tag{7.61}$$

and the orthogonality relation

$$\int d\hat{\mathbf{u}} Y_{\ell m}(\hat{\mathbf{u}}) Y_{\ell'm'}^*(\hat{\mathbf{u}}) = \delta_{\ell\ell'}\delta_{mm'}. \tag{7.62}$$

Substituting $\hat{u}_\alpha = (\hat{\mathbf{u}} \cdot \hat{\mathbf{r}}_\alpha)$ and $\hat{u}_{0\beta} = (\hat{\mathbf{u}}_0 \cdot \hat{\mathbf{r}}_\beta)$ (see Eq. (1.5)) to Eq. (7.60), one obtains after using the orthogonality relation

$$C_{\alpha\beta}(t) = \tfrac{1}{3}\delta_{\alpha\beta}e^{-2D_r t}. \tag{7.63}$$

The relaxation time τ_r of rotational diffusion of a particle with axial symmetry (e.g., a rod) is therefore given in terms of the rotation diffusion constant as

$$\boxed{\tau_r = (2D_r)^{-1}.} \tag{7.64}$$

7.8   *Ornstein-Uhlenbeck process

Ornstein-Uhlenbeck process describes diffusion in the harmonic potential

$$U = \tfrac{1}{2}kx^2, \qquad (7.65)$$

where k is the force constant of the harmonic spring. The corresponding overdamped Langevin equation (Eq. (7.46)) has the external force equal to $f = -dU/dx$. The equation of motion becomes

$$\dot{x} + \kappa x = \zeta^{-1}R \qquad (7.66)$$

with $\kappa = k/\zeta$ and R representing the stochastic force. One wants to find the transition probability (propagator) $p(x, t|x_0, 0)$ specifying the conditional probability for the particle to reach the coordinate x at time t given that the particle was at x_0 at $t = 0$. The solution of this problem is typically found by solving the Fokker-Planck equation [52] discussed in Sec. 7.6. An alternative approach of stochastic path integrals is discussed here to illustrate this theoretical technique [31] in application to this important stochastic process.

Equations (7.50) and (7.51) define overdamped stochastic dynamics in terms of the differential equation carrying significant similarity to the Schrödinger equation of quantum mechanics. In contrast to the wave function, it involves the probability, or transition probability, acted upon by the Hamiltonian defined in Eq. (7.51). The result of this action is equal to the first-order time derivative (Eq. (7.50)). This mathematical similarity implies that the method of path integrals developed for propagation of quantum-mechanical amplitudes (Sec. 4.17) affected by uncertainties of microscopic measurements can be equally applied to stochastic probabilities where uncertainty comes from the Brownian noise R on the right-hand side of Eq. (7.66). One important distinction from the quantum-mechanical transition amplitude (Eq. (4.171)) is that, unlike the Schrödinger equation, the Fokker-Planck equation does not have the imaginary unit i in the time derivative. This distinction implies that the imaginary unit does not appear in front of the action specified by the Lagrangian $L(x, \dot{x})$.

After dropping the imaginary unit in front of the action, the transition probability in Feynman's picture can be defined by the path integral

$$p(x, t|x_0, 0) = \int_{\{x, x_0\}} \mathcal{D}x(\tau) \exp\left[-\int_0^t d\tau L(x, \dot{x}) \right], \qquad (7.67)$$

in which one still needs to specify the Lagrangian $L(x, \dot{x})$. One can proceed by requiring that $x(t)$ satisfies the Langevin equation of motion (Eq. (7.66)) at each point of time τ for $0 \leq \tau \leq t$. In this derivation, we will be using proportionality sign to drop all integration constants while keeping in mind that the propagator needs to be normalized at the final step.

The condition of $x(\tau)$ satisfying the equation of motion is specified by the delta-function

$$p(x, t|x_0, 0) \propto \int_{\{x_0, x\}} \mathcal{D}x \left\langle \prod_{0 \leq \tau \leq t} \delta(\dot{x} + \kappa x - \zeta^{-1}R) \right\rangle_R, \tag{7.68}$$

where the product is over a large number N of intermediate points τ_n slicing the interval $0 \leq \tau \leq t$. The average $\langle \ldots \rangle_R$ is taken over the stochastic trajectory of random noice $R(\tau)$ and the trajectory $x(t)$ is constrained to start from x_0 and reach x after the travel time t (Fig. 7.8).

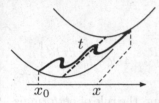

Fig. 7.8 Stochastic trajectory in a harmonic well connecting points x_0 and x. The travel time is t.

One can apply the Fourier integral for the delta-function to write

$$\prod_{n=1}^{N} \delta(\dot{x} + \kappa x - \zeta^{-1}R) = \prod_{n=1}^{N} \int \frac{d\xi_n}{2\pi} \exp\left[i\xi_n(\dot{x}(\tau_n) + \kappa x(\tau_n) - \zeta^{-1}R(\tau_n))\right].$$
$$\tag{7.69}$$

By making the slicing interval $\epsilon = \tau_n - \tau_{n-1}$ go to zero, the above equation can be written as a path integral of the scalar trajectory $\xi(\tau)$

$$p(x, t|x_0, 0) \propto \int_{\{x_0, x\}} \mathcal{D}x \int \mathcal{D}\xi$$
$$\left\langle \exp\left[i \int_0^t d\tau \, \xi(\dot{x} + \kappa x) - (i/\zeta) \int_0^t d\tau \xi(\tau)R(\tau)\right] \right\rangle_R.$$
$$\tag{7.70}$$

The random force $R(\tau)$ is a Gaussian stochastic variable at each point of time τ. The probability density of observing a given realization of the trajectory $R(\tau)$ is given by the Gaussian functional [32] (Sec. 5.14)

$$p[R] \propto \exp\left[-\tfrac{1}{2} \int_0^t d\tau \int_0^t d\tau' R(\tau)\chi_R^{-1}(\tau - \tau')R(\tau')\right]. \tag{7.71}$$

The function $\chi_R(\tau - \tau')$ measures how long in time the fluctuations of the random force stay correlated. For the white noise of Brownian motion, this function becomes delta-correlated

$$\chi_R(\tau - \tau') = \langle R(\tau)R(\tau')\rangle = 2D\zeta^2\delta(t - t'), \qquad (7.72)$$

where $D = k_BT/\zeta$ is the diffusion constant. The inverse of the correlation function $\chi_R^{-1}(\tau - \tau')$ in Eq. (7.71) is defined by the condition

$$\int d\tau'' \chi_R^{-1}(\tau - \tau'')\chi_R(\tau'' - \tau') = \delta(\tau - \tau'). \qquad (7.73)$$

Substituting Eq. (7.72), we get

$$\chi_R^{-1}(\tau - \tau') = (2D\zeta^2)^{-1}\delta(\tau - \tau'). \qquad (7.74)$$

One can now take the average over the random force by performing the path integral over $R(\tau)$

$$p(x,t|x_0,0) \propto \int_{\{x_0,x\}} \mathcal{D}x \int \mathcal{D}\xi \int \mathcal{D}R$$
$$\exp\left[i\int_0^t d\tau\ \xi(\dot{x} + \kappa x) - \frac{i}{\zeta}\int_0^t d\tau\xi R - \frac{1}{4D\zeta^2}\int_0^t d\tau R^2\right]. \qquad (7.75)$$

If we are not interested in the normalization constant, the rule of performing the Gaussian integral is to find the trajectory minimizing the action and then evaluate the action on that minimizing trajectory (Sec. 4.18). For the path $R(\tau)$, the corresponding functional is

$$\mathcal{F}[R] = -(i/\zeta)\int_0^t d\tau\xi R - \frac{1}{4D\zeta^2}\int_0^t d\tau R^2(\tau), \qquad (7.76)$$

and it is minimized by taking the functional derivative $\delta\mathcal{F}/\delta R|_{\tilde{R}} = 0$. The minimizing trajectory is

$$\tilde{R}(\tau) = -2i\zeta D\xi(\tau) \qquad (7.77)$$

and the functional taken on $\tilde{R}$ becomes

$$\tilde{\mathcal{F}} = -D\int_0^t d\tau\ \xi^2. \qquad (7.78)$$

One again gets a Gaussian path integral, now for the trajectory $\xi(\tau)$

$$p(x,t|x_0,0) \propto \int_{\{x_0,x\}} \mathcal{D}x \int \mathcal{D}\xi \exp\left[i\int_0^t d\tau\xi(\dot{x} + \kappa x) - D\int_0^t d\tau\xi^2\right]. \qquad (7.79)$$

Repeating the same steps for $\xi(\tau)$, one obtains

$$p(x,t|x_0,0) \propto \int_{\{x_0,x\}} \mathcal{D}x \exp\left[-\frac{1}{4D}\int_0^t d\tau(\dot{x} + \kappa x)^2\right]. \qquad (7.80)$$

Comparing this result with Eq. (7.67), we conclude that, within a normalization constant to be established, the Lagrangian $L(x, \dot{x})$ is given by the simple relation

$$L(x, \dot{x}) = (4D)^{-1}(\dot{x} + \kappa x)^2. \qquad (7.81)$$

It is instructive to notice a similarity between this expression and the Lagrangian of the particle in the electromagnetic field (Sec. 3.6). In both cases, the Lagrangian contains a term linear in the particle velocity. The calculation of the path integral is nevertheless similar to the steps already performed: one evaluates the action (Eq. (3.1)) on the trajectory minimizing the action functional. Fluctuations around this minimizing trajectory contribute only to the normalization factor.

The equation of motion for the trajectory minimizing the action along the path is found from the Euler-Lagrange equation (Eq. (3.7)), with the following result

$$\ddot{x} - \kappa^2 x = 0. \qquad (7.82)$$

The solution $\tilde{x}(\tau)$ has to satisfy the boundary conditions $\tilde{x}(0) = x_0$ and $\tilde{x}(t) = x$ (Fig. 7.8). The result is

$$\tilde{x}(\tau) = x \frac{\sinh \kappa \tau}{\sinh \kappa t} + x_0 \frac{\sinh \kappa(t - \tau)}{\sinh \kappa t}. \qquad (7.83)$$

The propagator is given by the exponent of the action calculated on the trajectory minimizing it

$$p(x, t | x_0, 0) \propto e^{-S_t[\tilde{x}(t)]}. \qquad (7.84)$$

The action on the minimizing trajectory is

$$\tilde{S}_t = \frac{x^2 - x_0^2}{4\sigma^2} + \frac{1}{4D} \dot{x} x \Big|_0^t = \frac{(x - x_0 \chi(t))^2}{2\sigma^2(1 - \chi^2(t))}, \qquad (7.85)$$

where $\sigma^2 = D/\kappa = \langle x^2 \rangle$ is the statistical variance of the displacement and

$$\chi(t) = \sigma^{-2} \langle x(t) x \rangle = e^{-\kappa t} \qquad (7.86)$$

is the normalized time autocorrelation function of the harmonic displacement. Since $\tilde{S}_t$ is quadratic in x, it describes a time-dependent Gaussian process with the variance given by $\sigma^2(1 - \chi^2(t))$. The normalization is obvious in this case and one gets the time-dependent propagator of the Ornstein-Uhlenbeck process

$$p(x, t | x_0, 0) = \frac{1}{\sqrt{2\pi\sigma^2(1 - \chi^2(t))}} \exp\left[-\frac{(x - x_0\chi(t))^2}{2\sigma^2(1 - \chi^2(t))} \right]. \qquad (7.87)$$

At short times, this equation describes free diffusion

$$p(x,t|x_0,0) \propto \exp\left[-\frac{(x-x_0)^2}{4Dt}\right]. \tag{7.88}$$

This is the propagator of a Brownian particle, which can be connected to the propagator of a free quantum particle in Eq. (4.190) upon replacing the diffusion constant with $i\hbar/2m$.

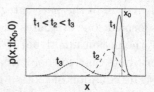

Fig. 7.9 Time dependence of $p(x,t|x_0,0)$ for the Ornstein-Uhlenbeck process.

In the opposite, long-time limit, $t \to \infty$, all information about the initial condition x_0 is lost, equilibrium is reached, and the propagator tends to the Gaussian distribution

$$p(x,t \to \infty|x_0,0) = \frac{1}{\sqrt{2\pi\sigma^2}}e^{-x^2/(2\sigma^2)}. \tag{7.89}$$

By adopting the equipartition theorem, $\sigma^2 = \langle x^2 \rangle = k_B T/k$, this equation becomes the equilibrium Boltzmann distribution. At intermediate times, the probability density $p(x,t|x_0,0)$ shifts and broadens between the delta-function at $t = 0$ and the equilibrium distribution at $t \to \infty$ (Fig. 7.9).

7.9   Mobility of ions

The Langevin equation for an ion diffusing in a polar liquid (e.g., water) can be made more explicit by including electrostatic interactions of the ion with the liquid, in addition to random collisions responsible for Brownian motion. The Langevin equation is particularly simple in the overdamped limit taken in Eq. (7.10)

$$\zeta\mathbf{v} = \mathbf{f}_{el} + \mathbf{R} = \mathbf{f}, \tag{7.90}$$

where $\mathbf{f}$ is the total force from electrostatic interactions, $\mathbf{f}_{el}$, and random collisions, $\mathbf{R}$. From this equation, one can obtain for the diffusion constant

$$D = \frac{1}{3}\int_0^\infty dt\langle\delta\mathbf{v}(t)\cdot\delta\mathbf{v}\rangle = \frac{1}{3\zeta^2}\int_0^\infty dt\langle\delta\mathbf{f}(t)\cdot\delta\mathbf{f}\rangle, \tag{7.91}$$

where $\delta\mathbf{f} = \mathbf{f} - \langle\mathbf{f}\rangle$ is the deviation from the average force. By applying Einstein's equation (Eq. (7.8)), one obtains the friction coefficient in terms of the force-force correlation function

$$\zeta = (\beta/3) \int_0^\infty dt \langle \delta\mathbf{f}(t) \cdot \delta\mathbf{f} \rangle. \tag{7.92}$$

One can next assume an exponential decay of electrostatic correlations with the relaxation time τ_E

$$\langle \delta\mathbf{f}_{el}(t) \cdot \delta\mathbf{f}_{el} \rangle = \langle (\delta\mathbf{f}_{el})^2 \rangle e^{-t/\tau_E}. \tag{7.93}$$

Substituting this into Eq. (7.92) and assuming that the random force of local molecular collisions is statistically uncorrelated from the long-range electrostatic forces, one obtains

$$\zeta = \zeta_H + \tfrac{1}{3}\beta \langle (\delta\mathbf{f}_{el})^2 \rangle \tau_E. \tag{7.94}$$

The hydrodynamic friction ζ_H can be expressed in terms of the Stokes hydrodynamic friction (Sec. 8.7)

$$\zeta_H = 6\pi\eta R \tag{7.95}$$

experienced by a spherical particle with the radius R in a liquid with the shear viscosity η. The second term in Eq. (7.94) is the dielectric friction due to the electric field of the surrounding liquid $\mathbf{E}_s$. This field defines the electrostatic force through the ionic charge q

$$\delta\mathbf{f}_{el} = q\delta\mathbf{E}_s. \tag{7.96}$$

The dielectric friction in Eq. (7.94) thus scales as q^2 with the ionic charge. One can express the dependence of the diffusion constant on the ionic charge with the equation

$$\frac{D(0)}{D(q)} = 1 + \frac{q^2}{3}\beta^2 \langle (\delta\mathbf{E}_s)^2 \rangle \tau_E D(0), \tag{7.97}$$

where $D(0) = k_B T/\zeta_H$ is the diffusion constant at zero charge.

Fluctuations of the electric field at the position of the ion can be related to solvation energy of a fictitious dipole $\mathbf{m}_{fic}$ placed at the position of the charge. The interaction energy of this dipole with the surrounding solvent is $-\mathbf{m}_{fic} \cdot \mathbf{E}_s$. The variance of this interaction allows one to determine the solvation free energy (Eq. (12.31)) as $\Delta\mu = -(\beta m_{fic}^2/6)\langle (\delta\mathbf{E}_s)^2 \rangle$ (see Sec. 12.3). One can therefore express the dependence of the diffusion constant on the ionic charge in terms of the free energy of dipole solvation

$$\frac{D(0)}{D(q)} = 1 - q^2 \tau_E D(0) \frac{2\beta\Delta\mu}{m_{fic}^2}. \tag{7.98}$$

Fig. 7.10 Delayed response of the liquid dipoles to the movement of the ion; τ_D is a characteristic time of dipole rotations.

In turn, $\Delta\mu$ can be expressed as the solvent-induced Stokes shift of an optical dye, as we discuss in more detail in Sec. 11.5.

The physical origin of dielectric friction is that the moving charge creates a time-dependent field in the surrounding polar liquid (Fig. 7.10). This field causes the dipoles of the liquid to rotate, but because their rotations are characterized by an intrinsic relaxation time, the response of the dipoles is delayed, leading to dielectric losses. The loss of energy to heating the liquid has the effect of slowing ion's motion down, i.e., to an effective friction. If the response of the dipoles was instantaneous, one would get $\tau_E = 0$ and no friction from electrostatic forces.

7.10   Memory function

Random forces acting on a Brownian particle and entering the Langevin equation are viewed as instantaneous collisions with very short time of correlation or memory. Each collision event happens very fast and is independent from the next collision event. This is reflected by the delta-function in the time correlation function of the random force in Eq. (7.18). This is obviously a simplification of actual interactions in the liquid which must happen on a finite, even if short, collision time. A generalization of the Brownian dynamics picture is achieved by the generalized Langevin equation, which is rigorously derived by the projection operators formalism [39, 41]. It leads to a nonlocal equation of motion which involves a time convolution integral for a dynamic variable $A(t)$. The generalized Langevin equation becomes

$$\dot{A}(t) + \int_0^t d\tau M(t - \tau) A(\tau) = R(t), \qquad (7.99)$$

where $\langle A \rangle = 0$ is assumed.

The random force $R(t)$ is statistically independent from the initial value $A = A(0)$

$$\langle AR(t) \rangle = 0. \qquad (7.100)$$

However, the time evolution of the random force involves some period of time when the memory of a random collision is preserved. The function which describes the decay of correlations between successive collisions is called the memory function

$$M(t) = \langle A^2 \rangle^{-1} \langle R(t)R \rangle. \tag{7.101}$$

It is exactly this function that enters the integral convolution in Eq. (7.99). This relation, which establishes a proportionality between the amplitude of the random noise and the memory function (also called the friction kernel), is often called the second fluctuation-dissipation theorem.

The normalized time autocorrelation function for the variable $A(t)$ is defined as

$$\Phi(t) = \langle A^2 \rangle^{-1} \langle A(t)A \rangle. \tag{7.102}$$

Given the lack of correlations between the initial value $A(0)$ and the random force (Eq. (7.100)), one can write a simple time evolution equation for the correlation function

$$\dot{\Phi} + \int_0^t d\tau M(t-\tau)\Phi(\tau) = 0. \tag{7.103}$$

This equation is known as the memory equation.

Taking the time derivative of Eq. (7.103) yields the memory function at $t = 0$

$$\ddot{\Phi}(t) + M(0)\Phi(t) + \int_0^t d\tau \dot{M}(t-\tau)\Phi(\tau) = 0. \tag{7.104}$$

By using the condition $\Phi(0) = 1$, one obtains at $t = 0$

$$\boxed{M(0) = \langle \dot{A}^2 \rangle / \langle A^2 \rangle.} \tag{7.105}$$

Here, we used the identity $\langle \dot{A}^2 \rangle = -\langle \ddot{A}A \rangle$.

Equation (7.104) can also be written as

$$M(0) = -\ddot{\Phi}(0)/\Phi(0) = \int_{-\infty}^{\infty} d\omega \omega^2 \Phi(\omega) / \int_{-\infty}^{\infty} d\omega \Phi(\omega), \tag{7.106}$$

where $\Phi(\omega)$ is the time Fourier transform of the correlation function. This relation thus establishes an important physical meaning of the memory function. Its initial value is equal to the normalized second frequency moment

$$\boxed{M(0) = \langle \omega^2 \rangle.} \tag{7.107}$$

The integral equation for the correlation function (Eq. (7.103)) is converted to an algebraic relation when the time convolution is eliminated by Fourier-Laplace transform

$$\tilde{\Phi}(\omega) = \left[-i\omega + \tilde{M}(\omega) \right]^{-1}. \tag{7.108}$$

This equation can be rewritten in the form that clarifies the reason for introducing the memory function. One can write

$$(1 + i\omega\tilde{\Phi})(1 - i\omega/\tilde{M}) = 1. \tag{7.109}$$

This equation is a time analog of the Ornstein-Zernike equation in the theory of liquid structure (Eq. (6.36)). Similarly to the propagation of chains of short-range spatial correlations (Fig. 6.8), Eq. (7.109) propagates chains of short-time memory functions to establish slower dynamics of $\Phi(t)$. The inverse of the memory function $\tilde{M}(\omega)^{-1}$ plays the role of the direct correlation function in reciprocal space.

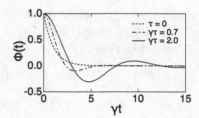

Fig. 7.11 Correlation function $\Phi(t)$ in the viscoelastic model at different values of $\gamma\tau$; $\tau = 0$ refers to the Langevin equation (Eq. (7.10)).

It is clear that the standard form of the Langevin equation (Eq. (7.10)) is recovered by adopting the delta-correlated memory function, $M(t) = \gamma\delta(t)$. For a more general description of molecular collisions, an exponential function is often adopted (viscoelastic model)

$$M(t) = (\gamma/\tau)e^{-t/\tau}. \tag{7.110}$$

With this memory function, the correlation function becomes

$$\tilde{\Phi}(\omega) = \left[-i\omega + \frac{\gamma}{1 - i\omega\tau}\right]^{-1}. \tag{7.111}$$

This function evolves with two eigenfrequencies, becoming oscillatory at $4\gamma\tau > 1$ (Fig. 7.11). This example illustrates the statement that a smooth temporal decay of the memory function $M(t)$ is projected by the generalized Langevin equation to a more complex dynamics of $\Phi(t)$. The transition from an oscillatory time correlation function to a smooth decay shown in Fig. 7.11 marks the transition in the dynamics from the underdamped to overdamped behavior.

At $\tau \to 0$, the memory function decays increasingly fast, approaching the limit of the Langevin equation for which

$$\Phi(t) = e^{-\gamma t}. \tag{7.112}$$

One notes that as $\tau \to 0$, the second frequency moment, which is often associated with the frequency of cage vibrations, is bound to diverge as $\langle \omega^2 \rangle \propto 1/\tau$ (Eqs. (7.107) and (7.110)) to allow a constant value of the friction coefficient γ and of the corresponding generalized diffusion constant

$$D = k_B T/(m\gamma) = k_B T/(m\tau \langle \omega^2 \rangle), \qquad (7.113)$$

where, from Eq. (7.106), $M(0) = \langle \omega^2 \rangle = \gamma/\tau$. Collisions become increasingly frequent as $\tau \to 0$, but the exponential decay of the memory function is still allowed.

If the dynamic variable $A = v_x$ is associated with one of the Cartesian projections of the particle velocity, D in Eq. (7.113) becomes that standard diffusion constant for translational motion of a particle. Equation (7.105) provides direct access to the friction coefficient γ for exponentially decaying memory functions

$$\gamma/\tau = \langle \omega^2 \rangle = \langle \dot{v}_x^2 \rangle / \langle v_x^2 \rangle = (\beta/m)\langle f_x^2 \rangle. \qquad (7.114)$$

The friction coefficient $\zeta = m\gamma$ becomes

$$\zeta = \beta\tau \langle (\delta f_x)^2 \rangle. \qquad (7.115)$$

This is Eq. (7.92) applied to the specific case of an exponentially relaxing memory function.

Chapter 8

Molecular Hydrodynamics

Molecular hydrodynamics describes the structure and dynamics of condensed materials in terms of spatially varying scalar and vector/tensor fields. The time evolution of these fields is described by the Liouville equation, which specifies the Hamiltonian dynamics of dynamic variables in the phase space of coordinates and momenta of the particles in the macroscopic sample. Conserved variables, which are properties not changing when the system evolves in time, are special and important cases of the Hamiltonian dynamics producing zero eigenvalues for the Liouville equation (these variables do not change). The total number of particles, mass and charge, and the total momentum and energy of the macroscopic sample are conserved variables for which hydrodynamic equations of motions can be written (such as the Navier-Stokes equation for the momentum density). Local variables, such as local density, are not conserved and vary in space. Their Fourier transforms produce corresponding scalar and vector fields in reciprocal space (Sec. 1.7), such as the time-dependent reciprocal-space density $\rho_{\mathbf{k}}(t)$ considered in this Chapter. These variables are conserved at $k \to 0$ and their dynamics satisfy in that limit the hydrodynamic equations for conserved variables. Molecular hydrodynamics [39, 41, 42, 55, 56] seeks to establish equations of motions for dynamic collective variables by combining both the correct hydrodynamic limit at $k \to 0$ and the microscopic properties of materials arising from local correlations between the molecules.

This Chapter starts with the Hamiltonian dynamics in the phase space of coordinates and momenta of an ensemble of particles (Liouville's equation). It then proceeds to the hydrodynamic Navier-Stokes equation and to the problem of hydrodynamic drag imposed by a viscous liquid on a moving rigid body. Still the main topic of the Chapter is to cover the ideas of molecular hydrodynamics operating in terms of collective variables defined at different spatial and temporal scales.

8.1   Liouville equation

Poisson bracket:

$$\{f, H\} = \sum_i \left(\frac{\partial f}{\partial x_i} \frac{\partial H}{\partial p_i} - \frac{\partial f}{\partial p_i} \frac{\partial H}{\partial x_i} \right)$$

Consider a system of N particles represented by their canonical variables $\mathbf{x} = \mathbf{x}_1 \ldots \mathbf{x}_N$ and $\mathbf{p} = \mathbf{p}_1 \ldots \mathbf{p}_N$. If we know the positions $\mathbf{x}^0$ and momenta $\mathbf{p}^0$ at $t = 0$, we can solve the Hamilton equations of motions and obtain with certainty the positions and momenta at time t. This program is difficult to realize for a realistic system since we cannot know the initial conditions at $t = 0$ with certainty. One, therefore, reduces the description to a statistical function $f(\mathbf{x}^0, \mathbf{p}^0)$ which specifies the probability density of obtaining a specific realization of initial conditions. As is true for any description involving probability, one requires that this function is nonnegative, $f(\mathbf{x}^0, \mathbf{p}^0) \geq 0$, and is normalized on the phase space of the system

$$\int f(\Gamma_0) d\Gamma_0 = 1, \quad d\Gamma_0 = \prod_i d\mathbf{x}_i^0 d\mathbf{p}_i^0. \tag{8.1}$$

It is important to realize that $f(\Gamma_0) = f(\mathbf{x}^0, \mathbf{p}^0)$ is not a description of the system, but a description of our ignorance about the state of the system at $t = 0$. One can ask if our ignorance is going to change as the particles of the system are allowed to move along the trajectories specified by the Hamilton equations of motion (Sec. 3.2).

To answer this question, one can chose a small volume of phase space $d\Gamma_0$ and the probability for this choice to realize

$$dp = f(\mathbf{x}^0, \mathbf{p}^0) d\Gamma_0. \tag{8.2}$$

As the system is allowed to evolve for time t, our ignorance will not increase or decrease since the Hamilton equations are deterministic and one can say with certainty that $\mathbf{x}^0$ will become $\mathbf{x}(t)$ and $\mathbf{p}^0$ will become $\mathbf{p}(t)$. Therefore, the probability dp will not change

$$dp = f(\mathbf{x}^0, \mathbf{p}^0) d\Gamma_0 = f(\mathbf{x}(t), \mathbf{p}(t)) d\Gamma_t, \tag{8.3}$$

where $d\Gamma_t$ is the element of the phase space evolved over time t. However, according to Liouville's theorem (Sec. 3.4) the volume of any given sub-ensemble chosen at $t = 0$ will not change at time t and $d\Gamma_t = d\Gamma_0$. One, therefore, obtains

$$f_t = f(\mathbf{x}(t), \mathbf{p}(t)) = f(\mathbf{x}^0, \mathbf{p}^0) = f_{t=0}. \tag{8.4}$$

This is the meaning of Liouville's equation which states that the total derivative of the phase space probability density is equal to zero

$$\boxed{\frac{df_t}{dt} = 0.} \tag{8.5}$$

In other words, this equation states that if the dynamic variables are allowed to evolve according to the Hamiltonian dynamics, the information content about the system (or the level of ignorance) does not change.

Liouville's equation is an extension of Newton's dynamics to a large ensemble of particles. We now provide a somewhat more formal derivation allowing us to introduce the Liouville operator. Let us define all time derivatives of $3N$ $\mathbf{x}$'s and $3N$ $\mathbf{p}$'s as the $6N$ vector $\mathbf{v} = \{\dot{\mathbf{x}}, \dot{\mathbf{p}}\}$. The normalization of probability density implies the continuity relation (also see Eqs. (3.34) and (7.47))

$$\partial_t f_t + \operatorname{div}(f_t \mathbf{v}) = 0. \tag{8.6}$$

We can now write

$$\operatorname{div}(f_t \mathbf{v}) = \nabla f_t \cdot \mathbf{v} + f_t \operatorname{div}(\mathbf{v}), \tag{8.7}$$

However, the phase-space "fluid" is incompressible, which means that $\operatorname{div}(\mathbf{v}) = 0$ (Liouville's theorem, Sec. 3.4).

The first summand in Eq. (8.7) can be expressed in terms of the Poisson bracket. Since the vector $\mathbf{v}$ includes both $\mathbf{x}$ and $\mathbf{p}$, one gets

$$\nabla f_t \cdot \mathbf{v} = \sum_i \left(\frac{\partial f_t}{\partial \mathbf{x}_i} \cdot \dot{\mathbf{x}}_i + \frac{\partial f_t}{\partial \mathbf{p}_i} \cdot \dot{\mathbf{p}}_i \right). \tag{8.8}$$

Upon substituting classical Hamilton equations of motion (Eq. (3.15)), one arrives at the time evolution for f_t given in terms of the Poisson bracket $\{H, f_t\}$

$$\boxed{\partial_t f_t = -\{f_t, H\} = \{H, f_t\}.} \tag{8.9}$$

From Eq. (3.23) specifying the total time derivative in terms of the Poisson bracket, Eqs. (8.5) and (8.9) are equivalent.

One next defines the Liouville operator

$$-i\hat{L}g = \{H, g\} \tag{8.10}$$

such that the Liouville equation becomes a linear operator equation, very much like the Schrödinger equation

$$\partial_t f_t = -i\hat{L}f_t \tag{8.11}$$

or symbolically

$$f_t = e^{-i\hat{L}t}f_{t=0}. \tag{8.12}$$

Since the probability density is inverse of the phase space occupied by the ensemble, the unitary operator $e^{i\hat{L}t}$ describes the evolution of the phase space Γ_t occupied by the ensemble with time (Fig. 8.1). This is expressed by the relation

$$\Gamma_t = e^{i\hat{L}t}\Gamma_0. \tag{8.13}$$

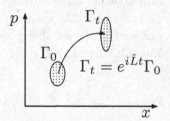

Fig. 8.1 Evolution of a dynamical ensemble.

The Liouville operator is Hermitian. This observation allows one to apply many results developed in quantum mechanics to problems of classical dynamics. For instance, one can derive the evolution equation for a dynamic variable $A_t = A(\Gamma_t) = A(\mathbf{x}(t), \mathbf{p}(t))$, which changes with time because the phase space $\mathbf{x}(t), \mathbf{p}(t)$ does.

Since the entire evolution of A_t is caused by the Newtonian dynamics of the particles in the system, one can write

$$\frac{dA_t}{dt} = \sum_i \left(\frac{\partial A_t}{\partial \mathbf{x}_i} \frac{\partial H}{\partial \mathbf{p}_i} - \frac{\partial A_t}{\partial \mathbf{p}_i} \frac{\partial H}{\partial \mathbf{x}_i} \right) = \{A_t, H\} = i\hat{L}A_t \tag{8.14}$$

and

$$A_t = A(\Gamma_t) = e^{i\hat{L}t}A(\Gamma_0) = e^{i\hat{L}t}A_{t=0}. \tag{8.15}$$

Note the opposite signs in front of the Poisson brackets defining the evolution of f_t in Eq. (8.9) and A_t in Eq. (8.14).

Equations (8.12) and (8.15) represent two alternative views on the evolution of dynamical variables and the calculation of statistical averages. In the first view (analogous to the Schrödinger picture in quantum mechanics), the dynamic variables depend on the coordinates in the phase space, $A(\Gamma_0)$, but do not evolve. The average is specified by the time-dependent probability density

$$\langle A \rangle_t = \int d\Gamma_0 A(\Gamma_0) f_t. \tag{8.16}$$

In an alternative representation (analog of the Heisenberg formulation in quantum mechanics), the dynamic variable is evolving due to evolving coordinates and momenta, but the average is taken over the distribution of initial coordinates

$$\langle A \rangle_t = \int d\Gamma_0 A_t f(\Gamma_0). \tag{8.17}$$

The Hermitian Liouville operator can act on the function either to the left or to the right of it (with the change of i to $-i$). One gets

$$\langle A \rangle_t = \int d\Gamma_0 f(\Gamma_0) e^{i\hat{L}t} A(\Gamma_0) = \int d\Gamma_0 e^{-i\hat{L}t} f(\Gamma_0) A(\Gamma_0) = \int d\Gamma_0 A(\Gamma_0) f_t. \tag{8.18}$$

Therefore, the two representations are equivalent in terms of producing identical statistical averages.

The time derivative of the average dynamic variable can be considered next

$$\frac{d\langle A \rangle}{dt} = \int d\Gamma_0 A(\Gamma_0) \partial_t f_t. \tag{8.19}$$

Note that the partial derivative of the probability density appears under the integral since no dynamics is attached to the integration variables $\mathbf{x}^0$ and $\mathbf{p}^0$. The time derivative of an ensemble average should be equal to zero in a closed mechanical system which reached equilibrium. This condition is satisfied if there is no explicit dependence on time for the probability density

$$\partial_t f_t = 0. \tag{8.20}$$

According to Eq. (8.9), this also implies $\{H, f_t\} = 0$. i.e., the phase probability density becomes a constant of motion. This observation allows one to suggest that the probability density can be viewed as a function of the Hamiltonian $f(H)$. Since the energy is conserved on each trajectory (and trajectories do not cross), the probability density for a given energy E is proportional to the delta function fixing the energy on the trajectory

$$f(\Gamma_t) \propto \delta\left(H(\Gamma_t) - E\right). \tag{8.21}$$

This is the microcanonical probability density.

These results describe the dynamics of a closed mechanical system. We next turn to the question of what happens when a weak external perturbation is applied and the system is not closed anymore. The effect of a weak perturbation on the dynamic variables is provided by the linear response theory developed by Kubo [57] and covered in the next section. One can skip the derivation presented below and proceed to the fluctuation-dissipation relations.

8.2   *Linear response theory

$$\boxed{\partial_t f_t = \{H, f_t\}}$$

The derivation of the linear response approximation starts with the assumption that the change in the system Hamiltonian is represented by the product of the external force $f(t)$, which in most cases of interest changes with time, and a conjugate displacement q. The product $H'(t) = -f(t)q$ is the energy supplied to the system externally. Neither the "force" nor the "displacement" need to carry physical meaning. The only restriction is that their product is an energy perturbation. The linear response theory seeks to establish the temporal variation of the generalized displacement $q(\Gamma_t)$ in response to the generalized force $f(t)$.

As an example, a chemical reaction can create charge $z(t)$ at the molecule. In that case $f(t) = z(t)$ is the external "force" and the negative of the electrostatic potential $-\phi(\Gamma_t)$, altered by the new charge, is the "displacement". The product $H'(t) = z(t)\phi(\Gamma_t)$ is obviously the electrostatic energy viewed as a perturbation.

We want to calculate the response $\Delta q(t)$ to an external perturbation $f(t)$ assuming that the system is at equilibrium at $t = 0$. This implies that $f_{t=0} = f_{eq} = Z^{-1}\exp[-\beta H]$, Z is the partition function. We need to calculate the following statistical average

$$\langle \Delta q \rangle_t = \int d\Gamma q(\Gamma)\Delta f_t, \tag{8.22}$$

where $f_t = f_{eq} + \Delta f_t$.

Both $H'(t)$ and the change of the probability density Δf_t are viewed as perturbations of the first order. By collecting only the terms of the first order in the energy perturbation, the Liouville equation becomes

$$\partial_t \Delta f_t = -i\hat{L}\Delta f_t - f(t)\{q, f_{eq}\}, \tag{8.23}$$

where $-i\hat{L}f_t = \{H, f_t\}$ is constructed from the unperturbed Hamiltonian H. Note that the term $\{q, \Delta f_t\}$ is omitted in the right-hand part of Eq. (8.23) because it is quadratic in the perturbation. Only terms of first order in the perturbation are included in the linear response theory. Further, for the last summand in Eq. (8.23) one can use Eq. (8.14) for the time derivative of a dynamic variable (the generalized displacement q here)

$$\{q, f_{eq}\} = -\beta\{q, H\}f_{eq} = -\beta\dot{q}f_{eq}. \tag{8.24}$$

Equation (8.23) is a linear differential equation with the initial condition $\Delta f_t(0) = 0$. It is solved by substituting Eq. (8.24) to the right-hand side

$$\Delta f_t = \beta \int_0^t d\tau f(\tau) e^{-i\hat{L}(t-\tau)}\dot{q}f_{eq}. \tag{8.25}$$

Correspondingly, one obtains for $\langle \Delta q \rangle_t$ in Eq. (8.22)

$$\boxed{\langle \Delta q \rangle_t = \int_0^t d\tau \chi_q(t-\tau)f(\tau),} \tag{8.26}$$

where we have defined the dynamic susceptibility or response function

$$\begin{aligned}
\chi_q(t-\tau) &= \beta \mathrm{Tr}\left[qe^{-i\hat{L}(t-\tau)}\dot{q}f_{eq}\right] \\
&= \beta \mathrm{Tr}\left[q(t-\tau)\dot{q}f_{eq}\right].
\end{aligned} \tag{8.27}$$

Replacing $\mathrm{Tr}\left[\ldots f_{eq}\right] = \int \ldots f_{eq}d\Gamma_0$ with $\langle \ldots \rangle$, one finally obtains

$$\boxed{\chi_q(t) = \beta\langle q(t)\dot{q}\rangle,} \tag{8.28}$$

where $\dot{q} = \dot{q}(0)$ and the angular brackets

$$\langle \ldots \rangle = \int d\Gamma_0 \ldots f_{eq}(\Gamma_0) \tag{8.29}$$

denote integration over the phase space of initial coordinates and velocities weighted with the equilibrium distribution function $f_{eq}(\Gamma_0)$.

Equations (8.26) and (8.28) are the main result of the linear response approximation for the dynamic variables evolving according to Newtonian dynamics. The linear response function represents a delayed change in the generalized displacement at time t due to the perturbation applied at an earlier time τ. The total displacement is the sum (integral) of all such delayed displacements.

One can write an alternative form of Eq. (8.28)

$$\chi_q(t) = -\beta\langle\dot{q}(t)q\rangle. \tag{8.30}$$

The reason for two alternative expressions is the invariance of the time correlation function to a uniform time shift

$$\frac{d}{ds}\langle q(t+s)q(s)\rangle = 0. \tag{8.31}$$

From this equation, one obtains by putting $s = 0$

$$\langle\dot{q}(t)q\rangle = -\langle q(t)\dot{q}\rangle. \tag{8.32}$$

8.3   Fluctuation-dissipation theorem

$$\boxed{q(t) = e^{i\hat{L}t}q}$$

The fluctuation-dissipation theorem (FDT) connects fluctuations caused by thermal agitation to dissipation of energy supplied by an external perturbation. From the qualitative perspective, it also points to the source of fluctuations in any given system: the part of the system responsible for dissipation is also the source of fluctuations [21].

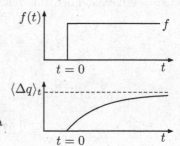

Fig. 8.2   Displacement $\langle\Delta q\rangle_t$ in response to a step of the force.

To appreciate different limits of the FDT, we first consider the simplest case of a step perturbation $f(t) = fh(t)$, where $h(t)$ is the Heaviside function. The perturbation Hamiltonian is the product of the generalized displacement q and the generalized force, $H'(t) = -qf(t)$. The linear change of the displacement is given in terms of the response function $\chi_q(t-\tau)$ by Eq. (8.26), which we can write now as

$$\langle\Delta q\rangle_t = -\beta f \int_0^t d\tau\langle\dot{q}(t-\tau)q\rangle = -\beta f \int_0^t d\tau\langle\dot{q}(\tau)q\rangle, \tag{8.33}$$

where the meaning of displacement implies $\langle q \rangle = 0$. This equation is integrated over τ to yield

$$\langle \Delta q \rangle_t = \beta \left[C_q(0) - C_q(t) \right] f, \tag{8.34}$$

where

$$C_q(t) = \langle q(t)q \rangle \tag{8.35}$$

is the time autocorrelation function of the stochastic variable $q(t)$ (Fig. 8.3). The variable $\langle \Delta q \rangle_t$ describes the evolution of displacement in response to the step of the force (Fig. 8.2). Since $C_q(t) \to 0$ at $t \to \infty$, the displacement saturates to the following value at the infinite time $t \to \infty$

$$\chi_q(\infty) = \frac{\langle \Delta q \rangle_\infty}{f} = \beta \langle q^2 \rangle. \tag{8.36}$$

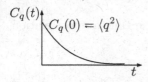

Fig. 8.3 Time autocorrelation function of the dynamic variable $q(t)$.

Equation (8.36) is the static limit of the FDT stating that the response to a weak perturbation is related to fluctuations of the same variable due to thermal agitation. Somewhat rearranged, this result was first formulated by Nyquist in an attempt to understand thermal noise of current in electrical circuits (Johnson noise [21]): the variance $\langle q^2 \rangle$ caused by thermal agitation is proportional to temperature

$$\boxed{\langle q^2 \rangle = k_{\mathrm{B}} T \chi_q(\infty).} \tag{8.37}$$

Higher temperature thus brings about proportionally stronger noise to macroscopic observables. The Einstein equation (7.6) for the diffusion constant is a specific realization of this general result.

The most celebrated form of the FDT is in the frequency domain. One assumes in this case that the external generalized force is oscillatory, $f(t) = f_\omega e^{-i\omega t}$. Additionally, we want to look at the stationary changes in the system and neglect transient effects of introducing the perturbation, just the opposite of what has been done when considering a step perturbation. To avoid transient effect, the point at which the perturbation is introduced is shifted to $t = -\infty$. In the stationary response, the displacement in response to an oscillating force is also oscillatory, $\langle \Delta q \rangle_t = q_\omega e^{-i\omega t}$. We obtain from Eq. (8.26)

$$q_\omega e^{-i\omega t} = f_\omega \int_{-\infty}^{t} d\tau \chi_q(t - \tau) e^{-i\omega\tau}. \tag{8.38}$$

This equation simplifies to

$$q_\omega = \tilde{\chi}_q(\omega) f_\omega, \tag{8.39}$$

where $\tilde{\chi}_q(\omega)$ is the Fourier-Laplace (one-sided Fourier) transform

$$\tilde{\chi}_q(\omega) = \int_0^{\infty} dt \chi_q(t) e^{i\omega t}. \tag{8.40}$$

From the linear response relation for the response function (Eq. (8.30)), we further obtain

$$\tilde{\chi}_q(\omega) = \beta \left[C_q(0) + i\omega \tilde{C}_q(\omega) \right], \tag{8.41}$$

where $\tilde{C}_q(\omega)$ is the Fourier-Laplace transform of the time autocorrelation function $C_q(t)$ (Eq. (8.35))

$$\tilde{C}_q(\omega) = \int_0^{\infty} dt C_q(t) e^{i\omega t}. \tag{8.42}$$

By taking the imaginary part of both sides of Eq. (8.41), we arrive at the frequency-domain formulation of the FDT

$$\boxed{2\tilde{\chi}_q''(\omega) = \beta \omega C_q(\omega),} \tag{8.43}$$

where now $C_q(\omega)$ is the full Fourier transform of $C_q(t)$ including integration over the entire time domain from $-\infty$ to $+\infty$:

$$C_q(\omega) = \int_{-\infty}^{\infty} dt C_q(t) e^{i\omega t}. \tag{8.44}$$

Note that $C_q(\omega)$, also known as spectral density, is real for autocorrelation functions.

Equation (8.43) is the celebrated FDT relation connecting the loss spectrum, $\tilde{\chi}_q''(\omega)$ to the spectral density $C_q(\omega)$. As mentioned above, the physical importance of this relation is the direct connection between dissipation of energy and the production of thermal fluctuations [39]. The notion of dissipation is related to the fact that the energy absorbed from the external field (force $f(t)$) and dissipated by the system into heat is proportional to $\omega \tilde{\chi}_q''(\omega)$.

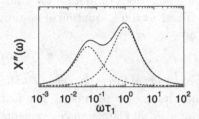

Fig. 8.4 Loss spectrum (Eq. (8.47)) of two exponential relaxation processes with relaxation times τ_1 and $\tau_2 = 20\tau_1$. The dashed lines indicate each Debye process separately.

It is instructive to consider a common case of dynamics in condensed materials when the time correlation function can be represented by a sum (finite or continuous) of decaying exponential functions

$$C_q(t) = C_q(0) \sum_i a_i e^{-t/\tau_i}, \quad \sum_i a_i = 1. \tag{8.45}$$

The one-sided Fourier-Laplace transform becomes

$$\tilde{C}_q(\omega) = C_q(0) \sum_i \frac{a_i \tau_i}{1 - i\omega\tau_i}. \tag{8.46}$$

The Fourier transform of the time correlation function is $2\mathrm{Re}[\tilde{C}_q(\omega)]$ and one gets for the loss spectrum

$$\tilde{\chi}_q''(\omega) = \beta C_q(0) \sum_i a_i \frac{\omega\tau_i}{1 + (\omega\tau_i)^2}. \tag{8.47}$$

The loss function of the form

$$\tilde{\chi}''(\omega) \propto \frac{\omega\tau}{1 + (\omega\tau)^2} \tag{8.48}$$

is known as the Debye spectrum. It is often encountered in dielectric spectroscopy of high-temperature polar liquids [58]. Figure 8.4 shows an example of the loss spectrum produced by two Debye processes. Note the logarithmic scale of frequencies in the plot.

8.4 Navier-Stokes equation

Viscosity:
$[\eta] = \mathrm{Pa} \times \mathrm{s}$
$\eta(\text{water}) \simeq 1 \ \mathrm{mPa} \times \mathrm{s}$
Kinematic viscosity:
$\nu = \eta/\rho_m, \ [\nu] = \mathrm{m}^2/\mathrm{s}$
Longitudinal and transverse fields:
$\nabla \times \mathbf{v}_L = 0, \ \nabla \cdot \mathbf{v}_T = 0$

Navier-Stokes equation is the Newtonian equations of motion for fluids with dissipation of energy due to friction. The Newton's equation of motion for a continuous medium can be written as

$$\rho_m \dot{v}_\alpha = -\rho \partial_\alpha U_{\text{ext}} + \partial_\beta \sigma_{\alpha\beta}. \qquad (8.49)$$

Here, U_{ext} represents the potential energy of external forces, $\rho_m = m\rho$ is the mass density, and summation over common indexes is assumed. The last term in the above equation implies the sum over all Cartesian components $\beta = x, y, x$

$$\partial_\beta \sigma_{\alpha\beta} = \sum_{\beta = x, y, z} \frac{\partial}{\partial r_\beta} \sigma_{\alpha\beta}. \qquad (8.50)$$

Further, $\sigma_{\alpha\beta}$ is the stress tensor related to pressure and viscous forces (see more on stress tensor in Secs. 9.6 and 9.7). The divergence of the two-rank stress tensor produces the α-component of the bulk force acting at a given point of the sample

$$f_\alpha = \partial_\beta \sigma_{\alpha\beta}. \qquad (8.51)$$

Here, f_α is the force per unit volume such that the force acting on the volume element dV is $dF_\alpha = f_\alpha dV$.

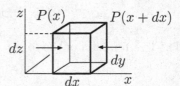

Fig. 8.5   Gradient of hydrostatic pressure.

The simplest model for the stress tensor is given by Euler's equation which assumes no dissipation in an isotropic liquid with the hydrostatic pressure $P(\mathbf{r})$. Figure 8.5 illustrates the connection between the force produced by the pressure gradient and the stress tensor. Consider the force acting on the cubic volume element $dV = dx dy dz$ caused by the pressure gradient along the x-axis. The pressure difference from the left (at x) and from the right (at $x + dx$) is multiplied with the area of the face $dy dz$

$$dF_x = [P(x) - P(x + dx)] \, dy dz \simeq -\partial_x P dV. \qquad (8.52)$$

One therefore obtains for the force per unit volume

$$f_x = -\partial_\beta \delta_{x\beta} P. \qquad (8.53)$$

The stress tensor is diagonal in this case becoming

$$\sigma_{\alpha\beta} = -\delta_{\alpha\beta} P. \qquad (8.54)$$

This is the isotropic part of the stress tensor. Substituting this result to Eq. (8.49), one arrives at Euler's equation

$$\dot{v}_\alpha = -m^{-1}\partial_\alpha U_\text{ext} - \rho_m^{-1}\partial_\alpha P. \tag{8.55}$$

This equation is usually used in hydrostatics when $\dot{v}_\alpha = 0$ and the change in pressure is related to the change in the potential energy: $\Delta P = -\rho\Delta U$. In the case of gravitational force, $\Delta U = -mgh$ and the change of pressure with the depth h of the liquid becomes $\Delta P = \rho_m gh$, where g is the acceleration of free fall.

In a more general case, the stress tensor producing linear shear deformation can be related to the time derivative of the strain tensor $\epsilon_{\alpha\beta}$ (see also Eq. (9.44)). Since shear is represented by a traceless tensor, it is given as [59]

$$\begin{aligned}
\sigma_{\alpha\beta} &= 2\eta \left[\dot{\epsilon}_{\alpha\beta} - \tfrac{1}{3}\delta_{\alpha\beta}\nabla \cdot \mathbf{v} \right] \\
&= \eta \left(\partial_\alpha v_\beta + \partial_\beta v_\alpha \right) - \tfrac{2}{3}\eta\delta_{\alpha\beta}\nabla \cdot \mathbf{v},
\end{aligned} \tag{8.56}$$

where η is the shear viscosity. One observes that shear is produced by the spatial variation of the velocity field.

The second component of stress is produced by bulk deformations (compressions) characterized by bulk viscosity η_B. This term contributes $\delta_{\alpha\beta}\eta_B\nabla \cdot \mathbf{v}$ to the stress tensor. The overall linear stress-strain relation adds the isotropic hydrostatic pressure P already accounted for in Euler's equation to the shear and bulk viscosities

$$\sigma_{\alpha\beta} = -\delta_{\alpha\beta}P + \eta \left(\partial_\alpha v_\beta + \partial_\beta v_\alpha \right) + \delta_{\alpha\beta}(\eta_B - \tfrac{2}{3}\eta)\nabla \cdot \mathbf{v}. \tag{8.57}$$

Substituting Eq. (8.57) into (8.49) and dropping the external forces, one obtains the Navier-Stokes equation

$$\boxed{\rho_m\dot{v}_\alpha = -\nabla_\alpha P + \eta\nabla^2 v_\alpha + \left(\eta_B + \tfrac{1}{3}\eta\right)\nabla_\alpha\nabla \cdot \mathbf{v}.} \tag{8.58}$$

Incompressible liquids have zero velocity divergence, $\nabla \cdot \mathbf{v} = 0$, and the Navier-Stokes equation is further simplified

$$\dot{\mathbf{v}} = -\rho_m^{-1}\nabla P + \nu\nabla^2\mathbf{v}, \tag{8.59}$$

where $\nu = \eta/\rho_m$ is the kinematic viscosity.

In a more general case of a compressible liquid, one can divide $\mathbf{v}$ into the longitudinal and transverse parts (Sec. 1.6): $\mathbf{v} = \mathbf{v}_L + \mathbf{v}_T$. The Navier-Stokes equation for the transverse velocity field reads ($\nabla \cdot \mathbf{v}_T = 0$)

$$\boxed{\dot{\mathbf{v}}_T = \nu\nabla^2\mathbf{v}_T.} \tag{8.60}$$

The time derivative of the velocity is separated into the partial time derivative $\partial/\partial t = \partial_t$ and the convection term

$$\frac{d}{dt} = \frac{\partial}{\partial t} + (\mathbf{v} \cdot \nabla). \tag{8.61}$$

If the convection terms is neglected, the transverse velocity field satisfies the diffusion equation

$$\partial_t \mathbf{v}_T = \nu \nabla^2 \mathbf{v}_T. \tag{8.62}$$

This equation of motion translates to the spectrum of hydrodynamic transverse-current fluctuations in the liquid considered below. Kinematic viscosity plays the role of the diffusion constant responsible for the diffusional relaxation of these fluctuations. The waves propagated by transverse velocity are known as shear waves, in contrast to sound waves related to the bulk compression.

For the longitudinal component of the velocity, one applies the identity from vector calculus (see Box in Sec. 2.8)

$$\nabla \times \nabla \times \mathbf{v} = \nabla\nabla \cdot \mathbf{v} - \nabla^2 \mathbf{v}. \tag{8.63}$$

Since $\nabla \times \mathbf{v}_L = 0$, one obtains $\nabla\nabla \cdot \mathbf{v}_L = \nabla^2 \mathbf{v}_L$ and

$$\boxed{\partial_t \mathbf{v}_L = -\rho_m^{-1} \nabla P + \nu_L \nabla^2 \mathbf{v}_L.} \tag{8.64}$$

Here, the convection term was dropped again and

$$\nu_L = \rho_m^{-1}\left(\eta_B + \tfrac{4}{3}\eta\right) \tag{8.65}$$

is known as the longitudinal kinematic viscosity.

The kinematic viscosity carries the units of $(\text{Length})^2/(\text{Time})$, which are the same units as for the diffusion constant. In fact, these two properties are equal for ideal gases, as we show in Sec. 8.12 below. From this analogy, ν_L is also often called the longitudinal diffusion constant [60].

8.5 Reynolds number

The importance of the temporal change of the velocity field is determined by the Reynolds number. It is established by introducing dimensionless time and coordinate variables: $t \to T\tau$ and $r_\alpha \to Lx_\alpha$. Equation (8.62) then reads

$$\mathcal{R}\partial_\tau \mathbf{v}_T = \nabla_x^2 \mathbf{v}_T. \tag{8.66}$$

The Reynolds number $\mathcal{R}$ here is

$$\mathcal{R} = \frac{L^2}{T\nu} \approx \frac{Lu}{\nu}, \tag{8.67}$$

where $u \simeq L/T$ is the average speed of the flow. For microscopic particles, $L^2/T \simeq D$ can be associated with the diffusion constant and one can write

$$\mathcal{R} \simeq \frac{D}{\nu}. \tag{8.68}$$

For a particle of the size of ~ 1 nm in a typical room-temperature dense liquid, one finds $D \simeq 10 \ \mu m^2/s$ and $\nu \simeq 10^6 \ \mu m^2/s$ (water at normal conditions). Therefore, while $D = \nu$ for ballistic diffusion in an ideal gas (Eq. (8.146)), $D/\nu \simeq 10^{-5}$ for Brownian diffusion of a nanometer-size particle in room-temperature water. Generally, the diffusive motion of a Brownian particle occurs in the regime of low Reynolds numbers.

These estimates translate to very low, $\mathcal{R} \simeq 10^{-5}$, Reynolds numbers for hydrodynamics at the length scale of a few nanometers typical for many chemical systems and biological macromolecules [61]. The low magnitude of $\mathcal{R}$ in Eq. (8.66) implies that the time derivative of the velocity on its left-hand side can be dropped, with the Poisson equation following for the velocity field: $\nabla^2 \mathbf{v}_T = 0$. Physically, this result implies that friction dominates in the flow dynamics and the dynamics of a target particle are well captured by the overdamped Langevin dynamics discussed in Sec. 7.6. When $\mathcal{R}$ is large, inertial effects determine the flow, friction is insignificant, and turbulent flow is produced.

8.6 Momentum density

Continuity relation:
$\partial_t \rho + \partial_\alpha(\rho v_\alpha) = 0$

The microscopic number density ρ_{mic} satisfies the continuity relation (Sec. 3.3) in terms of the microscopic current density $\mathbf{j}_{\mathrm{mic}}$

$$\dot{\rho}_{\mathrm{mic}} + \partial_\alpha j_{\mathrm{mic},\alpha} = 0. \tag{8.69}$$

Here we consider the density $\rho = \langle \rho_{\mathrm{mic}} \rangle$ coarse-grained over a small volume of the material. It changes in space over a length scale much greater than the length of molecular correlations. Correspondingly, the velocity $\mathbf{v}$ considered in hydrodynamics is the local average velocity defined as $\mathbf{j} = \langle \mathbf{j}_{\mathrm{mic}} \rangle = \rho \mathbf{v}$. Both $\rho = \rho(\mathbf{r}, t)$ and $\mathbf{v} = \mathbf{v}(\mathbf{r}, t)$ are spatially and temporally varying fields. Further, the coarse-grained density changes with time only through externally applied fields since the dynamic molecular variables have been integrated out in the statistical average. One can write $\langle \dot{\rho}_{\mathrm{mic}} \rangle = \partial_t \rho$ and arrive at the coarse-grained form of the continuity relation

$$\partial_t \rho + \partial_\alpha(\rho v_\alpha) = 0. \tag{8.70}$$

The conservation of the total momentum of the system requires a continuity relation for the momentum density

$$p_\alpha = m\rho v_\alpha = m j_\alpha. \tag{8.71}$$

We start by applying the continuity equation for the number density to derive a general mathematical identity for a vector field $\mathbf{a} = \mathbf{a}(\mathbf{r}, t)$

$$\rho \dot{a}_\alpha = \partial_t(a_\alpha \rho) + \partial_\beta (a_\alpha \rho v_\beta). \tag{8.72}$$

The following steps prove this result

$$\rho \dot{a}_\alpha = \rho \partial_t a_\alpha + \rho(v_\beta \partial_\beta)a_\alpha = \partial_t(\rho a_\alpha) + a_\alpha \partial_\beta(\rho v_\beta) + \rho(v_\beta \partial_\beta)a_\alpha, \tag{8.73}$$

where the density continuity relation (Eq. (8.70)) is used in the second step. Combining the two last terms in the above equation, one arrives at Eq. (8.72).

One can now apply Eq. (8.72) to the velocity field $\mathbf{a} = \mathbf{v}$ satisfying the Navier-Stokes equation in the absence of external fields

$$m\rho \dot{v}_\alpha = \partial_\beta \sigma_{\alpha\beta}. \tag{8.74}$$

By applying Eq. (8.72) it becomes the continuity equation for the momentum density

$$\partial_t p_\alpha + \partial_\beta \Pi_{\alpha\beta} = 0, \tag{8.75}$$

where the momentum flux density becomes

$$\boxed{\Pi_{\alpha\beta} = \rho_m v_\alpha v_\beta - \sigma_{\alpha\beta}.} \tag{8.76}$$

The tensor $\Pi_{\alpha\beta}$ has two components. The first term is the direct transfer of momentum due to liquid moving across the surface of a volume element. The second component, $-\sigma_{\alpha\beta}$, is the momentum transfer due to shear and compression (sound) waves and the pressure gradient.

8.7 *Stokes drag

A spherical particle with the radius R experiences a friction drag when driven through the liquid with the velocity $\mathbf{u}$. The hydrodynamic friction force $\mathbf{f}_H$ is proportional to the velocity with the friction coefficient ζ_H given by the Stokes law (Eq. (7.95))

$$\mathbf{f}_H = -\zeta_H \mathbf{u}, \quad \zeta_H = 6\pi\eta R. \tag{8.77}$$

The solution of this problem is achieved from the Navier-Stokes equation in the limit of the low Reynolds number

$$\eta \nabla^2 \mathbf{v} = \nabla P - \mathbf{f}_{\text{ext}}\delta(\mathbf{r}),$$
$$\nabla \cdot \mathbf{v} = 0, \tag{8.78}$$

where we have added a point source of the external force $\mathbf{f}_{\text{ext}} = \mathbf{f}$ at the origin of the coordinate system $\mathbf{r} = 0$ and adopted an incompressible liquid (second equation) [62]. These equations are known as creeping flow or Stokes equations.

Before proceeding to the solution, it is important to draw the connection between equations of hydrodynamics at $\mathcal{R} \ll 1$ and equations of electrostatics. Equation (8.78) is the Poisson equation with a source, applied to the vector field $\mathbf{v}$. In contrast, the Poisson equation is written for the scalar potential ϕ_q in electrostatics (Eq. (2.16)). Many of the results of hydrodynamics thus trace the corresponding results of electrostatics with the exception that the tensors involved are one step higher than for electrostatic quantities. For instance, one considers the scalar surface charge density σ_P in electrostatic problems, while the surface force $\sigma_{\alpha\beta}\hat{n}_\beta$ is a vector. The integral of this surface force over the surface of a rigid particle is the total force exerted from the particle on the liquid.

$$f_\alpha = \oint dS \sigma_{\alpha\beta}\hat{n}_\beta, \tag{8.79}$$

where $\hat{\mathbf{n}}$ is directed from the particle outward to the liquid. The force f_α is the analog of the monopole (the charge) in electrostatics.

Equation (8.78) is solved by applying the spatial Fourier transform to the pressure and velocity fields, which must be linear functions of the applied point force in the linear hydrodynamic equations

$$v_\alpha = G_{\alpha\beta}f_\beta, \quad P = g_\alpha f_\alpha. \tag{8.80}$$

We define Fourier transforms of the direct-space fields (Sec. 1.7)

$$\tilde{g}_\alpha = \int \frac{d\mathbf{k}}{(2\pi)^3} g_\alpha e^{-i\mathbf{k}\cdot\mathbf{r}}, \quad \tilde{G}_{\alpha\beta} = \int \frac{d\mathbf{k}}{(2\pi)^3} G_{\alpha\beta} e^{-i\mathbf{k}\cdot\mathbf{r}}. \tag{8.81}$$

Fourier transform of Eq. (8.78) becomes

$$ik_\alpha \tilde{g}_\beta - k^2 \eta \tilde{G}_{\alpha\beta} = -\delta_{\alpha\beta} \tag{8.82}$$

with the additional requirement $k_\alpha G_{\alpha\beta} = 0$ arising from the second equation in Eq. (8.78). Utilizing this constraint leads to the equation for $\tilde{g}_\beta$

$$\tilde{g}_\beta = ik_\beta/k^2. \tag{8.83}$$

Transforming this equation back to direct space one gets

$$g_\beta = -(4\pi)^{-1}\partial_\beta r^{-1}. \tag{8.84}$$

From this equation one obtains the field of the liquid pressure established in response to the applied point force

$$P(\mathbf{r}) = \frac{(\hat{\mathbf{r}} \cdot \mathbf{f})}{4\pi r^2}. \tag{8.85}$$

Note that this pressure adds to the hydrostatic pressure in the bulk at $r \to \infty$. This result is an analog of the electrostatic potential of a dipole (Fig. 8.6). The pressure in front of the applied force is positive, and it is negative on the opposite side of the applied force. The overall result of the pressure gradient is a drag directed opposite to the force.

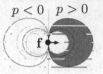

$$p < 0 \quad p > 0$$

Fig. 8.6 Distribution of the liquid pressure in response to the applied point force **f**.

One can next use Eq. (8.83) to solve for $\tilde{G}_{\alpha\beta}$ in Eq. (8.82)

$$\tilde{G}_{\alpha\beta} = \frac{\delta_{\alpha\beta}}{\eta k^2} - \frac{k_\alpha k_\beta}{\eta k^4}. \tag{8.86}$$

Transforming this equation to direct space leads to

$$G_{\alpha\beta} = \frac{\delta_{\alpha\beta}}{4\pi\eta r} + \eta^{-1}\partial_\alpha\partial_\beta \int \frac{d\mathbf{k}}{(2\pi)^3} \frac{e^{-i\mathbf{k}\cdot\mathbf{r}}}{k^4}. \tag{8.87}$$

One can show that the following identity holds

$$\int \frac{d\mathbf{k}}{(2\pi)^3} \frac{e^{-i\mathbf{k}\cdot\mathbf{r}}}{k^4} = -\frac{r}{8\pi}. \tag{8.88}$$

With this result used in Eq. (8.87), one finds

$$G_{\alpha\beta} = \frac{1}{8\pi\eta r} \left[\delta_{\alpha\beta} + \hat{r}_\alpha \hat{r}_\beta\right]. \tag{8.89}$$

The tensor $G_{\alpha\beta}$ provides the velocity field in response to a point source of force (Eq. (8.80)). For a force produced by the surface of a rigid sphere, one has to integrate over the sources of point forces

$$v_\alpha = \oint dS G_{\alpha\beta}(\mathbf{r} - \mathbf{x})\sigma_{\beta\gamma}(\mathbf{x})\hat{n}_\gamma(\mathbf{x}), \tag{8.90}$$

where the position $\mathbf{x}$ is at the surface of the sphere over which integration is performed. If the rigid body moves with the constant velocity $\mathbf{u}$, the stick boundary conditions require $\mathbf{v}(\mathbf{x}) = \mathbf{u}$.

At the distances far greater than the sphere radius R, one can replace $G_{\alpha\beta}(\mathbf{r} - \mathbf{x})$ with $G_{\alpha\beta}(\mathbf{r})$ with the result

$$v_\alpha = G_{\alpha\beta}f_\beta, \tag{8.91}$$

where **f** is the overall force exerted by the rigid body on the liquid (Eq. (8.79)).

It is natural to assume that $\mathbf{f}$ should be parallel to $\mathbf{u}$, $\mathbf{f} \propto \mathbf{u}$. This requires that at the surface of the sphere $v_\alpha(\mathbf{x}) = u_\alpha \propto \delta_{\alpha\beta} f_\beta$. It is clear that Eq. (8.91) does not provide such a solution since $G_{\alpha\beta}$ includes tensor components distinct from the Kronecker delta.

The solution of the problem is sought from the multipolar expansion of Eq. (8.90)

$$v_\alpha = \sum_n \frac{(-1)^n}{n!} M_\beta^n G_{\alpha\beta}, \quad M_\beta^n = \oint dS \sigma_{\beta\gamma}(\mathbf{x}) \hat{n}_\gamma(\mathbf{x}) x_{\alpha 1} \dots x_{\alpha n} \partial_{\alpha 1} \dots \partial_{\alpha n}.$$

(8.92)

Each $n > 0$ term in this expansion is a higher order multipolar correction to the monopole expression in Eq. (8.91). One finds that the degenerate quadrupole term has the correct structure

$$\nabla^2 G_{\alpha\beta} = \frac{1}{8\pi\eta} \left[\frac{2}{r^3} \delta_{\alpha\beta} - \frac{6}{r^3} \hat{r}_\alpha \hat{r}_\beta \right].$$

(8.93)

The sum of the monopole and the degenerate quadrupole terms yields

$$\left[1 + \tfrac{1}{6} R^2 \nabla^2 \right] G_{\alpha\beta} \big|_{r=R} = \frac{1}{6\pi\eta R} \delta_{\alpha\beta}.$$

(8.94)

One therefore arrives at the Stokes law of friction

$$\boxed{-f_{H\alpha} = f_\alpha = 6\pi\eta R u_\alpha.}$$

(8.95)

The Stokes flow around a rigid sphere requires a sum of the force monopole and a degenerate force quadrupole.

8.8 Dynamic structure factor

> Structure factor:
> $S(k) = N^{-1} \langle |\rho_\mathbf{k}|^2 \rangle$
> $S(0) = k_B T \rho \beta_T$
> Isothermal compressibility:
> $\beta_T = -V^{-1} (\partial V / \partial P)_T = \rho^{-1} (\partial \rho / \partial P)_T$
> Thermal velocity:
> $v_{\mathrm{rms}} = (3 k_B T / m)^{1/2}, v_T = (k_B T / m)^{1/2}$

Dynamics of soft materials are described by a number of correlation functions operating with the spatial Fourier transform of the time-dependent number density of the material. The instantaneous (microscopic) density $\rho(t) = \rho_{\mathrm{mic}}(t)$ can be defined as

$$\rho(t) = \sum_j \delta(\mathbf{r} - \mathbf{r}_j(t)),$$

(8.96)

where the sum runs over all molecules $j = 1, \ldots, N$ with center-of-mass coordinates $r_j(t)$ changing with time (also see Eq. (6.30)).

Scattering experiments (Sec. 11.10) usually provide information in reciprocal space. The density in reciprocal $\mathbf{k}$-space is

$$\rho_\mathbf{k}(t) = \sum_j e^{i\mathbf{k} \cdot \mathbf{r}_j(t)}. \tag{8.97}$$

The intermediate scattering function is the time correlation function of $\rho_\mathbf{k}(t)$

$$F(k,t) = N^{-1} \langle \rho_\mathbf{k}(t) \rho_{-\mathbf{k}} \rangle = N^{-1} \sum_{i,j} \langle e^{i\mathbf{k} \cdot \mathbf{r}_{ij}(t)} \rangle, \tag{8.98}$$

where $\mathbf{r}_{ij}(t) = \mathbf{r}_j(t) - \mathbf{r}_i$ and we put $\mathbf{r}_i = \mathbf{r}_i(0)$ and $\rho_{-\mathbf{k}} = \rho_{-\mathbf{k}}(0)$. The dynamic structure factor is the time Fourier transform of $F(k,t)$

$$\boxed{S(k,\omega) = \int_{-\infty}^{\infty} dt\, e^{-i\omega t} F(k,t).} \tag{8.99}$$

The intermediate scattering function is affected by cross-correlations in the coordinates of different particles i and j in the sample. In contrast, van Hove self-correlation function describes motions of a single particle. The self-correlation function in direct space determines the probability of a single particle to move by $\mathbf{r}$ during the time t. The probability is given by averaging the delta-function

$$G_s(r,t) = \langle \delta\left(\mathbf{r} - \Delta\mathbf{r}(t)\right)\rangle, \tag{8.100}$$

where $\Delta\mathbf{r}(t) = \mathbf{r}(t) - \mathbf{r}(0)$. This function is given in reciprocal space by the self-intermediate scattering function

$$F_s(k,t) = \langle e^{i\mathbf{k} \cdot \Delta\mathbf{r}(t)} \rangle. \tag{8.101}$$

One can establish a number of simple relations for the self-correlation functions in the long-time limit when hydrodynamic equations apply. In this limit, Fick's law of diffusion (Eq. (7.1)) can be applied to the number density of a single particle performing self-diffusion in the liquid. The target particle can be a single particle of the liquid or an impurity. In both cases, the single-particle number density

$$\rho_s(t) = e^{i\mathbf{k} \cdot \mathbf{r}(t)} \tag{8.102}$$

satisfies the diffusion equation

$$\partial_t \rho_s = -Dk^2 \rho_s. \tag{8.103}$$

This equation has a simple solution

$$\rho_s(t) = \rho_s e^{-k^2 Dt}. \tag{8.104}$$

The self-intermediate scattering function becomes

$$F_s(k, t) = e^{-k^2 Dt} \tag{8.105}$$

since $F_s(k, 0) = 1$ (Eq. (8.101)).

The transition to the self dynamic structure factor is accomplished by noting the connection between the Fourier transform and the one-sided Fourier-Laplace transform

$$\tilde{S}_s(k, \omega) = \int_0^\infty dt e^{i\omega t} F_s(k, t). \tag{8.106}$$

The self-dynamic structure factor is found as $S(k, \omega) = 2\mathrm{Re}[\tilde{S}(k, \omega)]$

$$S_s(k, \omega) = \frac{2Dk^2}{\omega^2 + (Dk^2)^2}. \tag{8.107}$$

One finds that $S_s(k, \omega)$ is a Lorentzian functions of ω in the hydrodynamic limit. This result applies to a limited range of wave vectors. At large k, the microscopic structure of the liquid at the length scale $2\pi/k$ is probed, and the hydrodynamic description is bound to fail. The wave-vector at which single-particle dynamics acquire ballistic features is found as $k \sim \omega_0/v_T$, where $v_T = (k_B T/m)^{1/2}$ is the thermal speed. Here, ω_0 is the Einstein frequency of harmonic motions of the particle within the cage formed by its neighbors. If the interparticle potential is $\phi(r)$, the frequency of cage vibrations becomes

$$\omega_0^2 = \kappa/m, \quad \kappa = \tfrac{1}{3}\langle \nabla^2 \phi \rangle. \tag{8.108}$$

In this equation, κ is the local force constant created by the interparticle potential. The self-structure factor becomes a Gaussian function at $k \gg \omega_0/v_T$ replacing the Lorentzian shape of the hydrodynamic limit

$$S_s(k, \omega) = \left[\frac{2\pi}{v_T k^2} \right]^{1/2} \exp\left[-\omega^2/(2v_T^2 k^2) \right]. \tag{8.109}$$

This is the limit of purely ballistic motion of the particle with the average thermal velocity v_T. In all cases, the self-structure factor is normalized

$$\int_{-\infty}^\infty \frac{d\omega}{2\pi} S_s(k, \omega) = F_s(k, 0) = 1. \tag{8.110}$$

The intermediate case between the ballistic and diffusion limits can be derived for the self-intermediate scattering function. If the displacement of a particle is a Gaussian variable, the statistical average in Eq. (8.101) is calculated directly

$$F_s(k,t) = \exp\left[-\tfrac{1}{6}\langle \Delta \mathbf{r}(t)^2 \rangle\right]. \tag{8.111}$$

Here, the displacement of the particle is related to the velocity autocorrelation function by Eq. (7.22). If the velocity autocorrelation function is given by an exponential decay

$$\langle \mathbf{v}(t) \cdot \mathbf{v} \rangle = \langle v^2 \rangle e^{-t/\tau_v}, \tag{8.112}$$

one obtains from Eq. (7.22)

$$\langle \Delta \mathbf{r}(t)^2 \rangle = 6D\left(t - \tau_v + \tau_v e^{-t/\tau_v}\right) \tag{8.113}$$

with the diffusion constant $D = \tau_v k_B T/m$ defined in terms of the velocity relaxation time τ_v. When $t \gg \tau$ is used in this equation, one obtains the hydrodynamic limit of Eq. (8.107). In the opposite limit, $t \ll \tau$, the ballistic result follows. The intermediate regime requires numerical evaluation of the Fourier-Laplace transform of Eq. (8.111).

8.9   Correlation functions of the current

In addition to the number density function (Eq. (8.96)), one can define the microscopic current density with $\mathbf{v}_j = \dot{\mathbf{r}}_j$

$$\mathbf{j}(t) = \sum_j \mathbf{v}_j \delta(\mathbf{r} - \mathbf{r}_j(t)). \tag{8.114}$$

The reciprocal-space current density becomes

$$\mathbf{j_k}(t) = \sum_j \mathbf{v}_j e^{i\mathbf{k}\cdot\mathbf{r}_j(t)}. \tag{8.115}$$

Since the current is a vector, the time correlation function of the current is a second-rank Cartesian tensor with the general form

$$\begin{aligned}
J_{\alpha\beta}(\mathbf{k},t) &= N^{-1}\langle j_{\mathbf{k}\alpha}(t) j_{\mathbf{k}\beta}\rangle \\
&= \hat{k}_\alpha \hat{k}_\beta J_L(k,t) + \left[\delta_{\alpha\beta} - \hat{k}_\alpha \hat{k}_\beta\right] J_T(k,t),
\end{aligned} \tag{8.116}$$

where $\hat{\mathbf{k}} = \mathbf{k}/k$ is the unit vector aligned with $\mathbf{k}$ and $j_{\mathbf{k}\beta} = j_{\mathbf{k}\beta}(0)$. The two scalar functions entering $J_{\alpha\beta}(\mathbf{k},t)$ are the longitudinal (L) and transverse (T) projections (Sec. 1.8) of the correlation function. The longitudinal projection takes molecular velocities along the $\mathbf{k}$-vector. It takes a simpler form if one aligns $\mathbf{k}$ with one of the Cartesian axes; usually the z-axis is taken.

One then gets for the longitudinal projection of the current correlation function

$$J_L(k,t) = N^{-1}\sum_{i,j}\langle v_{iz}(t)v_{jz}e^{ikz_{ij}(t)}\rangle.\qquad(8.117)$$

Correspondingly, the velocity projections perpendicular to the direction of the **k**-vector are taken (x is adopted here) in the transverse projection of the current correlation function

$$J_T(k,t) = N^{-1}\sum_{i,j}\langle v_{ix}(t)v_{jx}e^{ikz_{ij}(t)}\rangle.\qquad(8.118)$$

There is an important relation between the longitudinal current correlation function and the dynamic structure factor

$$\boxed{k^2 J_L(k,\omega) = \omega^2 S(k,\omega).}\qquad(8.119)$$

This equation is derived by noticing that

$$\frac{d^2 F(k,t)}{dt^2} = -N^{-1}\langle\dot\rho_{\mathbf{k}}(t)\dot\rho_{\mathbf{k}}\rangle = -k^2 J_L(k,t).\qquad(8.120)$$

Equation (8.119) follows from this relation after taking the time Fourier transform.

Equation (8.119) shows that at any value of the wave vector, $J_L(k,\omega)$ as a function of frequency starts from zero and has at least one peak before decaying to zero at higher frequencies (Fig. 8.7). Each curve at a given k-value has the same area under it since

$$\int_0^\infty \frac{d\omega}{2\pi} J_L(k,\omega) = J_L(k,t=0) = v_T^2.\qquad(8.121)$$

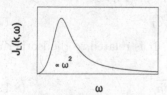

Fig. 8.7   Spectrum of longitudinal current.

Deriving the relation for the transverse current correlation function requires turning back to the transverse projection of the Navier-Stokes equation (Eq. (7.29))

$$\partial_t \mathbf{j}_T = \nu\nabla^2\mathbf{j}_T.\qquad(8.122)$$

Since $J_T(k,0) = \langle v^2 \rangle / 3$,

$$J_T(k,t) = \tfrac{1}{3}\langle v^2 \rangle e^{-k^2 \nu t}. \tag{8.123}$$

Correspondingly, the Lorentzian form follows for the frequency domain

$$\boxed{J_T(k,\omega) = \tfrac{2}{3}\langle v^2 \rangle \frac{k^2 \nu}{(k^2 \nu)^2 + \omega^2}.} \tag{8.124}$$

The transverse current spectrum appears to be a Lorentzian function in the hydrodynamic limit, with its width determined by the kinematic viscosity. This form is similar to the self-structure factor in the diffusion limit for a target particle dynamics. This mathematical form, which does not involve resonance peaks, implies that the liquid does not support propagation of transverse sound waves. This is a reflection of liquid's fluidity and inability to resist steady shear stress.

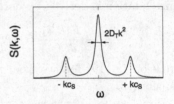

Fig. 8.8   Dynamic structure factor in the hydrodynamic limit.   Two Lorentzian lines at $\omega = \pm k c_S$ represent inelastic Brillouin scattering.   The central Lorentzian refers to elastic Rayleigh scattering.

The reason for a relatively simple form of the transverse current spectrum is that transverse current is decoupled from density fluctuations. In contrast, the spectrum of longitudinal current is directly related to the spectrum of density fluctuations (Eq. (8.119)). Since liquids support propagation of longitudinal (compression) sound waves, the dynamic structure factor $S(k,\omega)$ gains inelastic peaks at $\omega = \pm c_S k$ (Fig. 8.8), where

$$c_S = (c_P/c_V)^{1/2} c_T \tag{8.125}$$

is the adiabatic speed of sound (Eq. (6.53)). It is related to the isothermal speed of sound (Eq. (6.52))

$$c_T = 1/\sqrt{\beta_T \rho m} \tag{8.126}$$

through the square root of the ratio of constant-pressure, c_P, and constant-volume, c_V, heat capacities (per particle). The inelastic Brillouin peaks at $\omega = -k c_S$ and $\omega = +k c_S$ represent the Stokes $(-)$ and anti-Stokes $(+)$ inelastic resonances. In scattering experiments, the energy of the scattered radiation or neutrons is reduced by $\hbar k c_S$ for the Stokes resonance and is increased by $\hbar k c_S$ for the anti-Stokes line. This energy is transferred from radiation or incoming particles to the medium in the Stokes resonance. In opposite, the energy is scattered from particles which already possess an additional energy $\hbar k c_S$ in the anti-Stokes resonance and is thus added to the outcoming scattered wave. These inelastic resonances can be recorded by coherent neutron scattering discussed in Sec. 11.10.

The third, central line in the dynamic structure factor (Fig. 8.8) refers to elastic Rayleigh scattering. It is centered at $\omega = 0$ and is associated with temperature fluctuations in the liquid rising and decaying on the time-scale determined by thermal diffusivity

$$D_T = \lambda/(\rho m c_P), \tag{8.127}$$

where λ is the thermal conductivity. The Rayleigh elastic peak carries the generic diffusion functionality already seen for the self-structure factor and for the spectrum of transverse current (Eq. (8.124))

$$S_{\rm el}(k,\omega) \propto \frac{2D_T k^2}{\omega^2 + (D_T k^2)^2}. \tag{8.128}$$

8.10 Dispersion relation

The dispersion relation connects the characteristic frequency of collective excitations ω_k to the wave-vector k. One needs to calculate the normalized second moment of the frequency

$$\langle \omega_k^2 \rangle = \frac{\int_{-\infty}^{\infty} \omega^2 S(k,\omega) d\omega}{\int_{-\infty}^{\infty} S(k,\omega) d\omega}. \tag{8.129}$$

Since $\int S(k,\omega)d\omega/(2\pi) = S(k)$ and $\omega^2 S(k,\omega) = k^2 J_L(k,\omega)$ (Eq. (8.119)), one obtains

$$\langle \omega_k^2 \rangle = [S(k)]^{-1} k^2 J_L(k,0) = [3S(k)]^{-1} k^2 \langle v^2 \rangle, \tag{8.130}$$

where we used $\langle v^2 \rangle = 3v_T^2$. If the characteristic frequency is defined as $\omega_k = \sqrt{\langle \omega_k^2 \rangle}$, the dispersion relation becomes

$$\omega_k = k v_{\rm rms}/\sqrt{3S(k)}, \tag{8.131}$$

where $v_{\rm rms} = \sqrt{\langle v^2 \rangle}$. At $k \to 0$, the structure factor tends to $S(0) = k_B T \rho \beta_T$ given in terms of the isothermal compressibility β_T, which is also the inverse of the bulk modulus $K = \beta_T^{-1}$ (see also Eqs. (6.43) and (6.44)). One obtains a linear dispersion relation at $k \to 0$

$$\omega_k = k c_T, \tag{8.132}$$

Fig. 8.9 ω_k vs $k\sigma$ calculated from Eq. (8.131) for liquid argon with the atomic diameter σ. The dashed lines refer to linear asymptotes at small and large k-values. The arrow indicates de Gennes narrowing at $k = k_{max}$.

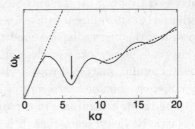

where the isothermal speed of sound c_T is given by Eq. (8.126). On the contrary, at high $k \to \infty$, $S(\infty) \to 1$ and $\omega_k \to k(v_{rms}/\sqrt{3}) = kv_T$.

In the intermediate range of k-values, the structure factor passes through a maximum at $k = k_{max}$ (Fig. 6.9). At this point, ω_k in Eq. (8.131) passes through a minimum as is illustrated in Fig. 8.9. The appearance of the minimum in the frequency of collective density excitations is known as de Gennes narrowing [41, 55]. Physically, it is viewed as slowing down of the relaxation time of a collective coordinate responsible for the local structure in the liquid. The strong peak of the reciprocal-space density structure factor at $k = k_{max}$ is a reflection of long-range oscillatory structure of the radial distribution function in real space. Slower relaxation at k_{max} is, therefore, assigned to a collective mode of density fluctuations in the liquid involving many particles. Generally, one anticipates that two values for the speed of sound should be reported for all non-crystalline materials (see two dashed lines in Fig. 8.9).

8.11 Shear viscosity

The spectrum of transverse current is determined by kinematic viscosity. One therefore can anticipate that shear viscosity can be derived from $J_T(k, \omega)$. This connection can indeed be directly established by taking the sequence of limits in Eq. (8.124) at $k \to 0$, followed with $\omega \to 0$

$$
\begin{aligned}
\eta &= (\beta\rho m^2/2) \lim_{\omega \to 0} \omega^2 \lim_{k \to 0} k^{-2} J_T(k, \omega) \\
&= \frac{\beta m^2}{V} \lim_{\omega \to 0} \lim_{k \to 0} k^{-2} \int_0^\infty dt \langle \dot{j}_k^x(t) \dot{j}_{-k}^x \rangle e^{i\omega t},
\end{aligned}
\tag{8.133}
$$

where $\dot{j}_k^x = dj_k^x/dt$. The time derivative of the number current comes from Eq. (8.75) for the momentum density as

$$
m\frac{d}{dt}j_x = m\partial_t j_x + m(v_\beta \partial_\beta)j_x = -\partial_\beta \Pi_{x\beta} + m(v_\beta \partial_\beta)j_x.
\tag{8.134}
$$

The term $(v_\beta \partial_\beta) j_x$, which is quadratic in the perturbed velocity, can be dropped for small perturbations with the result

$$m \frac{d}{dt} j_x(\mathbf{r}, t) = -\partial_\beta \Pi_{x\beta}(\mathbf{r}, t). \tag{8.135}$$

After performing the spatial Fourier transform, one gets

$$m\dot{j}_\mathbf{k}^x(t) = -ik\Pi_\mathbf{k}^{xz}(t), \tag{8.136}$$

where $\mathbf{k}$ is along the z-axis. Substituting this result into Eq. (8.133), one gets

$$\boxed{\eta = \frac{\beta}{V} \int_0^\infty dt \langle \Pi_0^{xz}(t) \Pi_0^{xz} \rangle,} \tag{8.137}$$

where $\Pi_0^{xz} = \Pi_{k=0}^{xz}$. Equation (8.137) is an important result of the linear response theory connecting the macroscopic shear viscosity to the time correlation function of the microscopic momentum flux density tensor. For a system of N particles interacting with the interaction potential $\phi(r)$, the microscopic momentum flux is [42]

$$\Pi_0^{\alpha\beta} = \sum_j v_{j\alpha} p_{j\beta} + \frac{1}{2} \sum_{j \neq k} r_{jk,\alpha} \left(\frac{d\phi(r_{jk})}{dr_{j,\beta}} \right), \tag{8.138}$$

where r_{jk} is the distance between molecules j and k.

8.12 Diffusion and viscosity in gases

The diffusion constant and shear viscosity are specific cases of a broad family of transport coefficients which can all be defined in terms of time correlation functions of molecular properties. In the case of the ideal gas, the diffusion constant and kinetic viscosity are in fact equal. Here we derive this equality.

The long-time decay of many correlation functions is exponential and this form, as employed in Eq. (8.112), is adopted here. The diffusion constant is then fully defined by the velocity relaxation time τ_v

$$D = \frac{1}{3} \int_0^\infty dt \langle \mathbf{v}(0) \cdot \mathbf{v}(t) \rangle = \frac{1}{3} \langle v^2 \rangle \tau_v. \tag{8.139}$$

In an ideal gas, one can estimate τ_v as the collision time $\tau_{\rm col} = l/v_{\rm rms}$, $v_{\rm rms}^2 = \langle v^2 \rangle$. The collision mean free path in gases is

$$l = \left[\sqrt{2} \pi \rho \sigma^2 \right]^{-1}, \tag{8.140}$$

where σ is the molecular diameter and $\sqrt{2}$ corrects for the fact that the relative velocities of colliding molecules, $\langle v_{rel} \rangle = \sqrt{2} v_{rms}$, enter the collision frequency. Collecting the terms, we obtain

$$D = \frac{v_{rms}}{3\sqrt{2}\pi\rho\sigma^2} \propto \rho^{-1}\sqrt{T/m}. \tag{8.141}$$

The diffusion coefficient in the gas is proportional to $\sqrt{T/m}$ and is inversely proportional to density.

As derived in Eq. (8.137), the shear viscosity is given by the time correlation function of the momentum flux density tensor, which, for an ideal gas, is given as the flux of the momentum (Eqs. (8.76) and (8.138))

$$\Pi_0^{xz} = \sum_j v_{jx}p_{jz}, \tag{8.142}$$

where the sum runs over all molecules in the gas $j = 1, \ldots, N$. We again assume that the correlation function decays with the correlation time equal to the collision time and obtain from Eq. (8.137) for the shear viscosity

$$\eta = \frac{\beta\tau_{col}}{V} \sum_{ij} \langle v_{xj}p_{zj}v_{xi}p_{zi} \rangle. \tag{8.143}$$

In the gas, only velocities of the same molecules are correlated and only the terms $i = j$ are non-zero. We also note that $\langle p_z^2 \rangle = mk_BT$ and one gets

$$\eta = \tfrac{1}{3}\rho m v_{rms}^2 \tau_{col}. \tag{8.144}$$

Upon substituting the collision time, one finds an equation in which the number density ρ cancels out

$$\eta = \frac{mv_{rms}}{3\sqrt{2}\pi\sigma^2} \propto \sqrt{mT}. \tag{8.145}$$

This result was first derived by Maxwell who was surprised to learn that the ideal gas viscosity does not depend on density. By converting η to kinematic viscosity $\nu = \eta/(m\rho)$ one arrives at the equality

$$\boxed{D = \nu.} \tag{8.146}$$

It is important to stress the distinction between this result and the Stokes-Einstein equation for self-diffusion in liquids (Eq. (7.8)), when one expects the product $D\eta \propto T$ to be fully determined by the geometry of the diffusing particle and temperature.

8.13 *Dielectric susceptibility

Polarization of a polar nonconducting material (dielectric) is a specific case of the linear response theory for the generalized external force taken as oscillating electrostatic potential $\phi_q(\mathbf{r}, t) = \phi_q(\mathbf{r})e^{-i\omega t}$. The potential is created by the oscillatory external charge $\rho_q(\mathbf{r}, t) = \rho_q(\mathbf{r})e^{-i\omega t}$. The two functions are connected by the Coulomb law (see Eq. (2.12))

$$\phi_q(\mathbf{r}) = \int \frac{d\mathbf{r}'}{|\mathbf{r} - \mathbf{r}'|} \rho_q(\mathbf{r}'), \tag{8.147}$$

which is easier to handle in reciprocal space converting the integral convolution to a linear algebraic equation (Sec. 1.7)

$$\boxed{\phi_\mathbf{k}^q = \frac{4\pi}{k^2} \rho_\mathbf{k}^q.} \tag{8.148}$$

The perturbation Hamiltonian is given as the integral in direct or reciprocal space of the external potential with the charge density of the material. We write it as the reciprocal-space sum over all possible values of the wave vector $\mathbf{k}$

$$H'(t) = \sum_\mathbf{k} \rho_\mathbf{k}^b \phi_\mathbf{k}^q e^{-i\omega t}. \tag{8.149}$$

The reciprocal space charge density in the material $\rho_\mathbf{k}^b$ is created by all partial atomic charges q_i with coordinates $\mathbf{r}_i$. This definition corresponds to the density of bound charge in standard theories of dielectrics (see Eq. (2.18)). The density of bound charge in reciprocal space is

$$\rho_\mathbf{k}^b = \sum_i q_i e^{i\mathbf{k}\cdot\mathbf{r}_i} = \int d\mathbf{r} \rho_b(\mathbf{r})e^{-i\mathbf{k}\cdot\mathbf{r}}. \tag{8.150}$$

Since the charge density and external potential enter the perturbation Hamiltonian as conjugate variables of generalized displacement (charge) and generalized external force (external potential), the linear-response theory offers a formalism to calculate a small perturbation of the charge density caused by the external potential

$$\delta\rho_\mathbf{k} = \delta\rho_\mathbf{k}^b = \chi_0(k, \omega)\phi_\mathbf{k}^q, \tag{8.151}$$

where $\chi_0(k, \omega)$ denotes the linear-response function to the perturbation caused by altering the potential of external charges.

The standard theory of dielectrics is typically formulated not in terms of the external potential, but in terms of the voltage applied to the plates of the plane capacitor in the dielectric experiment (see Sec. 2.9). One wonders what is the difference between the potential of external charges and the capacitor voltage? The capacitor voltage is the work required to drive a single unit charge from one plate to another. This charge, when driven between the plates, interacts not only with the free charges on the plates, but also with the molecular charges of the dielectric. The voltage thus incorporates the dielectric polarization.

The ratio of the capacitor voltage and the distance between the plates defines the combined field of the external charges ρ_q and of the bound charges ρ_b. This electric field is the Maxwell field $\mathbf{E}$ (Sec. 2.2). Dielectric measurements report the electric free energy stored in a plane capacitor $\frac{1}{2}C(\Delta\phi)^2$. It is expressed in terms of the voltage on the capacitor plates $\Delta\phi$ and the capacitance $C = A\epsilon_s/(4\pi d)$ for two plates with the area A separated by the distance d (Fig. 2.14). The uniform Maxwell field becomes $E = \Delta\phi/d$, and measurements of capacitance give access to the static dielectric constant ϵ_s.

The field of external charges $\mathbf{E}_{\mathbf{k}}^q = -i\mathbf{k}\phi_{\mathbf{k}}^q$ is obviously distinct from the Maxwell field, which, in reciprocal space, is related by a linear algebraic equation to the total charge density (Eqs. (2.29) and (2.30))

$$i\mathbf{k} \cdot \mathbf{E}_{\mathbf{k}} = 4\pi(\rho_{\mathbf{k}}^q + \delta\rho_{\mathbf{k}}). \tag{8.152}$$

On the contrary, the electric displacement field $\mathbf{D}_{\mathbf{k}}$ is caused by external charges only (Eq. (2.28))

$$i\mathbf{k} \cdot \mathbf{D}_{\mathbf{k}} = 4\pi\rho_{\mathbf{k}}^q. \tag{8.153}$$

The ratio of two fields defines the longitudinal dielectric function $\epsilon_L(k,\omega)$ depending on both the wave vector k and frequency ω

$$\epsilon_L(k,\omega)^{-1} = \frac{\mathbf{k} \cdot \mathbf{E}_{\mathbf{k}}}{\mathbf{k} \cdot \mathbf{D}_{\mathbf{k}}} = 1 + \frac{\delta\rho_{\mathbf{k}}}{\rho_{\mathbf{k}}^q}. \tag{8.154}$$

According to Eqs. (8.148) and (8.151),

$$\boxed{\epsilon_L(k,\omega)^{-1} = 1 + \frac{4\pi}{k^2}\chi_0(k,\omega).} \tag{8.155}$$

The response to the Maxwell field can be calculated by noting that both the external potential and the potential of induced charges need to be involved in the corresponding susceptibility

$$\delta\rho_{\mathbf{k}} = \chi(k,\omega)\left[\phi_{\mathbf{k}}^q + \delta\phi_{\mathbf{k}}\right], \tag{8.156}$$

where the potential of induced charges is given by the Poisson equation in reciprocal space

$$\delta\phi_{\mathbf{k}} = \frac{4\pi}{k^2}\delta\rho_{\mathbf{k}}. \tag{8.157}$$

One therefore obtains

$$\chi(k,\omega) = \frac{\chi_0(k,\omega)}{1 + (4\pi/k^2)\chi_0(k,\omega)} \tag{8.158}$$

and

$$\chi(k,\omega) = \epsilon_L(k,\omega)\chi_0(k,\omega). \tag{8.159}$$

This connection between the response to the external field and to the Maxwell field is known as screening of the field of external charges by the dielectric. One can also write $\epsilon_L(k,\omega)$ in terms of $\chi(k,\omega)$

$$\boxed{\epsilon_L(k,\omega) = 1 - \frac{4\pi}{k^2}\chi(k,\omega).} \tag{8.160}$$

Comparing this result to Eq. (8.155), one finds that the dielectric function represents the charge induced in the dielectric by the Maxwell field, while the dielectric modulus $\epsilon_L(k,\omega)^{-1}$ represents the induced charge due to the field of external charges.

Dielectric measurements do not give access to the non-local dielectric function $\epsilon_L(k,\omega)$. Frequency domain dielectric spectroscopy instead reports the $k \to 0$ limit of this function as the dielectric function depending only on the frequency of the applied voltage $\epsilon(\omega) = \epsilon_L(k \to 0, \omega)$ [58]. It is also important to keep in mind the origin of the dielectric function in the linear response approximation, which implies that this function is based on the Fourier-Laplace (from $t = 0$ to ∞) transform of the corresponding time susceptibility function.

Another response function reported experimentally is the longitudinal conductivity connecting longitudinal electric current $\mathbf{J_k} = \mathbf{j}_k^q + \delta\mathbf{j_k}$ to the Maxwell field

$$i\mathbf{k} \cdot \mathbf{J_k} = \sigma_L(k,\omega)i\mathbf{k} \cdot \mathbf{E_k}. \tag{8.161}$$

One can rewrite the above equation and Eq. (8.152) as

$$\rho_{\mathbf{k}}^q + \delta\rho_{\mathbf{k}} = \frac{1}{4\pi\sigma_L(k,\omega)}i\mathbf{k} \cdot \mathbf{J_k}. \tag{8.162}$$

Due to the continuity relation $\partial_t\delta\rho + \nabla \cdot \mathbf{J} = 0$ (Sec. 3.3),

$$-i\omega\delta\rho_{\mathbf{k}} + i\mathbf{k} \cdot \mathbf{J_k} = 0. \tag{8.163}$$

One therefore obtains from Eq. (8.154)

$$1 + \frac{\rho_{\mathbf{k}}^q}{\delta\rho_{\mathbf{k}}} = -\frac{\omega}{4\pi i\sigma_L(k,\omega)} = -\frac{1}{\epsilon_L(k,\omega) - 1}. \tag{8.164}$$

This expression implies that the longitudinal dielectric function and longitudinal conductivity measure the same property [39]

$$\boxed{\epsilon_L(k,\omega) = 1 + \frac{4\pi i\sigma_L(k,\omega)}{\omega}.} \tag{8.165}$$

8.14 Dielectric relaxation

Here we explore the dynamics of establishing the equilibrium polarization of the dielectric is response to a step of voltage $\Delta\phi_q$ in the plane capacitor. This derivation is a more detailed application of the step-function perturbation considered in Sec. 8.3. It serves as a realistic problem to practice with the application of the formalism of the linear response approximation.

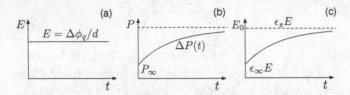

Fig. 8.10 Response of the dielectric to a step in the Maxwell field $E = \Delta\phi_q/d$ (a). The polarization density starts at $t = 0$ with the electronic polarization P_∞ and increases as $\Delta P(t)$ with the relaxation time τ_D (b). The field of external charges at the capacitor plates $E_0(t)$ starts with $\epsilon_\infty E$ and increases to $\epsilon_s E$ (c).

A step in the voltage $\Delta\phi$ initiates an instantaneous step in the Maxwell field $E(t) = Eh(t)$, where $E = \Delta\phi/d$ and d is the distance between the capacitor plates; $h(t)$ is the Heaviside step function (Fig. 8.10a, see the Box in Sec. 2.2). Dielectric polarization that follows the voltage step involves two components: the polarization of the electrons bound within the molecules and polarization caused by alignment of molecular multipoles along the field. The former polarization occurs on the electronic time scales, which can be viewed as instantaneous for a typical dielectric experiment. This electronic polarization is characterized by the high-frequency dielectric constant ϵ_∞ ($\epsilon_\infty \simeq 1.78$ for water at 298 K). In contrast, the total polarization of the dielectric, established when all degrees of freedom have sufficient time to relax, is characterized by the static dielectric constant ϵ_s ($\epsilon_s \simeq 78$ for water at 298 K). The frequency-dependent dielectric function $\epsilon(\omega)$ reported by dielectric spectroscopy changes between ϵ_s at $\omega = 0$ and ϵ_∞ at $\omega \to \infty$ (in practice, ϵ_∞ is often established by the highest instrumental frequency).

The uniform displacement field $D(t)$ in the plane capacitor is equal to the field $E_0(t) = 4\pi\sigma_0(t)$ of the external charges on the capacitor plates distributed with the surface charge density $\sigma_0(t)$. One, therefore, can write for the time-dependent fields

$$E_0(t) = D(t) = E(t) + 4\pi P_\infty(t) + 4\pi\Delta P(t), \tag{8.166}$$

where

$$4\pi P_\infty(t) = (\epsilon_\infty - 1)E(t) \tag{8.167}$$

is the instantaneous electronic polarization (Fig. 8.10b) and $\Delta P(t)$ is the time-dependent polarization of the dielectric due to the rearrangement of molecular multipoles (mostly reorientation of dipoles). One can convert Eq. (8.166) to the Fourier-Laplace transform with the result

$$\tilde{E}_0(\omega) = \epsilon_\infty \tilde{E}(\omega) + 4\pi \Delta \tilde{P}(\omega), \quad \tilde{E}(\omega) = iE/\omega. \tag{8.168}$$

We now turn to the response function connecting $\Delta P(t)$ with the field of external charges $E_0(t)$. While the time dependence $E_0(t)$ is unknown and needs to be determined, the linear response theory provides the connection

$$\Delta \tilde{P}(\omega) = \tilde{\chi}_0(\omega)\tilde{E}_0(\omega). \tag{8.169}$$

By using this equation in Eq. (8.168), one obtains

$$\tilde{E}_0(\omega) = (iE_0(0)/\omega)\left[1 - 4\pi\tilde{\chi}_0(\omega)\right]^{-1}, \tag{8.170}$$

where $E_0(0) = \epsilon_\infty E$ is the initial step in the field of external charges (Fig. 8.10c). The vacuum field $E_0(t)$ increases from the initial value $\epsilon_\infty E$ to the final value $\epsilon_s E$ because of the electron current from the power source maintaining the voltage to the capacitor plates. The increase in $E_0(t)$ thus reflects an increase in the surface charge $\sigma_0(t)$ at the plates.

The response function $\chi_0(\omega)$ can be calculated in terms of the Laplace-Fourier transform of the time correlation function in the absence of the external field (Eq. (8.41))

$$\tilde{\chi}_0(\omega) = \beta C_P(0)\left[1 + i\omega\tilde{\Phi}(\omega)\right], \tag{8.171}$$

where $C_P(t) = \langle \delta P(t)\delta P \rangle$ and $\tilde{\Phi}(\omega) = \tilde{C}_P(\omega)/C_P(0)$. We next assume that the time relaxation of the polarization correlation function is exponential with the longitudinal relaxation time τ_L

$$C_P(t) = C_P(0)e^{-t/\tau_L}. \tag{8.172}$$

The longitudinal character of relaxation is related to the fact that $P(t)$ is along the axial symmetry of the sample (z-axis in the notation scheme adopted above), which is usually associated with the direction of the wave vector corresponding to the external perturbation (also see below). With exponential relaxation for $C_P(t)$, one gets

$$1 + i\omega\tilde{\Phi}(\omega) = [1 - i\omega\tau_L]^{-1} \tag{8.173}$$

and

$$\tilde{\chi}_0(\omega) = \beta C_P(0)\left[1 - i\omega\tau_L\right]^{-1}. \tag{8.174}$$

One next notes that at $\omega \to 0$ (corresponding to $t \to \infty$), one expect the field of external charges to reach the limit $E_0 = \epsilon_s E$ (Fig. 8.10c). One obtains

$$\Delta P = \frac{\Delta\epsilon}{4\pi\epsilon_s}E_0, \quad \Delta\epsilon = \epsilon_s - \epsilon_\infty. \tag{8.175}$$

The limit $t \to \infty$ is the static limit of the FDT discussed in Sec. 8.3 and leading to Eq. (8.36). One obtains from that equation

$$\Delta P/E_0 = \beta C_P(0) = \frac{\Delta\epsilon}{4\pi\epsilon_s}. \tag{8.176}$$

One finally obtains for the sample polarization

$$\Delta\tilde{P}(\omega) = \tilde{\chi}_0(\omega)\tilde{E}_0(\omega) = \frac{i\Delta\epsilon E}{4\pi\omega}\frac{1}{1 - i\omega\tau_D}, \tag{8.177}$$

where

$$\boxed{\tau_D = (\epsilon_s/\epsilon_\infty)\tau_L} \tag{8.178}$$

is the Debye relaxation time.

The polarization of the sample in the time domain becomes

$$\Delta P(t) = \frac{\Delta\epsilon E}{4\pi}\left(1 - e^{-t/\tau_D}\right). \tag{8.179}$$

Similarly, the field of the plate charges $E_0(t)$ increases from the initial value $\epsilon_\infty E$ to the final value $\epsilon_s E$ on the time scale τ_D

$$E_0(t) = E\left[\epsilon_s - \Delta\epsilon e^{-t/\tau_D}\right]. \tag{8.180}$$

As already mentioned, the rise of the field $E_0(t)$ reflects the flow of charge to the capacitor plates from an external source. The relaxation time τ_D thus reflects the time of charge equilibration on the capacitor plates following a voltage step. The energy stored in the plane capacitor (Eq. (2.130)) will also increase on the time scale τ_D

$$w(t) = \frac{VE^2}{8\pi}\left[\epsilon_s - \Delta\epsilon e^{-t/\tau_D}\right], \tag{8.181}$$

where V is the capacitor volume.

Equations (8.178) and (8.179) show that the polarization of the dielectric sample in response to a step in the Maxwell field (capacitor voltage) occurs with the Debye relaxation time τ_D. This relaxation is much slower than the longitudinal relaxation for polar materials with $\epsilon_s/\epsilon_\infty \gg 1$. On

Fig. 8.11 Longitudinal (L) and transversal (T) polarization patterns in a polar liquid.

the contrary, since $\tilde{\chi}_0$ in Eq. (8.174) is given in terms of τ_L, polarization in response to the field E_0 should relax on the time of longitudinal relaxation τ_L. This relaxation occurs in response to altering the charge on the capacitor plates. The relaxation time τ_D, therefore, defines the Debye peak in the dielectric loss spectrum of $\epsilon(\omega)$ (Eqs. (8.48) and (8.201)), while τ_L gives the maximum of the loss spectrum of the dielectric modulus $\epsilon(\omega)^{-1}$.

The relaxation times τ_D and τ_L can be further associated with relaxation of, correspondingly, the divergence and curl of the vector field of polarization density $\mathbf{P}$ representing reorienting dipoles in the liquid. Equation (8.179) can be written as the time derivative

$$\tau_D \partial_t \mathbf{P} = \frac{\Delta\epsilon}{4\pi}\mathbf{E} - \mathbf{P}. \tag{8.182}$$

Given that $\nabla \times \mathbf{E} = 0$ (third Maxwell equation, Sec. 2.6), one obtains for the curl

$$\tau_D \partial_t \nabla \times \mathbf{P} = -\nabla \times \mathbf{P}. \tag{8.183}$$

When transformed to reciprocal space (Sec. 1.7), one obtains ($\hat{\mathbf{k}} = \mathbf{k}/k$ is the unit vector)

$$\tau_D \partial_t \tilde{\mathbf{P}}^T = -\tilde{\mathbf{P}}^T, \quad \tilde{\mathbf{P}}^T = \hat{\mathbf{k}} \times \tilde{\mathbf{P}}. \tag{8.184}$$

The Debye relaxation time τ_D thus describes the relaxation of the transverse polarization of the liquid.

To derive the relaxation equation for the longitudinal polarization, one applies the relation

$$\nabla \cdot \mathbf{D} = \nabla \cdot [\epsilon_\infty \mathbf{E} + 4\pi\mathbf{P}] = 0. \tag{8.185}$$

When used in Eq. (8.182), it yields

$$\tau_L \partial_t \nabla \cdot \mathbf{P} = -\nabla \cdot \mathbf{P}. \tag{8.186}$$

Transformation to reciprocal space leads to the relaxation equation for the longitudinal polarization

$$\tau_L \partial_t \tilde{\mathbf{P}}^L = -\tilde{\mathbf{P}}^L, \quad \tilde{\mathbf{P}}^L = \hat{\mathbf{k}}(\hat{\mathbf{k}} \cdot \tilde{\mathbf{P}}). \tag{8.187}$$

Longitudinal polarization excitations in the liquid produce head-to-head (or tail-to-tail) dipolar configurations with non-vanishing divergence of the polarization field (Fig. 8.11). In contrast, ring excitations with approximately closed circles of dipoles are responsible for the transverse polarization and the corresponding fluctuation (excitation) modes in the liquid.

Equation (8.178) is an analog of the Lyddane-Sachs-Teller equation for the ratio of longitudinal ω_{LO} and transverse ω_{TO} frequencies of optical vibrations in an ionic crystal [63]

$$\omega_{LO}^2/\omega_{TO}^2 = \epsilon_s/\epsilon_\infty. \qquad (8.188)$$

Similarly to polar liquids, where the longitudinal time is shorter, the frequency for longitudinal optical phonons is higher than the frequency of transverse phonons. However, in the case of overdamped relaxation of polar liquids (Sec. 7.6), one has to replace $\omega_{LO}^2/\omega_{TO}^2$ with τ_D/τ_L.

8.15 *Dynamic Kirkwood-Onsager equation

One can apply the linear response approximation to extend the static Kirkwood-Onsager equation for the dielectric constant ϵ_s (Sec. 2.19) to the frequency-dependent dielectric function measured by dielectric spectroscopy [58]. In contrast to the response of dielectrics to a voltage step worked out in the previous section, here an oscillatory electric field of external charges is considered. Following the main steps of the derivation of the static dielectric constant in Sec. 2.19, we consider the field first applied perpendicular to the plates of the capacitor and then parallel to them. In both cases, the time-dependent perturbation of the Hamiltonian is (see Eq. (2.185))

$$H'(t) = -M_\alpha E_{\alpha 0} e^{-i\omega t}, \qquad (8.189)$$

where either $\alpha = z$ or $\alpha = x$ are considered (Fig. 2.14). It is important to stress that $\mathbf{E}_0$ is the field of external charges, in contrast to a step in the Maxwell field considered in the previous section. It is this field that establishes the generalized force in the perturbation Hamiltonian in the framework of the linear-response theory.

From the linear response theory (Eq. (8.39)), the amplitude of the oscillatory dipole moment of the dielectric sample is

$$\langle M_\alpha(\omega)\rangle_E = \tilde{\chi}_\alpha(\omega)E_{\alpha 0}, \qquad (8.190)$$

where (see Eqs. (8.41) and (8.42))

$$\tilde{\chi}_\alpha(\omega) = \beta\left[\langle M_\alpha^2\rangle + i\omega\tilde{C}_\alpha(\omega)\right] \qquad (8.191)$$

and

$$\tilde{C}_\alpha(\omega) = \int_0^\infty dt \langle M_\alpha(t) M_\alpha \rangle e^{i\omega t}. \qquad (8.192)$$

To arrive at a practical relation for the dielectric function, one has to note that the response of a polar liquid has two components: polarization of the molecular electronic shells leading to induced electronic dipoles and polarization of permanent multipoles (mostly dipoles) in the liquid. We are concerned here with the latter and, therefore, an extra step is required to connect to the total permanent dipole of the sample $\langle M_\alpha \rangle_E$ to the dielectric function $\epsilon(\omega)$.

As is seen from Fig. 8.10b, the polarization P_∞ due to induced electronic dipoles of the molecules is created instantaneously in the capacitor. It needs, therefore, be subtracted from the total polarization perpendicular to the capacitor plates. The remaining permanent dipole is proportional to $\epsilon(\omega) - \epsilon_\infty$

$$\langle M_z(\omega) \rangle_E = V \frac{\epsilon(\omega) - \epsilon_\infty}{4\pi\epsilon(\omega)} E_{z0}, \qquad (8.193)$$

where V is the sample volume.

A field parallel to the dielectric boundary cannot be created in a capacitor since there is no field inside a metal. However, the response to the parallel field E_{x0} is experimentally realized in absorption of radiation by dielectric materials when electromagnetic wave propagates perpendicular to the dielectric slab. The direction of the oscillating electric field lies in the plane of the slab surface. To separate instantaneous (on the time-scale of the dielectric experiment) electronic polarization from retarded polarization of permanent dipoles, one can consider the configuration in which the electromagnetic wave propagates in the medium with the dielectric constant ϵ_∞ in contact with the dielectric. Since dielectric problems depend only on the ratio of dielectric constants in contacting media, one gets

$$\langle M_x(\omega) \rangle_E = V \frac{\epsilon(\omega) - \epsilon_\infty}{4\pi\epsilon_\infty} E_{x0}. \qquad (8.194)$$

It is clear that the solution depends only on the ratio $\epsilon(\omega)/\epsilon_\infty$ and not on each dielectric constant separately. The same observation applies to Eq. (8.193).

For isotropic dielectrics, the dipole induced in the dielectric slab in the y-direction is equal to that in the x-direction and one can construct the sum of all three dipolar projections $\sum_\alpha \langle M_\alpha(\omega) \rangle_E$, which allows to produce the result in terms of $\langle \mathbf{M}^2 \rangle = \langle M_\alpha M_\alpha \rangle$. This scalar product is invariant to rotations of the reference laboratory frame and, therefore, results expressed in terms of $\langle \mathbf{M}^2 \rangle$ are applicable to an arbitrary dielectric sample and do not need to rely anymore on the specific slab geometry considered here.

By taking the sum $\sum_\alpha \langle M_\alpha(\omega) \rangle_E$ in Eq. (8.190) and (8.191), one obtains the right-hand side of the dynamic version of the Kirwood-Onsager equation

$$\frac{(\epsilon(\omega) - \epsilon_\infty)(2\epsilon(\omega) + \epsilon_\infty)}{9\epsilon_\infty \epsilon(\omega)} = y g_K \left[1 + i\omega \tilde{\Phi}_M(\omega) \right]. \tag{8.195}$$

The left-hand side of this equation is the result of evaluating the sum $\sum_\alpha \langle M_\alpha(\omega) \rangle_E$ from Eqs. (8.193) and (8.194). Further, $y = (4\pi/9)\beta m^2 \rho$ on the right-hand side of Eq. (8.195) is the effective density of dipoles which appears also in the static Kirkwood-Onsager equation (Eq. (2.190)). In condensed materials, m is assigned the meaning of the condensed-phase dipole moment m', which is usually higher than the dipole moment in the gas phase due to molecular polarizability. Further, the Kirkwood factor g_K here is the same static Kirkwood factor which appeared in Eqs. (2.191) and (2.192). It is related to the variance of the total dipole moment of the sample containing N dipoles as

$$\langle \mathbf{M}^2 \rangle = N m^2 g_K. \tag{8.196}$$

Finally, $\tilde{\Phi}_M(\omega)$ in Eq. (8.195) is the Fourier-Laplace transform of the normalized time correlation function of the sample dipole

$$\Phi_M(t) = C_M(t)/C_M(0), \quad C_M(t) = \langle \mathbf{M}(t) \cdot \mathbf{M}(0) \rangle. \tag{8.197}$$

It can be viewed as the definition of the frequency-dependent Kirkwood factor $g_K(\omega)$ since one can write

$$\tilde{\Phi}_M(\omega) = \frac{g_K(\omega)}{g_K}. \tag{8.198}$$

Here, $g_K(\omega)$ is specified through mutual time correlations of a target dipole with the unit direction $\hat{\mathbf{u}}_i = \hat{\mathbf{u}}_i(0)$ at $t = 0$ with all dipoles $\hat{\mathbf{u}}_j(t)$ at a later time t

$$g_K(\omega) = \int_0^\infty dt \sum_j \langle \hat{\mathbf{u}}_j(t) \cdot \hat{\mathbf{u}}_i \rangle e^{i\omega t}. \tag{8.199}$$

Note that the sum over j here includes $j = i$, which corresponds to single-particle orientational dynamics of the target molecule i.

Equation (8.199) points to the main difficulty of theories of dielectric relaxation. The static Kirkwood factor in Eq. (8.195) can be determined from the static Kirkwood-Onsager equation. Orientational dynamics of a single particle $\langle \hat{u}(t) \cdot \hat{u} \rangle$ can also be modeled or measured in the laboratory. It is the multiparticle dynamics involving time-dependent cross-correlations between different dipoles $j \neq i$ in the condensed phase that is the most challenging part of this problem.

Equation (8.195) leads to the standard Debye relaxation spectrum (Eq. (8.48)) when two approximations are adopted. One first assumes that the relaxation function $\Phi_M(t)$ is single-exponential with the collective relaxation time τ_D. This assumption leads to the expression

$$1 + i\omega \tilde{\Phi}_M(\omega) = \frac{1}{1 - i\omega\tau_D} \tag{8.200}$$

on the right-hand side of Eq. (8.195) (also see Eq. (8.173)): Further, one assumes that $\epsilon(\omega) \gg \epsilon_\infty$ in the range of circular frequencies ω of interest. This assumption allows one to reduce the left-hand side of Eq. (8.195) to $2(\epsilon(\omega) - \epsilon_\infty)/(9\epsilon_\infty)$. The same approximation obviously applies to ϵ_s, which leads in the $\omega \to 0$ limit to $yg_K = 2\Delta\epsilon/(9\epsilon_\infty)$, $\Delta\epsilon = \epsilon_s - \epsilon_\infty$. Combining these approximations, one obtains the Debye equation for the frequency-dependent dielectric function

$$\epsilon(\omega) - \epsilon_\infty = \frac{\Delta\epsilon}{1 - i\omega\tau_D}. \tag{8.201}$$

Deriving the Debye equation as a specific limit of the Kirkwood-Onsager equation shows clearly where the underlying approximations are introduced. For instance, a popular generalization of the Debye equation is the Cole-Davidson function

$$\epsilon(\omega) - \epsilon_\infty = \frac{\Delta\epsilon}{(1 - i\omega\tau_D)^\gamma}, \tag{8.202}$$

where the exponent $0 < \gamma \leq 1$ is used to empirically describe a multitude of relaxation processes contributing to dielectric relaxation. This distribution of dielectric relaxation times is also known as dielectric dispersion. If one applies $\epsilon_s > \epsilon(\omega) \gg \epsilon_\infty$ to the Kirkwood-Onsager equation (8.195), one obtains

$$\epsilon(\omega) - \epsilon_\infty = \Delta\epsilon \left(1 + i\omega \frac{g_K(\omega)}{g_K} \right). \tag{8.203}$$

Comparing this result to Eq. (8.202), it is clear that complex dynamics of correlations between orientations of distinct dipoles in the liquid, expressed by $g_K(\omega)$, are responsible for dielectric dispersion. The equation for the dynamic Kirkwood factor can be written in a more compact form as

$$\boxed{\frac{\epsilon_s - \epsilon(\omega)}{\epsilon_s - \epsilon_\infty} = -i\omega \frac{g_K(\omega)}{g_K}.} \tag{8.204}$$

Chapter 9

Elasticity

Solids of macroscopic (size of a sample) and mesoscopic (hundreds of molecular sizes) dimensions can be viewed as continuous media with no gaps or voids. The properties of such materials can be represented by continuous tensor fields, which, for time-independent phenomena, are functions of position only. This is the realm of *continuum mechanics* [3, 4, 59]. Here we focus on linear elastic deformations of solids and touch on viscoelasticity applicable to liquids and soft matter [38, 64, 65].

9.1 Elastic deformation

Solids can change shape and volume under the action of forces. This simple observation suggests that at least two elastic parameters, or elastic moduli, are needed to describe deformation. We will see that, indeed, bulk and shear moduli are required for isotropic materials.

The position of each point in a solid is displaced by deformation. If the displacement vector $\mathbf{u}(\mathbf{r})$ is assigned to the point $\mathbf{r}$ in the material, the entire deformation can be characterized by the continuous vector field of displacements $\mathbf{u}(\mathbf{r})$.

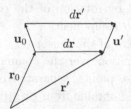

Fig. 9.1 Continuous deformation of the medium.

Consider two close points, $\mathbf{r}_0$ and $\mathbf{r}' = \mathbf{r}_0 + d\mathbf{r}$ [59] (Fig. 9.1). The distance between them is $dl^2 = dr_\alpha dr_\alpha$. Upon deformation these two points become $\mathbf{r}_0 + \mathbf{u}(\mathbf{r}_0)$ and $\mathbf{r}' + \mathbf{u}(\mathbf{r}')$. We get $dr' = dr + du$ and $(dl')^2 = dr'_\alpha dr'_\alpha$. By expanding du_α in dr_β as $du_\alpha = (\partial u_\alpha / \partial r_\beta) dr_\beta$ one gets

$$\begin{aligned}(dl')^2 &= dl^2 + 2\partial_\beta u_\alpha dr_\alpha dr_\beta + \partial_\beta u_\alpha \partial_\gamma u_\alpha dr_\beta dr_\gamma \\ &\simeq dl^2 + 2\epsilon_{\alpha\beta} dr_\alpha dr_\beta,\end{aligned} \tag{9.1}$$

where $\partial_\beta = \partial / \partial r_\beta$ and

$$\boxed{\epsilon_{\alpha\beta} = \tfrac{1}{2}\left(\partial_\beta u_\alpha + \partial_\alpha u_\beta\right).} \tag{9.2}$$

In the second line of Eq. (9.1), the second-order terms in u_α are dropped for a linear deformation.

The second-rank tensor $\epsilon_{\alpha\beta}$ is the strain tensor. The strain tensor is not affected by rigid-body rotations (Sec. 3.5) since distance is invariant in respect to rotations. The derivation also shows that deformation is characterized by a spatial variation of displacement, but not by displacement itself. This is particularly clear when one considers a finite translation of a rigid body, which creates displacement but no deformation.

From its definition, the strain tensor is symmetric

$$\epsilon_{\alpha\beta} = \epsilon_{\beta\alpha}. \tag{9.3}$$

This means that $\epsilon_{\alpha\beta}$ has only six independent components. This simplification is the basis of the Voigt convention for numbering the tensor components: $xx = 1$, $yy = 2$, $zz = 3$, $xy = yx = 4$, $yz = zy = 5$, $xz = zx = 6$.

Fig. 9.2 Surface element ΔA with the unit normal $\hat{n}$ and the force $\Delta\mathbf{f}$ applied from outside of the selected volume.

Any symmetric tensor can be diagonalized by rotation of the coordinate system, which leaves only three non-zero components $\epsilon^\alpha = \epsilon_{\alpha\alpha}$ in that specific coordinates known as principal axes of the tensor ($\epsilon_{\alpha\alpha}$ is the diagonal element of the tensor here; no summation over the common indices is taken). One can now choose two close points, separated by dx, on the principal x-axis. Upon deformation, one obtains from Eq. (9.1) at $dr_\alpha = dr_\beta = dx$: $dx' = dx\sqrt{1 + 2\epsilon^x} \simeq dx(1 + \epsilon^x)$. Repeating these steps for two other axes, one obtains the volume element upon deformation

$$dV' = dV(1 + \epsilon^x)(1 + \epsilon^y)(1 + \epsilon^z) \simeq dV(1 + \epsilon^x + \epsilon^y + \epsilon^z). \tag{9.4}$$

The trace of the tensor $\text{Tr}[\epsilon] = \epsilon_{\alpha\alpha}$ is invariant under coordinate transformations and the sum of the diagonal components of the strain tensor determines the relative volume change in an arbitrary system of coordinates

$$(dV' - dV)/dV = \epsilon_{\alpha\alpha}. \tag{9.5}$$

Since the trace of a tensor is invariant under rotations, this result is preserved when the axes of the laboratory frame are rotated from the principal axes (Sec. 1.2) and the strain tensor is not diagonal anymore.

When the body is deformed, it develops internal stresses expressed as the vector force density $\mathbf{f}^{\text{int}}(\mathbf{r})$ acting on each volume element of the body $d\mathbf{r}$. The total force is $\int d\mathbf{r} \mathbf{f}^{\text{int}}$. In equilibrium, it must be balanced by the external forces applied to the body to deform it. These forces can be viewed as surface forces.

Consider an element of the surface of the body ΔA with the surface normal $\hat{\mathbf{n}}$ (Fig. 9.2). The force applied to the this surface element by the surrounding parts or external forces is $\Delta \mathbf{f}$. The second-rank stress tensor $\sigma_{\alpha\beta}$ is defined to produce the vector field of the force when contracted with the vector of the normal to the surface

$$\Delta f_\alpha = \Delta A \sigma_{\alpha\beta} \hat{n}_\beta. \tag{9.6}$$

In this equation, the product $\Delta A n_\beta$ can be viewed as the component of the area ΔA perpendicular to the axis β (Fig. 9.3). The diagonal components of the stress tensor (e.g., σ_{xx}) are called its normal components, while off-diagonal components (e.g., σ_{xy}) are shearing components.

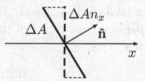

Fig. 9.3 Component of the surface element ΔA perpendicular to the x-axis. Also shown is the unit normal $\hat{\mathbf{n}}$.

The sum of internal forces must be balanced by the total surface force. One gets

$$\int_V d\mathbf{r} f_\alpha^{\text{int}} = \oint_A dA \sigma_{\alpha\beta} \hat{n}_\beta = \int_V d\mathbf{r} \partial_\beta \sigma_{\alpha\beta}, \tag{9.7}$$

where the Gauss's theorem (Eq. (1.40)) was used in the last equality. One therefore obtains the connection between the bulk forces and the stress tensor

$$\boxed{f_\alpha^{\text{int}} = \partial_\beta \sigma_{\alpha\beta}.} \tag{9.8}$$

9.2   Hooke's law

For small deformations, the stress tensor is proportional to the strain tensor, which is the meaning of the Hooke's law. By using the Voigt convention, the linear relation is given in terms of the 6×6 matrix of elastic constants C_{ij}, $i, j = 1, \ldots, 6$

$$\sigma_i = C_{ij}\epsilon_j. \tag{9.9}$$

Both the stress and strain tensors are symmetric, and the matrix C_{ij} is symmetric as well. The elastic constants carry the units of energy/volume. Because of symmetry, there are $6(6 + 1)/2 = 21$ independent components.

Specific symmetries reduce the number of independent elastic constants. Only two are required for isotropic materials. Isotropic materials are liquids, amorphous solids, and mixtures of polymers. For such materials $\lambda = C_{12}$ and $\mu = (C_{11} - C_{12})/2$ are sufficient to define the elastic deformation; λ and μ are called the Lamé coefficients. They can be related to the experimentally accessible bulk modulus K and the shear modulus G equal to μ. Additionally, Young's modulus E and Poisson's ratio ν are reported. Any pair of them can be used for isotropic materials to find the rest of the moduli.

The Lamé coefficients appear in the quadratic expansion of the free energy of the deformed body in the strain tensor. The Hooke's law applied to a linear spring states that the free energy of deformation is quadratic in the spring displacement. A similar statement applies to a linear elastic deformation. Since the free energy is a scalar, one has to form scalars quadratic in $\epsilon_{\alpha\beta}$. Only two such combinations can be formed, $\epsilon_{\alpha\alpha}^2 = \mathrm{Tr}[\epsilon]^2$ and $\epsilon_{\alpha\beta}^2 = \epsilon_{\alpha\beta}\epsilon_{\alpha\beta}$. The Lamé coefficients are expansion coefficients of the deformation free energy ΔF in terms of these two scalars

$$\Delta F = \tfrac{1}{2}\lambda\epsilon_{\alpha\alpha}^2 + \mu\epsilon_{\alpha\beta}^2. \tag{9.10}$$

To decompose an elastic deformation into a pure shear and pure hydrostatic compression, $\epsilon_{\alpha\beta}$ is split into a spherical (proportional to the identity matrix) tensor and a traceless (deviatoric) tensor $\bar\epsilon_{\alpha\beta}$ ($\bar\epsilon_{\alpha\alpha} = 0$)

$$\epsilon_{\alpha\beta} = \tfrac{1}{3}\delta_{\alpha\beta}\epsilon_{\alpha\alpha} + \bar\epsilon_{\alpha\beta}, \quad \epsilon_{\alpha\alpha} = \mathrm{Tr}[\epsilon]. \tag{9.11}$$

When this decomposition is applied to the free energy of elastic deformation, one obtains

$$\Delta F = \tfrac{1}{2}K\epsilon_{\alpha\alpha}^2 + \mu\bar\epsilon_{\alpha\beta}^2. \tag{9.12}$$

Here, $K = \lambda + (2/3)\mu$ is the bulk modulus. The inverse of K is the isothermal compressibility β_T (Eq. (6.44)). Since $\sigma_{\alpha\beta} = \partial \Delta F / \partial \epsilon_{\alpha\beta}$ [59], one obtains

$$\boxed{\sigma_{\alpha\beta} = K\epsilon_{\alpha\alpha}\delta_{\alpha\beta} + 2\mu\bar{\epsilon}_{\alpha\beta}.} \tag{9.13}$$

This is the isotropic form of the general Hooke's law in Eq. (9.9). From this equation, $\sigma_{\alpha\alpha} = 3K\epsilon_{\alpha\alpha}$ and

$$\epsilon_{\alpha\beta} = (9K)^{-1}\sigma_{\alpha\alpha}\delta_{\alpha\beta} + (2\mu)^{-1}\bar{\sigma}_{\alpha\beta}, \quad \bar{\sigma}_{\alpha\beta} = \sigma_{\alpha\beta} - \tfrac{1}{3}\delta_{\alpha\beta}\sigma_{\alpha\alpha}. \tag{9.14}$$

9.3 Deformation of a rod

Consider a rod along the z-axis which is stretched in both direction by the force pA, where A is the area of the surface perpendicular to the z-axis (Fig. 5.11). We assume that $\epsilon_{\alpha\beta}$ is constant in the rod, which means that $\sigma_{\alpha\beta}$ is also constant. Since the force is applied only along the z direction, the only non-zero stress component is $\sigma_{zz} = p$. Further, from Eq. (9.14) one gets

$$\epsilon_{zz} = \frac{p}{9K} + \frac{p}{3\mu}, \quad \epsilon_{xx} = \epsilon_{yy} = \frac{p}{9K} - \frac{p}{6\mu}. \tag{9.15}$$

The coefficient connecting p with ϵ_{zz} is the Young's modulus: $p = E\epsilon_{zz}$. It becomes

$$E = \frac{9K\mu}{3K + \mu}. \tag{9.16}$$

The ratio of the transverse compression and longitudinal extension is Poisson's ratio

$$-\frac{\epsilon_{xx}}{\epsilon_{zz}} = \nu = \frac{1}{2}\frac{3K - 2\mu}{3K + \mu}. \tag{9.17}$$

Since K and μ are positive, ν varies between -1 for $K = 0$ and $1/2$ for $\mu = 0$

$$-1 \leq \nu \leq 1/2. \tag{9.18}$$

Alternatively, one can write K and μ in terms of E and ν

$$K = \frac{1}{3}\frac{E}{1 - 2\nu},$$
$$\mu = \frac{1}{2}\frac{E}{1 + \nu}. \tag{9.19}$$

From these equations, one also obtains the relative change of the volume of the stretched rod (Eq. (9.5))

$$\frac{\Delta V}{V} = \frac{p}{E}(1 - 2\nu). \tag{9.20}$$

As expected, $\nu = 1/2$ corresponds to an incompressible material discussed below for the problem of rubber elasticity. The result in Eq. (9.20) is obviously different from the relative change of the volume under the hydrostatic pressure since the pulling force is applied in one direction only: to two end of the rod. It is instructive to consider the case of hydrostatic compression separately, as we do next.

9.4 Hydrostatic compression of a block

Consider a rectangular block with the length L_z, width L_x, and height L_y. If equal hydrostatic pressure p is applied to all its sides, the strain in each given direction is affected by strains in perpendicular directions through Poisson's ratio. Consider the block's length. Pressure applied to two ends of the block leads to

$$\Delta L_z^{(1)}/L_z = -p/E. \tag{9.21}$$

On the other hand, pressure applied to two sides also makes the strain $\Delta L_x^{(1)}/L_x = -p/E$. The sideways strain is transformed, through Poisson's ratio, to a change in the length along z-axis. The x- and y-components add up to

$$- \nu \left(\Delta L_x^{(1)}/L_x + \Delta L_y^{(1)}/L_y \right) = 2\nu p/E. \tag{9.22}$$

The overall change in the length is

$$\Delta L_z/L_z = -(1 - 2\nu)p/E. \tag{9.23}$$

The same considerations apply to other sides of the block and one gets

$$\Delta L_x/L_x = \Delta L_y/L_y = -(1 - 2\nu)p/E. \tag{9.24}$$

The change in the volume $V = L_x L_y L_z$ is

$$\Delta V = \Delta L_x L_y L_z + L_x \Delta L_y L_z + L_x L_y \Delta L_z. \tag{9.25}$$

The relative volume change becomes

$$\frac{\Delta V}{V} = \frac{\Delta L_z}{L_z} + \frac{\Delta L_x}{L_x} + \frac{\Delta L_y}{L_y} = -3(1 - 2\nu)\frac{p}{E}. \tag{9.26}$$

In this case, the relative volume change is three times the change obtained by applying pressure to the ends of the rods in Eq. (9.20) (the negative sign specifies compression in contrast to expansion considered for the linear rod). With account for Eq. (9.19), one also obtains

$$\Delta V/V = -p/K. \tag{9.27}$$

This result is expected since the bulk modulus is the inverse of the isothermal compressibility (Eq. (6.44)).

9.5 Rubber elasticity

Rubbers are made of cross-linked polymer chains. In contrast to deformation of solids, which mostly involves energy penalty, elastic properties of rubbers are controlled by the entropy of chain extension between the cross-linking points. The ideal rubber is an idealized material in which the interaction energy between the chains is neglected and only the entropic contribution to the free energy of the chain is considered. More specifically, one starts with the free energy of the material under the action of externally applied force f_{ext}. The change in the free energy is

$$dF = -SdT - PdV + f_{ext}dL, \tag{9.28}$$

where dL is the elongation of the material caused by the force. From this equation, one finds the force as

$$f_{ext} = \left(\frac{\partial F}{\partial L}\right)_{T,V} = \left(\frac{\partial U}{\partial L}\right)_{T,V} - T\left(\frac{\partial S}{\partial L}\right)_{T,V}. \tag{9.29}$$

For the ideal rubber, $U = 0$ and one gets the entropic force $f_{ext} = f_S$

$$\boxed{f_S = -T\left(\partial S/\partial L\right)_{T,V}.} \tag{9.30}$$

Rubber elasticity can be viewed from the standpoint of the affine network model of elasticity, which assumes that deformation of each chain follows the deformation of the entire macroscopic sample. This simplification allows one to relate the macroscopic deformation of the sample to the microscopic deformation of an individual chain. As we found in Sec. 6.12, polymer chains loose entropy when stretched and $\partial S/\partial L < 0$. Therefore, the entropic component of the force increases with increasing temperature, which is the consequence of the entropic origin of rubber elasticity. In other words, rubber gets stiffer when heated.

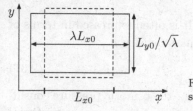

Fig. 9.4 Deformation of a rectangular block preserving its volume.

We now turn to calculating the shear modulus of rubber by adopting the affine network model. Consider a macroscopic rectangular block (Fig. 9.4) which is stretched by λ along the x-axis: the original length L_{x0} changes to $L_x = \lambda L_{x0}$. One anticipates that the length in y- and z-directions also change: $L_\alpha = \lambda' L_{\alpha0}$. If one requires that the total volume of the sample stays constant, $L_x L_y L_z = L_{x0} L_{y0} L_{z0}$, this allows to connect λ' to λ: $\lambda' = 1/\sqrt{\lambda}$. This is the case of Poisson's ratio equal to $1/2$ (Eq. (9.18)).

One can now consider a part of the chain between cross-linking points. The initial end-to-end distance is R_0. In the affine network, the deformation of the sample will cause equally scaled changes in the x, y, z-components of R_0

$$R_0^2 = X_0^2 + Y_0^2 + Z_0^2 \to \lambda^2 X_0^2 + \lambda^{-1}(Y_0^2 + Z_0^2). \qquad (9.31)$$

Therefore, one can write the change in the free energy of the chain (see Eq. (5.198))

$$\Delta F_{\text{chain}} = \frac{3k_{\mathrm{B}}T}{2Nb^2} \left[(\lambda^2 - 1)\langle X_0^2 \rangle + (\lambda^{-1} - 1)\langle Y_0^2 \rangle + (\lambda^{-1} - 1)\langle Z_0^2 \rangle \right]. \qquad (9.32)$$

Since $\langle X_0^2 \rangle = \langle Y_0^2 \rangle = \langle Z_0^2 \rangle = \langle \mathbf{R}_0^2 \rangle/3 = Nb^2/3$, one can write the free energy of the sample by multiplying ΔF_{chain} with the number n of strands between the cross-linking points

$$\Delta F = n\Delta F_{\text{chain}} = \tfrac{1}{2} n k_{\mathrm{B}} T \left(\lambda^2 + 2/\lambda - 3 \right). \qquad (9.33)$$

The external entropic force along the x-axis becomes

$$f_{Sx} = L_{0x}^{-1} \left(\partial \Delta F / \partial \lambda \right)_{T,V} = \frac{n k_{\mathrm{B}} T}{L_{x0}} \left(\lambda - \lambda^{-2} \right). \qquad (9.34)$$

The xx-component of the stress tensor σ_{xx} is obtained by dividing the force by the area $L_y L_z$ perpendicular to the direction of applied force

$$\sigma_{xx} = \frac{f_{Sx}}{L_y L_z} = \frac{\lambda n k_{\mathrm{B}} T}{V_0} \left(\lambda - \lambda^{-2} \right) = G \left(\lambda^2 - \lambda^{-1} \right), \qquad (9.35)$$

where $V = V_0 = L_{x0} L_{y0} L_{z0}$ is the sample volume, which stays unchanged before and after the deformation. Further,

$$G = \frac{n k_{\mathrm{B}} T}{V_0} \qquad (9.36)$$

carries the meaning of the shear modulus. It is often expressed in terms of the molar mass of the strand M_s and the mass density ρ_m

$$G = \frac{\rho_m R T}{M_s}. \qquad (9.37)$$

As is the case with other entropic forces (e.g., the osmotic pressure considered in Sec. 10.2), the stress tensor for the rubber extension is proportional to temperature. At small deformations, $\lambda - 1 \ll 1$, the stress is linear in the strain according to Hooke's law, $\sigma_{xx} \simeq 3G(\lambda - 1)$, and $E = 3G = 3\mu$ in accordance with Eq. (9.19) at $\nu = 1/2$.

9.6 Simple shear

Consider a block of solid material deformed by applying the force parallel to its upper face (Fig. 9.5). The lower face is glued to the substrate and only the upper face is shifted horizontally by displacement u_x. This displacement makes the side face tilt by the angle $\chi = u_x/d$, where d is the height of the block. We therefore obtain the strain

$$\epsilon_{xy} = \tfrac{1}{2}\left(\partial_y u_x + \partial_x u_y\right) = \tfrac{1}{2}\chi. \tag{9.38}$$

The stress follows from Hooke's law (Eq. (9.13))

$$\sigma_{xy} = 2\mu\epsilon_{xy} = \mu(u_x/d). \tag{9.39}$$

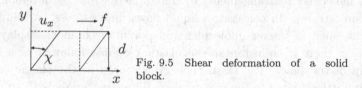

Fig. 9.5 Shear deformation of a solid block.

An alternative to elastic shear is the shear stress in a viscous liquid (Sec. 8.4). Consider a liquid of thickness d sandwiched between two plates. The force f is applied to the upper plate, which moves with the velocity v (Fig. 9.6). If the area of the upper plate is A, this is the area perpendicular to the y-axis (Fig. 9.3). According to Eq. (9.6), we obtain $f_x = A\sigma_{xy}\hat{n}_y$. Since $\hat{n}_y = 1$ and $f_x = f$, the shear stress is $\sigma_{xy} = f/A$. The shear viscosity η is defined through the linear relation between the shear stress and the rate of strain

$$\sigma_{xy} = 2\eta\dot{\epsilon}_{xy}. \tag{9.40}$$

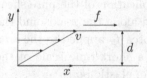

Fig. 9.6 Shear of a liquid between two plates.

In the configuration shown in Fig. 9.6, only the x-projection of the velocity exists and one finds

$$\dot{\epsilon}_{xy} = \tfrac{1}{2}\left(\partial_y v_x + \partial_x v_y\right) = \tfrac{1}{2}\partial_y v_x, \tag{9.41}$$

where $v_x = \dot{u}_x$ and $v_y = \dot{u}_y$. If the liquid velocity changes linearly from zero at the lower plate to v at the upper plate (assuming sticking of the liquid's upper layer to the plate), one gets $\partial_y v_x = v/d$ and

$$\eta = \frac{f}{A}\frac{d}{v}. \tag{9.42}$$

From this equation, the units of shear viscosity η is pressure$\times$time, which becomes Pa$\times$s in SI units. For instance, the viscosity of water at normal conditions is $\eta \sim 10^{-3}$ Pa$\times$s. We next combine the concepts of elastic and viscous shear into a single property of viscoelasticity.

9.7 Viscoelasticity

An elastic solid responds instantaneously (on the time scale far exceeding time of molecular rearrangement) to the applied stress by developing deformation (strain). In contrast, a liquid flows under stress. A number of materials, such as viscous molecular and polymeric liquids, display both properties: there is an instantaneous elastic response followed by a slower build-up of the flow. The elastic energy is stored in the elastic components and is dissipated in the viscous flow. These materials both store and dissipate elastic energy and display the viscoelastic response to stress [64]. Both solids and liquids can produce the viscoelastic response. The difference between these two materials is that a liquid, when exposed to pure shear deformation, will dissipate the stress to zero after transient effects. In contrast, a solid will maintain a residual shear stress.

Consider a viscoelastic material exposed to two consecutive stresses. The first stress creates an instantaneous elastic response and a slow viscous response. If the second stress is applied, the response will depend on how far has the material evolved as a viscous medium. In other words, response of a viscoelastic material to stress involves memory effects.

Before introducing these memory effects, it is instructive to look at two limiting cases of very fast and very slow application of the pure shear stress to the material. Consider the force applied along the x-axis and the corresponding stress σ_{xy}, which is equal to the force f_x applied to the area A_y perpendicular to the y-axis (Fig. 9.5). On a short time scale of the quickly applied shear, the material responds as an elastic body

$$\bar{\sigma}_{xy} = 2\mu\bar{\epsilon}_{xy}, \tag{9.43}$$

where $\bar{\epsilon}_{\alpha\beta}$ is the traceless tensor defined by Eq. (9.11). One can introduce a characteristic relaxation time τ_M (Maxwell relaxation time) of shear relaxation. Fast shear implies that the frequency ω of an oscillatory applied force satisfies the condition $\omega\tau_M \gg 1$, which means that the external force changes faster than the viscous response.

In the opposite limit of a nearly constant force applied to the material, $\omega\tau_M \ll 1$, the material will flow as a Newtonian (incompressible) fluid. In that case one gets the newtonian stress tensor

$$\sigma_{xy} = 2\eta\dot{\epsilon}_{xy} = 2\eta\dot{\bar{\epsilon}}_{xy}. \tag{9.44}$$

The last equality is the result of liquid's incompressibility, $\epsilon_{\alpha\alpha} = 0$ (Eq. (9.5)), which also means $\sigma_{\alpha\alpha} = 0$ and $\sigma_{\alpha\beta} = \bar{\sigma}_{\alpha\beta}$ (shear only).

In order to combine two limits in one formula, one can consider a frequency-dependent friction $\eta(\omega)$. If $\epsilon_{xy}(t)$ is given by an oscillatory function

$$\epsilon_{xy}(t) = \epsilon_0 e^{i\omega t}, \tag{9.45}$$

one gets, from Eq. (9.44), the generalized form of the viscous flow limit

$$\bar{\sigma}_{xy}(\omega) = 2\eta(\omega)i\omega\bar{\epsilon}_{xy}(\omega). \tag{9.46}$$

In order to satisfy the elastic limit, one has to require

$$i\omega\eta(\omega) \to G, \quad \omega\tau_M \gg 1. \tag{9.47}$$

This requirement is satisfied by the function

$$\boxed{\eta(\omega) = \frac{G\tau_M}{1 + i\omega\tau_M},} \tag{9.48}$$

where $G = \mu$ is the high-frequency shear modulus. This equation also requires that the zero-frequency shear viscosity is connected to G through the Maxwell relaxation time

$$\eta = G\tau_M. \tag{9.49}$$

Equation (9.48) establishes the Maxwell model of viscoelasticity. The shear modulus $i\omega\eta(\omega)$ vanishes at $\omega \to 0$ representing a viscoelastic fluid.

Instead of interpolating between two limits of fast and slow response, one can introduce a linear integral connection between the shear stress and rate of strain in terms of the memory function $G(t - \tau)$

$$\bar{\sigma}_{\alpha\beta} = 2\int_{-\infty}^{t} d\tau\, G(t - \tau)\dot{\bar{\epsilon}}_{\alpha\beta}(\tau). \tag{9.50}$$

If one assumes oscillatory strain $\bar{\epsilon}_{\alpha\beta}(\tau) = \bar{\epsilon}_{\alpha\beta}(\omega)e^{i\omega t}$, this equation can be written as a linear relation between the oscillation amplitudes $\bar{\sigma}_{\alpha\beta}(\omega)$ and $\bar{\epsilon}_{\alpha\beta}(\omega)$

$$\bar{\sigma}_{\alpha\beta}(\omega) = 2\tilde{G}(-\omega)i\omega\bar{\epsilon}_{\alpha\beta}(\omega). \tag{9.51}$$

Here, $\tilde{G}(\omega)$ is the Fourier-Laplace transform (one-sided, $0 \leq t \leq \infty$, Fourier transform) of the memory function

$$\tilde{G}(\omega) = \int_0^\infty dt e^{i\omega t} G(t). \tag{9.52}$$

The most common form adopted for $G(t)$ is a simple exponential decay

$$G(t) = G e^{-t/\tau_M}. \tag{9.53}$$

From this equation, one obtains

$$\tilde{G}(\omega) = \frac{G\tau_M}{1 - i\omega\tau_M} \tag{9.54}$$

and

$$\bar{\sigma}_{\alpha\beta}(\omega) = \frac{2Gi\omega\tau_M}{1 + i\omega\tau_M}\bar{\epsilon}_{\alpha\beta}(\omega) = 2\eta(\omega)i\omega\bar{\epsilon}_{\alpha\beta}(\omega). \tag{9.55}$$

The Maxwell form of the frequency-dependent shear viscosity in Eq. (9.48) corresponds to a single-exponential decay of the memory function, Eq. (9.53).

9.8 *Ponderomotive forces in dielectrics

Ponderomotive forces are volume forces experienced by a deformable dielectric. Volume forces, derived first, are used, in the second step, to arrive at the Maxwell stress tensor in dielectrics.

The work done on charges at constant temperature and volume is equal to the change of the electrostatic free energy F_{el} (Eq. (2.123))

$$\delta F_{el} = \frac{1}{4\pi} \int d\mathbf{r} \mathbf{E} \cdot \delta \mathbf{D}. \tag{9.56}$$

This equation indicates that $F_{el} = F_{el}(\mathbf{D})$ is a thermodynamic potential with respect to $\mathbf{D}$. The Maxwell electric field (Sec. 2.2) can by found as the derivative

$$\mathbf{E} = 4\pi \left(\partial F_{el}/\partial \mathbf{D} \right)_{\rho, T}. \tag{9.57}$$

If the condition of a constant electric field, in contrast to the constant electric displacement, is required, one applies the Legendre transformation (Sec. 5.11) by subtracting the product of two conjugate variables, $\mathbf{E}$ and $\mathbf{D}$ in this case. One obtains the electrostatic free energy $\tilde{F}_{el}$ for which $\mathbf{E}$, instead of $\mathbf{D}$, is the independent variable [8]

$$\tilde{F}_{el} = F_{el} - \frac{1}{4\pi} \int d\mathbf{r} \mathbf{E} \cdot \mathbf{D}. \tag{9.58}$$

The change of the electrostatic free energy becomes

$$\delta \tilde{F}_{\text{el}} = -\frac{1}{4\pi} \int d\mathbf{r} \mathbf{D} \cdot \delta \mathbf{E}. \tag{9.59}$$

If $\mathbf{E}$ and $\mathbf{D}$ are coupled by a linear constitutive relation, integration over $\mathbf{E}$ from zero to the final value yields

$$\tilde{F}_{\text{el}} = -\frac{1}{8\pi} \int d\mathbf{r} \mathbf{E} \cdot \mathbf{D}. \tag{9.60}$$

We now consider a plane capacitor held at a constant voltage on the plates, i.e., at the constant value of the Maxwell field E. The question addressed here is to estimate the work done to produce a deformation in the dielectric represented by a nonuniform displacement vector $\mathbf{u}(\mathbf{r})$. The displacement is assumed to be parallel to the plates at the surfaces bounding the dielectric. Since the deformation is done at constant E the work done is described as

$$\delta \tilde{F}_{\text{el}} = -\frac{1}{8\pi} \int d\mathbf{r} \delta \epsilon_s \mathbf{E}^2. \tag{9.61}$$

For generality, we assume that the dielectric constant is inhomogeneous, $\epsilon_s = \epsilon_s(\mathbf{r})$, and that the deformation alters the density of the dielectric. There will be two components, $\delta\epsilon_1$ and $\delta\epsilon_2$, to the overall change of the dielectric constant.

A small element of the volume dV_1 before deformation becomes $dV_2 = (1 + \nabla \cdot \mathbf{u})dV_1$ after the deformation (Eq. (9.5)). The number of molecules in a small volume is conserved and one obtains

$$\rho_2 dV_2 = \rho_1 dV_1 = \rho_2(1 + \nabla \cdot \mathbf{u})dV_1. \tag{9.62}$$

From this equation, one find the density change related to the change in the volume

$$\delta\rho = \rho_2 - \rho_1 = -\rho \nabla \cdot \mathbf{u}, \tag{9.63}$$

where ρ is the number density before the deformation was applied. If the dielectric constant depends on density, the related change in the dielectric constant is

$$\delta\epsilon_1 = -\rho(\partial\epsilon_s/\partial\rho)\nabla \cdot \mathbf{u}. \tag{9.64}$$

The second source of changing dielectric constant is the deformation itself. If ϵ_1 is $\epsilon_s(\mathbf{r})$ fore the deformation, the element of space originally at $\mathbf{r} - \mathbf{u}$ and with $\epsilon_s(\mathbf{r} - \mathbf{u})$ will be brought in place of ϵ_1 after the deformation

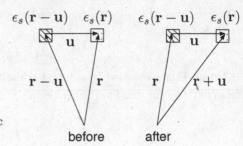

Fig. 9.7 Change in the dielectric constant due to deformation.

(Fig. 9.7). One obtains the second component of the change in the dielectric constant

$$\delta\epsilon_2 = \epsilon_s(\mathbf{r} - \mathbf{u}) - \epsilon_s(\mathbf{r}) = -\mathbf{u} \cdot \nabla\epsilon_s. \tag{9.65}$$

The sum of two contributions $\delta\epsilon_s = \delta\epsilon_1 + \delta\epsilon_2$ in Eq. (9.61) yields

$$\delta\tilde{F}_{\text{el}} = \frac{1}{8\pi} \int d\mathbf{r} E^2 \left[\mathbf{u} \cdot \nabla\epsilon_s + \rho(\partial\epsilon_s/\partial\rho)\nabla \cdot \mathbf{u} \right]. \tag{9.66}$$

One can next use the identity

$$E^2\rho(\partial\epsilon_s/\partial\rho)\nabla \cdot \mathbf{u} = \nabla \cdot \left(E^2\rho(\partial\epsilon_s/\partial\rho)\mathbf{u} \right) - \mathbf{u} \cdot \nabla\left(E^2\rho(\partial\epsilon_s/\partial\rho) \right). \tag{9.67}$$

The first term is the divergence, and it converts by Gauss's theorem to the surface integral

$$\oint_S E^2\rho(\partial\epsilon_s/\partial\rho)(\mathbf{u} \cdot d\mathbf{S}). \tag{9.68}$$

Given that we assumed that the displacements are parallel to the plates at the surface of the dielectric in contact with the plates, this integral vanishes.

One can next relate the work done to change the electrical free energy to the volume force $\mathbf{f}_{\text{el}}$. The sum of the electrical work and the work by this volume force $\int d\mathbf{r}\, \mathbf{f}_{\text{el}} \cdot \mathbf{u}$ should cancel at equilibrium, which allows us to write

$$\boxed{\mathbf{f}_{\text{el}} = -\frac{1}{8\pi}E^2\nabla\epsilon_s + \frac{1}{8\pi}\nabla\left(E^2\rho\frac{\partial\epsilon_s}{\partial\rho} \right).} \tag{9.69}$$

This the volume force acting on the dielectric. The first term is due to an inhomogeneous dielectric constant of the medium. The second term, known as electrostriction, appears in an inhomogeneous electric field which allows a nonzero gradient of the electric field energy per unit volume $\propto \nabla E^2$.

When the dielectric constant is homogeneous and temperature is held constant, the gradient of the dielectric constant becomes

$$\nabla\epsilon_s = (\partial\epsilon_s/\partial\rho)_T\nabla\rho. \tag{9.70}$$

The equation for the force is simplified to

$$\mathbf{f}_{el} = \frac{\rho}{8\pi} \nabla \left[E^2 \left(\frac{\partial \epsilon_s}{\partial \rho} \right)_T \right]. \qquad (9.71)$$

For a liquid dielectric, one can add the hydrostatic pressure to the electrostatic force and rewrite this equation in terms of the total force per unit volume

$$\mathbf{f} = -\rho \nabla \left[\mu_0 - \frac{1}{8\pi} E^2 \left(\frac{\partial \epsilon_s}{\partial \rho} \right)_T \right], \qquad (9.72)$$

where μ_0 is the chemical potential in the absence of the field. The property in the brackets is the chemical potential in the presence of the electric field

$$\boxed{\mu_{el} = \mu_0 - \frac{1}{8\pi} E^2 \left(\frac{\partial \epsilon_s}{\partial \rho} \right)_T.} \qquad (9.73)$$

The difference in chemical potentials between the liquid within the plane capacitor and outside of it can be used to estimate the pressure difference. To compensate for the negative contribution to the chemical potential from polarizing the dielectric, the liquid inside the capacitor is under pressure p exceeding the pressure p_0 outside of the capacitor. The equilibrium is established at

$$\mu_0(p) - \frac{1}{8\pi} E^2 \left(\frac{\partial \epsilon_s}{\partial \rho} \right)_T = \mu_0(p_0). \qquad (9.74)$$

Since $\mu_0(p) - \mu_0(p_0) \simeq \rho^{-1} \Delta p$, $\Delta p = p - p_0$, one gets

$$\Delta p = \frac{E^2}{8\pi} \rho \left(\frac{\partial \epsilon_s}{\partial \rho} \right)_T. \qquad (9.75)$$

For water at room temperature, $\rho (\partial \epsilon_s / \partial \rho)_T \simeq 81$ [16] (Table 2.2) and one gets $\Delta p \simeq 0.4$ atm at $E = 100$ kV/cm.

9.9 *Maxwell tensor in dielectrics

We next proceed to derive the electric Maxwell stress tensor in the presence of dielectrics. One can assume that a body carrying the charge density ρ_q is inserted in the dielectric. The force acting on the dielectric is modified compared to Eq. (9.69) to include the force $\rho_q \mathbf{E}$ acting on the charges. The total force acting on the dielectric becomes

$$F_\alpha = \int_\Omega d\mathbf{r} \left[\rho_q E_\alpha - \frac{1}{8\pi} E^2 \partial_\alpha \epsilon_s + \frac{1}{8\pi} \partial_\alpha \left(E^2 \rho \frac{\partial \epsilon_s}{\partial \rho} \right) \right], \qquad (9.76)$$

where the integration is extended over the volume of the dielectric Ω and we use the Cartesian coordinates $\alpha = x, y, z$ to make the differential more explicit.

By using the first Maxwell's equation, one can write $\rho_q E_\alpha = (4\pi)^{-1} E_\alpha \partial_\beta D_\beta$, where summation over common indecies is assumed. One next writes

$$E^2 \partial_\alpha \epsilon_s = \partial_\alpha(\epsilon_s E^2) - 2\epsilon_s E_\beta \partial_\alpha E_\beta. \tag{9.77}$$

By using $D_\beta = \epsilon_s E_\beta$ and $\partial_\alpha E_\beta = \partial_\beta E_\alpha$ (which is true because of $\nabla \times \mathbf{E} = 0$, Eq. (1.34)), one gets

$$F_\alpha = \int_\Omega d\mathbf{r} \partial_\beta T_{\alpha\beta} = \oint_S T_{\alpha\beta} \hat{n}_\beta dS. \tag{9.78}$$

Here, $T_{\alpha\beta}$ is the Maxwell tensor [8]

$$\boxed{T_{\alpha\beta} = \frac{1}{4\pi} \left[\epsilon_s E_\alpha E_\beta - \frac{1}{2}\delta_{\alpha\beta} E^2 \left(\epsilon_s - \rho \frac{\partial \epsilon_s}{\partial \rho} \right) \right].} \tag{9.79}$$

It converts to Eq. (2.112) for the Maxwell tensor in vacuum when $\epsilon_s = 1$.

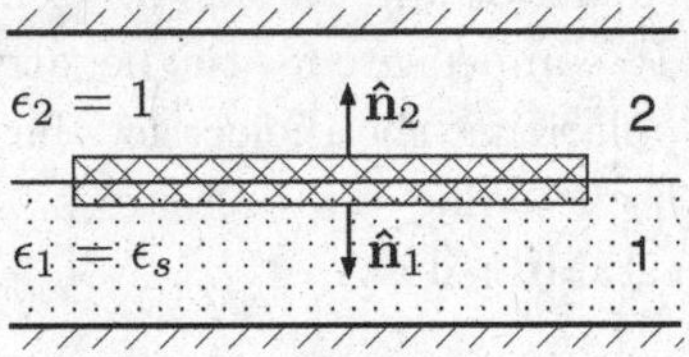

Fig. 9.8 Plane capacitor half-filled with a polar liquid.

The stress tensor can be used to calculate the force acting on the liquid placed in the electric field. As a specific example, the force acting on the liquid in a half filled capacitor is considered here (Fig. 9.8). To calculate the force acting on the unit area of the surface, one can choose a small surface element with two normal unit vectors, $\hat{\mathbf{n}}_1$ and $\hat{\mathbf{n}}_2$, pointing into corresponding media with the dielectric constants $\epsilon_1 = \epsilon_s$ (liquid) and $\epsilon_2 = 1$ (air). From the problem geometry, one has $\hat{\mathbf{n}} = \hat{\mathbf{n}}_2 = -\hat{\mathbf{n}}_1$ and $E_2 = D = \epsilon_s E_1$. The projection of the force per unit area on the normal to the liquid is obtained from Eq. (9.78) by taking the contraction, $\hat{\mathbf{n}} \cdot \mathbf{T} \cdot \hat{\mathbf{n}}_i$, $i = 1, 2$ of the stress tensor in Eq. (9.79) on two surfaces of the slab shown by the hatched area in Fig. 9.8

$$4\pi(\hat{\mathbf{n}} \cdot \mathbf{F}) = \epsilon_2(\hat{\mathbf{n}} \cdot \mathbf{E}_2)^2 - \epsilon_1(\hat{\mathbf{n}} \cdot \mathbf{E}_1)^2 - \gamma_2 E_2^2 + \gamma_1 E_1^2, \tag{9.80}$$

where $2\gamma_i = \epsilon_i - \rho(\partial\epsilon_i/\partial\rho)$. This equation can be simplified in terms of the electric displacement D in the capacitor to produce a positive force pulling the liquid upward to fill up the capacitor

$$\hat{\mathbf{n}} \cdot \mathbf{F} = \frac{D^2}{8\pi} \left(1 - \frac{1}{\epsilon_s} - \frac{\rho}{\epsilon_s^2} \frac{\partial \epsilon_s}{\partial \rho} \right). \tag{9.81}$$

Chapter 10

Solutions and Electrolytes

This Chapter describes ions dissolved in polar liquids. An ensemble of charges which are not prohibited from moving toward each other by constraining forces is mechanically unstable [6, 8]. *Electrolytes* in the bulk and in interfaces [66] still exist in an equilibrium configuration because they are thermodynamic and not mechanical systems. A finite temperature creates thermal agitation and is responsible for an entropic drive to separate the ions despite electrostatic attractions favoring condensation. The balance between entropic and energetic (enthalpic) forces leads to a mean-field screening of charges as considered already for ionic liquids in Sec. 6.12. In those dense systems, most of the resistance to electrostatic attractions is achieved by repulsive cores of the molecules. For dilute electrolytes, the balance is between the entropic free energy of the ideal gas and the free energy of Coulomb attractions and repulsions. The long-range Coulomb interactions are captured by the mean-field formalism of the Debye-Hückel theory of weak electrolytes [36, 67]. The long-range character of Coulomb interactions also leads to an unusual virial expansion for the equation of state and long-range fluctuations of the charge density. Those are described by dynamic structure factors (Sec. 8.8), which take a particularly simple form for weak electrolytes and can be characterized by two parameters: the plasmon frequency and the Debye screening length [40].

10.1   Solutions

Equation (5.73) for the chemical potential of an ideal gas can be applied to each component of a solution assuming that individual particles dissolved in a liquid do not interact with each other (ideal solution). If one assumes that N_i particles of the component i occupy the solution volume

V, the number density becomes $\rho_i = N_i/V$. One additionally needs the thermal de Broglie wavelength λ_i (Eq. (5.38)), which will be different for each component because of the molecular mass m_i involved

$$\lambda_i = h/\left(2\pi m_i k_{\mathrm{B}} T\right)^{1/2}. \tag{10.1}$$

The chemical potential of the component i in solution becomes

$$\beta\mu_i = \ln[\rho_i \lambda_i^3]. \tag{10.2}$$

The number density ρ_i can be expressed in terms of the molar concentration c_i carrying the units of $\mathrm{M} = \mathrm{mol/L}$

$$\rho_i = N_A c_i \times 10^3, \tag{10.3}$$

where the factor 10^3 transforms L^{-1} to m^{-3} and N_A is the Avogadro number. One can next separate the term carrying the logarithm of the concentration to write the chemical potential in the form commonly adopted in physical chemistry of solutions (also see Eq. (5.81))

$$\boxed{\beta\mu_i = \beta\mu_i^0 + \ln[c_i/c^0].} \tag{10.4}$$

Here, $c^0 = 1$ M is introduced to eliminate units under the logarithm and

$$\beta\mu_i^0 = \ln[N_A c_0 \lambda_i^3 \times 10^3]. \tag{10.5}$$

In this equation, λ_i caries the units of meters.

Deviations from the ideal-solution equation are usually described in terms of the activity coefficient γ_i. It allows one to write the chemical potential of a nonideal solution in the form maintaining the mathematical structure established for the ideal solution

$$\beta\mu_i = \beta\mu_i^0 + \ln[\gamma_i(c_i/c^0)]. \tag{10.6}$$

The goal of the equilibrium theory of solutions is to calculate γ_i.

10.2   Osmotic pressure

Osmotic pressure develops at the contact between the solution and the bulk solvent to compensate for the difference in chemical potentials between them. According to thermodynamics,

$$(\partial\mu/\partial P)_T = V/N = \rho^{-1}. \tag{10.7}$$

From this equation, a small variation of the chemical potential can be related to a small change of pressure

$$\beta\Delta\mu = \rho^{-1}\beta\Delta P = \rho^{-1}\beta\Pi, \tag{10.8}$$

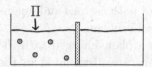

Fig. 10.1   Osmotic pressure Π applied to the solution.

where $\Delta P = \Pi$ is the pressure that needs to be applied from the solution side to compensate for a higher chemical potential of the solvent in the bulk compared to the chemical potential of the solvent (e.g., water for aqueous solutions) in solution (Fig. 10.1).

One can apply Eq. (10.2) either to the chemical potential of a given component in the solution or to the solvent itself. When applied to the solvent, one writes

$$\beta \Delta \mu = \ln[\rho \lambda^3] - \ln[\rho(1 - x_0)\lambda^3], \qquad (10.9)$$

where x_0 is the mole fraction of all solutes in the solution. Assuming that x_0 is small, $x_0 \ll 1$, one obtains from the expansion of the logarithm $\ln[1 - x] \simeq -x - x^2/2$

$$\beta \Delta \mu \simeq x_0. \qquad (10.10)$$

Combining Eqs. (10.8) and (10.10) for $\Delta \mu$, one arrives at the van't Hoff equation for the osmotic pressure

$$\beta \Pi = x_0 \rho. \qquad (10.11)$$

Assuming $V = 1$ L and $x_0 \rho = N_0/V = (N_0/N_A)(N_A/V)$, one arrives at the commonly used form

$$\boxed{\Pi_{\text{id}} = c_0 RT,} \qquad (10.12)$$

where $c_0 = (N_0/N_A)V^{-1}$ is the molar concentration and $R = k_B N_A$ is the gas constant. The subscript "id" in this equation refers to the assumption that solutes do not interact and Eq. (10.4) for the chemical potential of an ideal solution can be applied. This assumption needs revision described below for electrolytes. In that case, the osmotic pressure is reduced due to Coulomb interactions between the ions.

10.3   Gouy-Chapman screening

To introduce screening in electrolytes, we consider first the screening of a negatively charged substrate by its counterions (cations) in solution [61,66] (Fig. 10.2). There are no anions in solution to make the discussion easier to follow. The screening by an electrolyte made of both cations and anions in considered next in relation to the Debye-Hückel problem.

The configuration considered here consists of a plane charged with the negative surface charge density σ_-. The cations are distributed in the solution according to the Boltzmann (Gibbs) distribution

$$\rho_+(x) = \rho_0 e^{-\beta e\phi(x)}, \tag{10.13}$$

where $\phi(x)$ is the electrostatic potential acting on the positive charges carrying the unit charge e. The potential is shifted to be equal to zero at $x = 0$, ρ_0 becomes the surface concentration of cations.

By applying the Maxwell electrostatic equation $d^2\phi/dx^2 = -4\pi\rho_+/\epsilon_s$ for the scaled dimensionless potential $\bar{\phi} = \beta e\phi$, one arrives at the Poisson-Boltzmann equation

$$d^2\bar{\phi}/dx^2 = -(4\pi\beta e\rho_0/\epsilon_s)e^{-\bar{\phi}}, \tag{10.14}$$

where ϵ_s is the static dielectric constant of the solvent. This differential equation needs to be supplemented with the boundary conditions

$$\begin{aligned} d\bar{\phi}/dx\big|_{x=0} &= 4\pi\beta e|\sigma_-|/\epsilon_s, \\ d\bar{\phi}/dx\big|_{x=\infty} &= 0, \quad \bar{\phi}\big|_{x=0} = 0. \end{aligned} \tag{10.15}$$

These conditions state that the electric field at $x = 0$, $E_0 = 4\pi\sigma_- = -4\pi|\sigma_-|$ is produced by the surface charge density at the substrate. The electric field decays to zero at infinity. In addition, the potential is zero at $x = 0$ by the choice of parameters in the Boltzmann distribution.

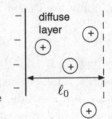

Fig. 10.2 Screening of the negatively charge substrate by the diffuse layer of cations

The solution of the second-order differential equation with the boundary conditions is

$$\bar{\phi} = 2\ln(1 + x/\ell_0), \tag{10.16}$$

where the Gouy-Chapman length

$$\ell_0 = \frac{\epsilon_s k_{\mathrm{B}} T}{2\pi e|\sigma_-|} \tag{10.17}$$

is obtained to satisfy the boundary conditions. One gets an additional condition for the density of solution ions at the substrate from the Poisson-Boltzmann equation

$$\frac{\rho_0}{e} = \frac{\epsilon_s k_B T}{2\pi e^2 \ell_0^2} = \frac{2\pi |\sigma_-|^2}{\epsilon_s k_B T}. \tag{10.18}$$

The fact that $\ell_0 \propto k_B T$ is an important physical result. It tells us that the diffuse layer of ions screening the charged substrate is caused by their thermal motion, which compensates, with entropy, for the pull of the negative charge of the substrate applied to the positive charge in solution. This is a general observation applicable to all finite-temperature electrolytes. The length ℓ_0 becomes equal to zero at $T = 0$, when the counterions condense on the negatively charged substrate thus fully compensating its charge. The surface charge density of counterions diverges in this limit, $\rho_0 \to \infty$. In addition, increasing ϵ_s makes the dielectric screening stronger, thus reducing the pull of the substrate and broadening the diffuse layer.

10.4 Debye-Hückel screening

Here we consider screening of the electrostatic potential around an ion in the bulk electrolyte. This is a three-dimensional problem with the radial symmetry. This means that the electrostatic potential $\phi(r)$ depends only on the distance r from a target spherical ion. The Poisson equation in 3D now involves the Laplacian operator $\Delta = \nabla^2$ in place of the second derivative: $\nabla^2 \phi = -4\pi \rho_q / \epsilon_s$. In addition, the charge density ρ_q is a sum of charge densities of cations and anions. We naturally expect that if we consider the potential around a positively charged cation, we find more anions on average it its vicinity (counterions) than cations (coions). These counterions form the diffuse screening layer with the radius λ_D known as the Debye-Hückel length. The equation for it is

$$\boxed{\lambda_D = \left(\epsilon_s k_B T / (8\pi e^2 \rho_0)\right)^{1/2},} \tag{10.19}$$

where now ρ_0 is the concentration of each ion (cation or anion) in the electrolyte solution (we assume that cations and anions carry charges of equal magnitude e). As one sees from this equation, one obtains perfect ion pairing in solution when $T \to 0$ and $\lambda_D \to 0$. The diffuse Debye-Hückel screening layer is thus held by the balance between the entropy forcing the ions to separate and the attraction energy between them forcing condensation. Attractions win over the thermal motion when $T \to 0$.

For water at 300 K, one obtains

$$\lambda_D \simeq \frac{3.0}{\sqrt{c_i}} \text{ Å}, \tag{10.20}$$

where c_i is the molar concentration of either cations or anions in the 1:1 electrolyte. One can next introduce the ionic strength

$$I = \tfrac{1}{2} \sum_k z_k^2 (c_k/c^0), \tag{10.21}$$

which is the sum over all ions with charges z_k present in solution with the molar concentrations c_k normalized to $c^0 = 1$ M. For a 1:1 electrolyte, $I = c_i/c^0$. The Debye screening length can then be given in terms of the ionic strength, the ratio T/T_0 with $T_0 = 300$ K, and the dielectric constant ϵ_s as follows

$$\lambda_D \simeq 0.344 \sqrt{\frac{\epsilon_s T}{I T_0}} \text{ Å}. \tag{10.22}$$

We now turn to the solution of the Debye-Hückel problem. The Poisson-Boltzmann equation reads in this case

$$\nabla^2 \bar{\phi} = -(4\pi\beta\rho_0 e^2/\epsilon_s)\left[e^{-\bar{\phi}} - e^{\bar{\phi}}\right]. \tag{10.23}$$

Two terms in the brackets represent the Boltzmann distributions for the cations ("-", depletion) and anions ("+", augmentation). By comparing with the definition of the Debye-Hückel length in Eq. (10.19), one can rewrite the Poisson-Boltzmann equation in a compact form

$$\boxed{\nabla^2 \bar{\phi} = \lambda_D^{-2} \sinh \bar{\phi}.} \tag{10.24}$$

10.5 Weak electrolytes

We consider a model of the restricted primitive electrolyte made of equal-size ions carrying charges $+e$ (cations) and $-e$ (anions) (Fig. 10.3). The "restricted" model implies equal sizes of the anions and cations for which we assume the diameter σ_i. The meaning of "primitive" implies that no molecular structure is assigned to the liquid in which ions are dissolved, and it is viewed as a dielectric with the dielectric constant ϵ_s.

The Debye-Hückel theory [67] seeks to solve the Poisson-Boltzmann equation by linearizing it, $\sinh(\bar{\phi}) \simeq \bar{\phi}$. The linearized Eq. (10.24) becomes

$$\nabla^2 \bar{\phi} = \kappa^2 \bar{\phi}, \tag{10.25}$$

where $\kappa^2 = \lambda_D^{-2}$ is the Debye-Hückel screening parameter. The solution of this equation is separated into the region of the bulk $r > \sigma_i$ and the region

$0 < r < \sigma_i$ inaccessible to the ions. The electrostatic potential in the bulk decays to zero at $r \to \infty$ and is given by the solution (cf. to Eq. (6.108))

$$\bar{\phi}(r) = \frac{A}{r} e^{-\kappa r}. \tag{10.26}$$

Inside the repulsion core, the potential is a sum of the potential of the ion $q = \pm e$ and a constant offset B

$$\bar{\phi}_0(r) = B + \frac{\beta e q}{\epsilon_s r}. \tag{10.27}$$

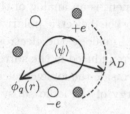

Fig. 10.3 Electrostatic potential of a weak 1:1 electrolyte. The target ion in the center is shown not to scale.

The two solutions for the electrostatic potential should satisfy the continuation boundary conditions

$$\bar{\phi}(\sigma_i) = \bar{\phi}_0(\sigma_i), \quad \bar{\phi}'(\sigma_i) = \bar{\phi}_0'(\sigma_i). \tag{10.28}$$

Since the dielectric constant is continuous across the boundary at $r = \sigma_i$, the second condition guarantees the continuity of the electric displacement across the boundary. These conditions are satisfied by the following solution

$$\bar{\phi} = \frac{\beta e q}{r \epsilon_s (1 + \kappa \sigma_i)} e^{-\kappa(r - \sigma_i)}, \quad \bar{\phi}_0 = \frac{\beta q e}{\epsilon_s r} + \langle \bar{\psi} \rangle. \tag{10.29}$$

The second term in $\bar{\phi}_0$ is the average potential created by the ionic atmosphere at the center of the target ion (Fig. 10.3)

$$\langle \bar{\psi} \rangle = \beta e \langle \psi \rangle = -\frac{\beta q e}{\epsilon_s} \frac{\kappa}{1 + \kappa \sigma_i}. \tag{10.30}$$

The electrostatic potential of a target ion in the liquid becomes

$$\phi = \frac{q}{\epsilon_s r} \frac{1}{1 + \kappa \sigma_i} e^{-\kappa(r - \sigma_i)}. \tag{10.31}$$

It is a product of the bare potential of the ion $q/(\epsilon_s r)$ and the screening factor $\propto \exp[-(r/\lambda_D)]$ exponentially damping the electrostatic potential on the characteristic distance of the Debye-Hückel screening length λ_D. This result is usually associated with the ionic atmosphere predominantly made of counterions and surrounding a target ion carrying the charge q. One has to be warned agains a mechanistic understanding of the ionic atmosphere as a static structure. On the contrary, it has to be understood in the dynamic terms. If the potential is averaged over some observation time much longer than the time of Brownian motion of the ions (picoseconds in liquids at normal conditions), the counterions moving nearly freely in solution will spend more time around a target ion than ions with the same charge. This time average will display an average opposite charge building up around each ion in solution.

The electrostatic potential inside the ion is composed of the electrostatic potential of the ion itself and an additional electrostatic potential $\langle \bar{\psi} \rangle = \beta e \langle \psi \rangle$ (Eq. (10.29)). This second component is an analog of the reaction field inside a spherical dipole produced by the dielectric polarized by the same dipole (Sec. 2.18). In a direct physical analogy, the electrostatic potential $\langle \psi \rangle$ is produced by the ionic atmosphere induced in the surrounding electrolyte by the target ion. The potential inside the ionic core can be used to calculate the thermodynamic free energy of the ionic subsystem. We will proceed in two steps.

The first step is to calculate the electrostatic free energy, which is the reversible work of creating the ionic atmosphere around each target ion. This can be done by continuously charging the target ion by changing its charge from zero to $q = \pm e$. Since $\langle \psi \rangle \propto q$, integration over q in the calculation of the work to assemble the charges leads to a factor of $1/2$ in the following expression (see also Eq. (2.153))

$$w_{\text{el}} = \pm \tfrac{1}{2} e \langle \psi_\pm \rangle, \tag{10.32}$$

where $e\langle \psi_+ \rangle$ is used for cations ($q = +e$) and $-e\langle \psi_- \rangle$ applies to anions ($q = -e$). Given that there are N_0 ions of each kind with the number density $\rho_0 = N_0/V$, one obtains the electrostatic energy of the ionic subsystem

$$\frac{\beta W_{\text{el}}}{V} = -\frac{1}{8\pi} \frac{\kappa^3}{1 + \kappa \sigma_i}, \tag{10.33}$$

where

$$\kappa = \left(\frac{8\pi \beta e^2 \rho_0}{\epsilon_s} \right)^{1/2}. \tag{10.34}$$

Even though W_{el} is the work done to assemble the charges, this electrostatic work carries the meaning of the internal energy from the perspective of thermodynamics. The reason is that this work is performed isoentropically, without heat exchange with the surrounding thermal bath. It, therefore, carries the usual meaning of the potential energy of charges.

The Helmholtz free energy is obtained by integrating the electrostatic potential energy in terms of the inverse temperature according to the equation

$$\frac{\beta F_{el}}{V} = \frac{1}{V} \int_0^\beta W_{el}(\beta')d\beta' = -\frac{1}{8\pi} \int_0^\beta \frac{d\beta'}{\beta'} \frac{\kappa^3}{1 + \kappa\sigma_i}. \tag{10.35}$$

One can next change the integration variable from β to κ by using the identity $d\beta/(2\beta) = d\kappa/\kappa$ with the result

$$\frac{\beta F_{el}}{V} = -\frac{1}{4\pi} \int_0^\kappa \frac{d\kappa' \kappa'^2}{1 + \kappa'\sigma_i}. \tag{10.36}$$

The result of integration is

$$\boxed{\frac{\beta F_{el}}{V} = -\frac{1}{4\pi\sigma_i^3} \left[\ln(1 + \kappa\sigma_i) + \tfrac{1}{2}(\kappa\sigma_i)^2 - \kappa\sigma_i \right].} \tag{10.37}$$

In the limit of small electrolyte concentration satisfying the condition $\kappa\sigma_i \ll 1$, one obtains

$$\frac{\beta F_{el}}{V} \simeq -\frac{\kappa^3}{12\pi}. \tag{10.38}$$

Since $\kappa \propto T^{-1/2}$, $F_{el} \propto T^{-1/2}$. The free energy of ions becomes less negative with increasing temperature. This scaling implies that ions produce a negative contribution to the system entropy

$$S_{el} = -(\partial F_{el}/\partial T)_{NV} < 0. \tag{10.39}$$

From the solution for the free energy, one finds the contribution to pressure from electrostatic interactions

$$\beta P_{el} = -\left(\frac{\partial \beta F_{el}}{\partial V} \right)_{NT} = \frac{\kappa^3}{12\pi} + \frac{\kappa^2 V}{4\pi} \left(\frac{\partial \kappa}{\partial V} \right)_{NT}. \tag{10.40}$$

Given that $d\kappa/\kappa = -dV/(2V)$, one obtains

$$\beta P_{el} = -\kappa^3/(24\pi). \tag{10.41}$$

One finds that the lowest-order contribution to the pressure arising from Coulomb interactions is negative (on average, attractions between the ions). If the osmotic pressure $\Pi_{id} = 2c_0 RT$ (Eq. (10.12)) is associated with the ideal pressure of both cations and anions (the factor of two), the electrostatic pressure P_{el} contributes to the reduction of the osmotic pressure due to ions' attractions

$$\boxed{\Pi = \Pi_{id} - k_B T \kappa^3/(24\pi).} \tag{10.42}$$

The attraction pressure due to ions is not given in the form of the mean-field van der Waals attraction component $-\beta a \rho^2$ entering the compression factor (Eq. (6.24)). Instead, one finds that electrolyte's contribution to pressure scales as $\rho_0^{3/2}$. This scaling of pressure with the density of ions is a signature of long-range Coulomb interactions, which distinguishes this case from the term $B(T)\rho^2$ appearing in the virial expansion due to short-range interactions (Sec. 6.3).

10.6   Activity coefficients

Thermodynamic equations derived here are known as the extended Debye-Hückel theory. An outcome of this theory that finds most applications is the result for the chemical potential of an ion. It is given as the interaction of the ionic charge with the average electrolyte potential at the position of the charge

$$\beta\mu_{\pm} = \pm\langle\bar{\psi}_{\pm}\rangle = -\frac{\beta e^2}{\epsilon_s}\frac{\kappa}{1+\kappa\sigma_i}. \tag{10.43}$$

This equation can be proved directly by taking the derivative of the free energy over the number of particles for a given component in Eq. (10.36)

$$\frac{\beta\mu_+}{V} = \frac{1}{V}\left(\frac{\partial\beta F_{el}}{\partial N_0}\right)_{VT} = -\frac{1}{4\pi}\left(\frac{\partial\kappa}{\partial N_0}\right)_{VT}\frac{\partial}{\partial\kappa}\int_0^\kappa\frac{d\kappa'\kappa'^2}{1+\kappa'\sigma_i}. \tag{10.44}$$

The result is obtained by noting that $d\kappa/\kappa = dN_0/(2N_0)$ if T, V are held constant

$$\beta\mu_+ = -\frac{1}{8\pi\rho_0}\frac{\kappa^3}{1+\kappa\sigma_i} = \beta e\langle\psi_+\rangle. \tag{10.45}$$

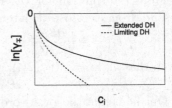

Fig. 10.4   Extended Debye-Hückel (DH) theory for the activity coefficient vs the limiting DH theory based on $\sigma_i \to 0$.

The chemical potential arising from interactions between the ions specifies the activity coefficient, which applies to both cations and anions

$$\beta\mu_{\pm} = \ln\gamma_{\pm}. \tag{10.46}$$

Since $\kappa \propto \sqrt{I}$, this relation leads to the commonly listed result for the activity coefficient $\gamma_{\pm}$ in terms of the ionic strength of the solution (Fig. 10.4)

$$\ln\gamma_{\pm} = -\frac{A\sqrt{I}}{1+B\sigma_i\sqrt{I}}. \tag{10.47}$$

A more physically motivated results can be obtained in terms of three length scales characterizing the physics of the problem.

The limiting Debye-Hückel theory corresponds to $\kappa\sigma_i \ll 1$ (Fig. 10.4). In this case, the $\kappa\sigma_i$ term in the denominator of Eq. (10.43) can be dropped. The activity coefficient is then given as the ratio of two characteristic lengths describing the range of electrostatic interactions in solutions: the Bjerrum length ℓ_B and the Debye-Hückel length λ_D

$$\ln\gamma_\pm = -\ell_B/\lambda_D. \tag{10.48}$$

The Bjerrum length establishes the distance at which binary Coulomb interaction energies between pairs of ions become equal to the thermal energy $k_B T$

$$\ell_B = \beta e^2/\epsilon_s. \tag{10.49}$$

In contrast, λ_D establishes the characteristic length of collective excitations of the density of ions in solution. This latter interpretation of λ_D follows from the structure factor of weak electrolytes considered in the next section. For water at normal conditions $\ell_B \simeq 7$ Å, while, according to Eq. (10.20), $\lambda_D \simeq 9.5$ Å at $c_i = 0.1$ M. The two lengths are comparable at these conditions, but one gets $\lambda_D \gg \ell_B$ at $c_i \to 0$. This is the limit of an ideal solution when one can neglect the activity coefficient in the chemical potential.

The limiting Debye-Hückel theory applies to the limit of small ions $\kappa\sigma_i \ll 1$ and contains two characteristic lengths, ℓ_B and λ_D. In contrast, the extended Debye-Hückel theory does not neglect the size of the ions and adds one more characteristic length, the ionic diameter σ_i. In terms of these three lengths, Eq. (10.43) yields for the activity coefficient

$$\boxed{\ln\gamma_\pm = -\frac{\ell_B}{\lambda_D + \sigma_i}.} \tag{10.50}$$

10.7 Plasmons

Fluctuations of the charge density in plasmas and electrolytes are called plasmons. Those are propagating (non-dissipative) waves. The dynamic structure factor of the electrolyte involves the δ-function of the plasmon frequency $\omega_p(k)$

$$S(k,\omega) = S(k)\delta(\omega - \omega_p(k)). \tag{10.51}$$

Here, we consider an example of the one-component plasma, i.e., an ensemble of ions immersed in a uniform neutralizing background charge also characterized by the dielectric constant ϵ_s. For a one-component plasma, the charge density is directly related to the number density and there is no need to distinguish between the structure factors of charge density and number density.

The connection between the dynamic structure factor and the longitudinal current correlation function (Eq. (7.55)) implies the sum rule

$$\int_{-\infty}^{\infty} \frac{d\omega}{(2\pi)} \omega^2 S(k,\omega) = k^2 J_L(k,0) = (v_T k)^2, \qquad (10.52)$$

where $v_T^2 = (\beta m)^{-1}$ is the thermal speed and m is the mass of an ion. The plasmon dispersion relation therefore becomes

$$\omega_p^2(k) = (v_T k)^2 / S(k). \qquad (10.53)$$

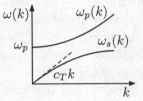

Fig. 10.5   Dispersion of propagating optical plasmons with the frequency $\omega_p(k)$ compared to acoustic waves with the dispersion relation $\omega_s(k)$; c_T is the isothermal speed of sound (Eq. (6.52)).

One can apply the Ornstein-Zernike equation (Sec. 6.5),

$$S(k) = [1 - \rho_0 c(k)]^{-1} \qquad (10.54)$$

for the static structure factor; ρ_0 is the number density of ions in the one-component plasma. The Fourier transform of the direct correlation function $c(k)$ can be taken for dilute electrolytes in the random-phase approximation (Eq. (6.34)) as the negative interaction potential energy divided by $k_B T$: $c(k) = -\beta v(k)$. The Coulomb interaction between the ions in the one-component plasma, which are assumed to carry the unit charge e, in given in reciprocal space as

$$v(k) = \frac{4\pi e^2}{\epsilon_s k^2}. \qquad (10.55)$$

From this equation, the static structure factor of the one-component plasma is

$$\boxed{S(k) = \frac{k^2}{k^2 + \kappa^2},} \qquad (10.56)$$

where $\kappa = \lambda_D^{-1}$ is the Debye-Hückel screening parameter, $\kappa^2 = 4\pi\beta e^2 \rho_0 / \epsilon_s$. This implies the following dispersion law for the plasmon excitations

$$\omega_p^2(k) = v_T^2(\kappa^2 + k^2). \qquad (10.57)$$

Plasmons are therefore optical excitations (frequency is nonzero at $k = 0$, Fig. 10.5). The $k = 0$ value is known as the plasma frequency satisfying the equation

$$\omega_p^2 = (v_T \kappa)^2. \tag{10.58}$$

The definition of the plasma frequency can be extended to the 1:1 electrolyte, in which case it becomes

$$\boxed{\omega_p^2 = \frac{8\pi e^2 \rho_0}{\epsilon_s m}.} \tag{10.59}$$

Note that ω_p is independent of temperature. The temperature dependence cancels between the thermal velocity v_T and λ_D. Equation (10.58) shows that the inverse plasmon frequency is equal to the inverse time required for a particle with thermal velocity to travel the screening length.

10.8 Electrolyte conductivity

Here we illustrate the application of the memory function formalism introduced in Sec. 7.10 to calculate the electrolyte conductivity. The arguments are given in terms of a phenomenological exponential relaxation time of the memory function. This is followed by a more microscopic arguments in the next section, where this relaxation time is expressed in terms of the diffusion constants of the ions and the Debye-Hückel screening length.

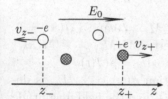

Fig. 10.6 Electrolyte in the electric field of external charges E_0.

Consider an electrolyte placed in the oscillatory external field $E_0(t) = E_\omega e^{-i\omega t}$ directed along the z-axis of the laboratory frame (Fig. 10.6). The total dipole moment of the sample is

$$M_z = e \sum_{i=1}^{N_0} (z_{i,+} - z_{i,-}), \tag{10.60}$$

where $z_{i,\pm}$ are the z-coordinates of N_0 cations ($+$) and anions ($-$). The external electric field is responsible for the perturbation of the system of charges with the perturbation Hamiltonian $H'(t) = -M_z E_\omega e^{-i\omega t}$. This field induces an oscillatory dipole due to electrolyte ions changing their positions. The dipole moment is oriented along the z-axis and has the amplitude M_ω. The latter is given by the linear response theory (Sec. 8.2)

$$M_\omega = \tilde{\chi}_M(\omega)E_\omega. \tag{10.61}$$

The Fourier-Laplace transform of the linear response function in this equation is found from Eq. (8.28)

$$\tilde{\chi}_M(\omega) = \beta \int_0^\infty dt e^{i\omega t} \langle M_z(t)\dot{M}_z \rangle. \tag{10.62}$$

The time-dependent susceptibility function here is

$$\chi_M(t) = \beta \langle M_z(t)\dot{M}_z \rangle. \tag{10.63}$$

The total electric current through the sample is the time derivative of the dipole moment

$$J(t) = \dot{M}_z(t) \tag{10.64}$$

and one finds $J_\omega = -i\omega M_\omega$. Substituting this relation to Eq. (10.61), one obtains

$$J_\omega = -E_\omega \int_0^\infty \chi_M(t)de^{i\omega t} = \beta E_\omega \int_0^\infty dt e^{i\omega t} \langle J(t)J \rangle. \tag{10.65}$$

Arriving at the last equation requires integration by parts

$$\int_0^\infty \chi_M(t)de^{i\omega t} = \chi_M(t)e^{i\omega t}\Big|_0^\infty - \int_0^\infty dt\dot{\chi}_M(t)e^{i\omega t}. \tag{10.66}$$

One notes that, from Eq. (10.63), $\chi_M(0) = \beta\langle M_z\dot{M}_z \rangle = 0$ and $\chi_M(\infty) = 0$. The first result is the consequence of the property $\langle M_z\dot{M}_z \rangle = -\langle \dot{M}_z M_z \rangle$ (Eq. (8.32)). Finally, one has from Eq. (10.63) $\dot{\chi}_M(t) = \beta\langle J(t)J \rangle$.

The frequency-dependent conductivity function $\sigma(\omega) = j_\omega/E_\omega$ is defined by the ratio of the current density $j_\omega = J_\omega/V$ and the field magnitude

$$\sigma(\omega) = (\beta/V) \int_0^\infty dt e^{i\omega t} \langle J(t)J \rangle. \tag{10.67}$$

In turn, the zero-frequency conductivity $\sigma = \sigma(\omega = 0)$ can be written as

$$\boxed{\sigma = \frac{\beta}{3V} \int_0^\infty dt\langle \mathbf{J}(t) \cdot \mathbf{J} \rangle,} \tag{10.68}$$

where now all three equivalent Cartesian components of the current $\mathbf{J}(t)$ in an isotropic material are included in the current-current time correlation function.

One can view the total electric current $J(t)$ through the sample as a dynamic variable and write the memory equation for it (Eq. (7.99))

$$\dot{J} + \int_0^t d\tau M_J(t-\tau)J(\tau) = R_J(t), \qquad (10.69)$$

where R_J is the random force producing fluctuations of the dynamic variable J. The corresponding equation for the time correlation function (Eq. (7.103)) $C_J(t) = \langle J(t)J \rangle$, $\langle J \rangle = 0$ reads

$$\dot{C}_J + \int_0^t d\tau M_J(t-\tau)C_J(\tau) = 0. \qquad (10.70)$$

One can perform Fourier-Laplace transform of this equation to obtain

$$\sigma(\omega) = (\beta/V)\tilde{C}_J(\omega) = \frac{\omega_p^2}{4\pi} \frac{1}{-i\omega + \tilde{M}_J(\omega)}, \qquad (10.71)$$

where $C_J(0) = 2e^2 v_T^2 N_0$ is used. For the memory function given by an exponential decay

$$M_J(t) = \omega_0^2 e^{-t/\tau_q}, \qquad (10.72)$$

one arrives at the final result

$$\sigma(\omega) = \frac{\omega_p^2}{4\pi}\left[-i\omega + \frac{\omega_0^2 \tau_q}{1 - i\omega\tau_q} \right]^{-1}. \qquad (10.73)$$

This expression results in a finite zero-frequency conductivity $\sigma = \omega_p^2/(4\pi\omega_0^2\tau_q)$, which can be used to determine ω_0.

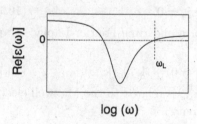

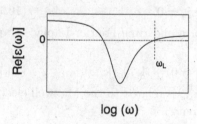

Fig. 10.7 Real part of the dielectric function from Eq. (10.74).

The frequency-dependent dielectric function of the electrolyte can be found from Eq. (8.165)

$$\epsilon(\omega) = 1 + \frac{4\pi i\sigma(\omega)}{\omega} = 1 + \frac{i\omega_p^2}{\omega}\left[-i\omega + \frac{\omega_0^2\tau_q}{1 - i\omega\tau_q} \right]^{-1}. \qquad (10.74)$$

This dielectric function diverges at $\omega \to 0$, as expected for a conductor.

The dielectric function passes through zero, $\epsilon(\omega_L) = 0$, at the longitudinal optical frequency ω_L (Fig. 10.7). This point corresponds to the divergence of the longitudinal dielectric susceptibility $\propto (1 - \epsilon^{-1}(\omega))$ (Sec. 8.13). The transverse dielectric susceptibility scales as $\propto (\epsilon(\omega) - 1)$, and its divergence is reached at the transverse optical frequency ω_T: $\epsilon(\omega_T) \to \infty$. At $\omega_p \tau_q \ll 1$ and $\omega_0 \tau_q \ll 1$, Eq. (10.74) predicts

$$\omega_L \simeq \sqrt{\omega_0^2 + \omega_p^2}, \quad \omega_T \simeq \omega_0. \tag{10.75}$$

One gets $\omega_L > \omega_T$ in accord with the Lyddane-Sachs-Teller equation (Eq. (8.188)) discussed in Sec. 8.14. The range of frequencies where $\mathrm{Re}[\epsilon(\omega)] < 0$ (Fig. 10.7) corresponds to full reflection of electromagnetic radiation by the material.

10.9   *Dynamics of electrolytes

Charge density:
$\rho_q(t) = \sum_{j,\nu} q_\nu \delta(\mathbf{r} - \mathbf{r}_{j,\nu}(t)), \; \nu = 1, 2$
Current density:
$\mathbf{j}_q(t) = \sum_{j,\nu} q_\nu \mathbf{v}_{j,\nu} \delta(\mathbf{r} - \mathbf{r}_{j,\nu}(t))$
Conductivity tensor:
$\boldsymbol{\sigma}(\mathbf{k}) = \hat{\mathbf{k}}\hat{\mathbf{k}}\sigma_L(k) + (\mathbf{1} - \hat{\mathbf{k}}\hat{\mathbf{k}})\sigma_T(k).$

The total charge of an electrolyte has to be conserved. The differential form of the conservation law is the continuity relation between the charge density derivative and the divergence of the current density

$$\partial_t \rho_q + \nabla \cdot \mathbf{j}_q = 0. \tag{10.76}$$

Performing the spatial Fourier transform and time Fourier-Laplace transform, one obtains (Eq. (1.50))

$$- i\omega \tilde{\rho}_q(\mathbf{k}, \omega) = \rho_q(\mathbf{k}, t = 0) + i\mathbf{k} \cdot \tilde{\mathbf{j}}_q(\mathbf{k}, \omega). \tag{10.77}$$

The longitudinal current density $\hat{\mathbf{k}} \cdot \tilde{\mathbf{j}}_q$ is related to the Maxwell electric field through the longitudinal conductivity

$$- i\mathbf{k} \cdot \tilde{\mathbf{j}}_q(\mathbf{k}, \omega) = -\sigma_L(\mathbf{k}, \omega)i\mathbf{k} \cdot \tilde{\mathbf{E}}(\mathbf{k}, \omega) = (4\pi\sigma_L(\mathbf{k}, \omega)/\epsilon(\omega))\tilde{\rho}_q(\mathbf{k}, \omega), \tag{10.78}$$

where the Maxwell equation (Eq. (2.71)) was used in the second equality. The dielectric function $\epsilon(\omega)$ accounts for dielectric screening by the solvent. The equation for the charge density becomes

$$\tilde{\rho}_q(\mathbf{k}, \omega) = \frac{\rho_q(\mathbf{k}, t = 0)}{-i\omega + 4\pi\sigma_L(k, \omega)/\epsilon(\omega)}. \tag{10.79}$$

By multiplying this equation with the complex conjugate charge density $\rho_q(-\mathbf{k}, t = 0) = \rho_q(-\mathbf{k}, 0)$ and taking the statistical average $\langle \ldots \rangle$, one arrives at the equation for the Fourier-Laplace transform of the intermediate scattering function

$$F_q(\mathbf{k}, t) = N^{-1} \langle \rho_q(\mathbf{k}, t) \rho_q(-\mathbf{k}, 0) \rangle. \tag{10.80}$$

It is given by the following equation

$$\tilde{F}_q(k, \omega) = \frac{S_q(k)}{-i\omega + 4\pi\sigma_L(k, \omega)/\epsilon(\omega)}. \tag{10.81}$$

Here, the static structure factor of the charge density is

$$\boxed{S_q(k) = N^{-1} \langle |\rho_q(\mathbf{k})|^2 \rangle.} \tag{10.82}$$

It describes correlated fluctuations of the charge density, in contrast to correlated fluctuations of the number density described by the density structure factor.

The dynamic structure factor $S_q(k, \omega) = 2\text{Re}[\tilde{F}_q(k, \omega)]$ follows next

$$S_q(k, \omega) = S_q(k) \frac{8\pi\text{Re}[\sigma_L(k, \omega)/\epsilon(\omega)]}{(\omega - 4\pi\text{Im}[\sigma_L(k, \omega)/\epsilon(\omega)])^2 + (4\pi\text{Re}[\sigma_L(k, \omega)/\epsilon(\omega)])^2}. \tag{10.83}$$

This is a Lorentzian function with the relaxation time of the electrolyte at $k \to 0$ equal to

$$\tau_q = \epsilon_s/(4\pi\sigma), \tag{10.84}$$

where $\epsilon_s = \epsilon(\omega = 0)$ is the static dielectric constant of the medium and $\sigma = \sigma_L(k = 0, \omega = 0)$ is the zero-frequency conductivity of the electrolyte.

The conductivity follows from the linear-response approximation as given by Eq. (10.68) (cf. to Eqs. (7.16) and (8.137)). The time correlation function for the total current is

$$\langle \mathbf{J}(t) \cdot \mathbf{J} \rangle = \sum_{\nu, \mu = 1, 2} \sum_{i, j} q_\nu q_\mu \langle \mathbf{v}_{i, \nu}(t) \cdot \mathbf{v}_{j, \mu} \rangle. \tag{10.85}$$

Assuming that velocities of different ions are uncorrelated

$$\langle \mathbf{v}_{i, \nu}(t) \cdot \mathbf{v}_{j, \mu} \rangle = \delta_{\nu\mu} \delta_{ij} \langle \mathbf{v}_i(t) \cdot \mathbf{v}_i \rangle, \tag{10.86}$$

one arrives at the Nernst-Einstein equation

$$\sigma = \beta\rho(x_1 q_1^2 D_1 + x_2 q_2^2 D_2), \quad \rho = 2\rho_0, \tag{10.87}$$

where $x_{1,2}$ and $D_{1,2}$ are the ions' mole fractions and diffusion constants, respectively. In the simple case of $x_1 = x_2 = 1/2$, $q_1 = -q_2 = e$, the above equation simplifies to

$$\sigma = \bar{D}\beta e^2 \rho, \tag{10.88}$$

where $\bar{D} = (D_1 + D_2)/2$ is the average diffusion constant of the ions. The relaxation time of the ionic atmosphere becomes

$$\tau_q = \left(\kappa^2 \bar{D}\right)^{-1} = \lambda_D^2/\bar{D}. \qquad (10.89)$$

At the physiological 0.1 M electrolyte concentration by putting $D_1 = D_2 \simeq 10^{-5}$ cm^2/s one arrives at $\tau_q \approx 1$ ns in water.

Equation (10.89) is an appealing physical result. It shows that relaxation of an electrolyte is caused by diffusion of ions, with the average diffusion constant $\bar{D}$, through the Debye-Hückel screening length.

Chapter 11

Spectra

Spectroscopy provides us with the tool to learn about physics of microscopic dimensions by observing the interaction of light and particles with atoms and molecules [68,69]. Transition energies for light absorption and emission are affected by local fields acting on the light absorber [70]. The transition line shape is affected either by average static fields (solvatochromism) or by field dynamics and fluctuations (line broadening function) [71]. The energy absorbed by the molecule in an electronic transition is transferred through radiationless transitions to nonequilibrium populations of molecular vibrations [72]. Relaxation of these hot vibrational states is driven by forces exerted by the medium on molecular oscillators and can be expressed as the friction spectrum experienced by the molecule from the surrounding thermal bath. Finally, thermal agitation of scattering centers affects scattering cross-section allowing one to learn about microscopic dynamics from scattering of radiation and subatomic particles (neutrons and electrons) [51,73–75]. Many of the theoretical concepts considered in previous sections find direct applications in spectroscopy. This Chapter aims at establishing these connections. Spectroscopy of molecules and condensed materials is a true integrator of many areas of physics and chemistry requiring input from a number of theoretical formalisms: from quantum mechanics to the theory of correlation functions and from linear response theories to molecular hydrodynamics and solvation. One of the goals of this book is to demonstrate how these often distinct areas of study intertwine in molecular science. Spectroscopy becomes a prime example of such an integration.

11.1   Transition probability

Fourier integral:
$$\delta(x) = \int_{-\infty}^{\infty} e^{ixt} dt/(2\pi)$$

Most of the results for molecular spectroscopy can be obtained in the quasi-classical approximation, when the molecule is viewed as a quantum system interacting with the classical electromagnetic field. The dipolar approximation for the molecule-light interaction is considered here. When not prohibited by selection rules, it yields the highest intensities for absorption and emission.

The interaction potential energy which perturbs the system to cause a quantum transition is between the dipole moment operator $\hat{\mu}$ and the oscillating classical electric field $\mathbf{E}(t) = \mathbf{E}_0 \exp[-i\omega t]$ of the electromagnetic wave (external electric field)

$$H'(t) = -\chi_c \hat{\mu} \cdot \mathbf{E}(t). \tag{11.1}$$

In this equation, χ_c is the cavity field susceptibility (Sec. 2.16) correcting the local field acting on a given molecule compared to the external field $\mathbf{E}(t)$. For transitions at optical frequencies, $\chi_c = \chi_c(n_D)$ is a function of the medium's refractive index n_D.

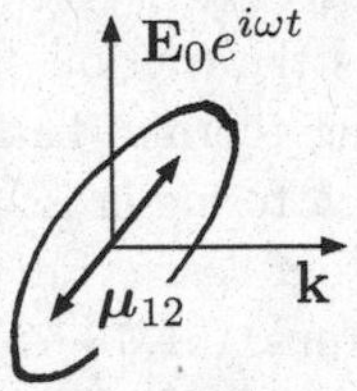

Fig. 11.1   Interaction of the electromagnetic wave with the transition dipole μ_{12} of the molecule.

The Fermi's golden rule can be applied to obtain the transition probability w between the initial state 1 and the final state 2 with the absorption of the photon carrying the energy $\hbar\omega$. The result is (Eq. (4.90))

$$w = \frac{2\pi}{\hbar} \chi_c^2 |\mu_{12}|^2 (\hat{\mathbf{e}} \cdot \mathbf{E}_0)^2 \delta(E_2 - E_1 - \hbar\omega), \tag{11.2}$$

where $\delta(x)$ is the delta function. Further, $\mu_{12} = \langle 1|\hat{\mu}|2\rangle$ is the transition dipole between the states 1 and 2. Finally, $\hat{\mathbf{e}}$ determines the direction of the transition dipole: the angle between the transition dipole and the polarization of the electric field affects the transition probability. Since all observables in quantum mechanics are not affected by the phase shift of the wave function, the transformation $\hat{\mathbf{e}} \to -\hat{\mathbf{e}}$ should not affect the observables;

the transition dipole is often shown with a double arrow pointing in opposite directions to reflect this symmetry (Fig. 11.1).

The direction of propagation of the electromagnetic field is determined by the wave vector $\mathbf{k}$

$$\mathbf{k} = \hat{\mathbf{k}}(2\pi/\lambda) = \hat{\mathbf{k}}2\pi\bar{\nu}, \tag{11.3}$$

where λ is the wave length and $\bar{\nu} = 1/\lambda$ is the wavenumber.

Assuming that transitions to all possible final states E_f of the molecule can occur and different initial states E_i can be involved, Eq.(11.2) can be altered to include all those states. The final states are included by summing over them, but the initial states E_i should come with their respective probabilities ρ_i (statistical weights). For the equilibrium distribution, one has $\rho_i = Z^{-1}\exp[-\beta E_i]$, where Z is the partition function. One therefore obtains

$$\boxed{w = \frac{2\pi}{\hbar}\chi_c^2(\hat{\mathbf{e}} \cdot \mathbf{E}_0)^2 \sum_{i,f} |\mu_{if}|^2 \delta(E_f - E_i - \hbar\omega)\rho_i.} \tag{11.4}$$

We now consider separately the term under the sum and apply the Fourier integral for the delta-function to obtain

$$I(\omega) = \int_{-\infty}^{\infty} \frac{dt}{2\pi\hbar} I(t)e^{-i\omega t}, \tag{11.5}$$

where

$$I(t) = \sum_{i,f} \langle i|\hat{\mu}^*|f\rangle\langle f|\hat{\mu}|i\rangle e^{i(E_f - E_i)t/\hbar}\rho_i. \tag{11.6}$$

Since $|i\rangle$ and $|f\rangle$ are the eigenstates of the unperturbed Hamiltonian $\hat{H}|i,f\rangle = E_{i,f}|i,f\rangle$, one can bring the exponents carrying energies under the quantum bracket by using the identity $\exp[i\hat{H}t/\hbar]|i\rangle = \exp[iE_i t/\hbar]|i\rangle$

$$I(t) = \sum_{i,f} \langle i|e^{-i\hat{H}t/\hbar}\hat{\mu}^* e^{i\hat{H}t/\hbar}|f\rangle\langle f|\hat{\mu}|i\rangle\rho_i. \tag{11.7}$$

Since one has the identity $\sum_f |f\rangle\langle f| = 1$, the final result becomes

$$I(t) = \mathrm{Tr}\left[\hat{\mu}^*(t)\hat{\mu}\hat{\rho}\right], \tag{11.8}$$

where $\mathrm{Tr}[\ldots] = \sum_i \langle i|\ldots|i\rangle$ and $\hat{\mu}(t)$ denotes the Heisenberg representation of the quantum operator (see Eq. (4.31))

$$\hat{\mu}(t) = e^{i\hat{H}t/\hbar}\hat{\mu}e^{-i\hat{H}t/\hbar} \tag{11.9}$$

and

$$\hat{\rho} = Z^{-1}\exp[-\beta\hat{H}] \tag{11.10}$$

is the density matrix operator.

11.2   Oscillator strength

The oscillator strength is often used in spectroscopy to characterize the intensity of an optical line. As we show below, it is directly related to the extinction coefficient. The oscillator strength for the transition between quantum states 1 and 2 is defined by the following equation

$$f_{12} = (2m_e/3\hbar)\omega_{12}|\mathbf{r}_{12}|^2 = (2m_e/\hbar)\omega_{12}|x_{12}|^2, \qquad (11.11)$$

where $\omega_{12} = (E_2 - E_1)/\hbar$ is the transition frequency between energy levels E_1 and E_2 and m_e is the electron mass. The second relation here is obtained with the account of isotropy of space which demands that all three Cartesian coordinates of the electron should be equivalent and $|\mathbf{r}_{12}|^2 = 3|x_{12}|^2$. Here, $x_{12} = \langle 1|\hat{x}|2\rangle$ is the bra-ket of the displacement operator taken between the quantum states $|1\rangle$ and $|2\rangle$.

The advantage of using f_{12} instead of the transition dipole μ_{12} is that the former is dimensionless and the latter has to carry the units of the electric dipole moment. It is instructive to realize where this definition is coming from and why do we call it the oscillator strength. This perspective is provided by looking at the x_{12} for the harmonic oscillator.

One can define the operator of displacement of the harmonic oscillator in terms of raising, $a^\dagger$, and lowering, a, operators (Sec. 4.9)

$$\hat{x} = \sqrt{\frac{\hbar}{2m_e\omega}}\left(a^\dagger + a\right), \qquad (11.12)$$

where it is assumed that the response of an electron in the atom or molecule to the field of electromagnetic radiation can be represented by the displacement of a harmonic oscillator with the frequency ω. The bra-ket x_{12} between the ground, $|0\rangle$, and first excited, $|1\rangle$, states of the harmonic oscillator is given by the expression (see Eq. (4.105))

$$\langle 1|\hat{x}|0\rangle = \sqrt{\hbar/(2m_e\omega)}. \qquad (11.13)$$

One therefore anticipates that each electron, represented by a harmonic oscillator, contributes the value

$$f_{01} = (2m_e/\hbar)\omega\langle 1|\hat{x}|0\rangle^2 = 1 \qquad (11.14)$$

to the total absorption. For a molecule with N_e electrons, one automatically satisfies the Thomas-Reiche-Kuhn sum rule

$$\sum_i f_{01}(i) = N_e, \qquad (11.15)$$

where the sum runs over all electrons in the molecule. The actual transition dipoles in atoms and molecules, $\boldsymbol{\mu}_{12} = -e\sum_i\langle 1|\hat{\mathbf{r}}_i|2\rangle$ ($\mathbf{r}_i$ is the position of electron i), deviate from the simple result for the harmonic oscillator. The definition of f_{12} according to Eq. (11.11) quantifies the extent of this deviation.

11.3 Absorption of light

We consider here absorption of radiation leading to transition $1 \rightarrow 2$ between two states in the molecule. The summation over states of the molecule is eliminated in Eq. (11.4), but the summation over the states of the photons in the cubic volume L^3 still needs to be included if the molecules undergoing quantum transitions stay in equilibrium with electromagnetic radiation (Fig. 11.2).

Fig. 11.2 A two-state system absorbing radiation in equilibrium with an ensemble of photons.

Given a macroscopic number of photons, the summation over the photon states implies integration over the photon frequencies with the density of states determined for the back-body radiation (Sec. 4.1). Rewriting Eq. (4.11) in terms of circular frequencies $\omega = 2\pi c/(n_D\lambda)$, the number of photons with frequency ω is

$$N(\omega) = \frac{8\pi n_D^3 L^3}{3(2\pi c)^3}\omega^3. \tag{11.16}$$

The factor of n_D^3 appears from the speed of light c/n_D in the medium with the refractive index n_D. One can rewrite this equation in terms of the number of photons per energy interval $dE = \hbar d\omega$ and in the solid angle $d\Omega = \sin\theta d\theta d\phi$

$$\rho(E) = dN/dE = 2\left(\frac{n_D L}{2\pi c\hbar}\right)^3 E^2 d\Omega. \tag{11.17}$$

We will now use the real oscillating electric field $\mathbf{E}(t) = \mathbf{E}_0\cos\omega t = \mathbf{E}_0(e^{i\omega t} + e^{-i\omega t})/2$ to connect it with the electromagnetic energy. The perturbation theory in Eq. (11.2) can still be maintained with the replacement $\mathbf{E}_0 \rightarrow \mathbf{E}_0/2$ (rotating wave approximation, see Sec. 4.8). The summation over the initial states i in Eq. (11.4) is replaced with the integral over the photon energies $E = \hbar\omega$ as

$$w_{\text{abs}} = \int w(E)\rho(E)dE \qquad (11.18)$$

with the result

$$w_{\text{abs}} = \frac{\pi n_D^3 L^3 \omega^2}{(2\pi c)^3 \hbar^2} \chi_c^2 |\mu_{12}|^2 E_0^2 (\hat{\mathbf{e}} \cdot \hat{\mathbf{e}}_0)^2 d\Omega, \qquad (11.19)$$

where $\hat{\mathbf{e}}_0$ is the unit vector along the electric field vector $\mathbf{E}_0$ and $\omega_{12} = \omega$ is the transition frequency.

The energy of the electromagnetic field in the volume L^3 can be related to the number of photons N_{ph} through the relation

$$N_{\text{ph}}\hbar\omega = \frac{n_D^2 L^3}{8\pi}\left[\overline{\mathbf{E}^2 + \mathbf{B}^2}\right] = \frac{n_D^2 L^3}{4\pi}\overline{\mathbf{E}^2}. \qquad (11.20)$$

The bar here refers to a time average over the period T of the field oscillations

$$\overline{\mathbf{E}^2} = \frac{1}{T}\int_0^T dt E_0^2 \cos^2 \omega t = E_0^2/2. \qquad (11.21)$$

Further, n_D^2 in the nominator is the high-frequency dielectric constant of the medium at the optical radiation frequency (it replaces the static dielectric constant in Eq. (2.141)). One has to replace n_D^2 with the real part of the dielectric function $\epsilon'(\omega)$ for transitions at lower frequencies.

One therefore obtains

$$E_0^2 = 8\pi N_{\text{ph}}\hbar\omega/(n_D^2 L^3). \qquad (11.22)$$

The absorption intensity can be integrated over the angles of the incoming radiation. Assuming random orientations of the molecule, one obtains

$$\int (\hat{\mathbf{e}} \cdot \hat{\mathbf{e}}_0)^2 d\Omega = 4\pi/3 \qquad (11.23)$$

and the overall absorption rate

$$\boxed{w_{\text{abs}} = \frac{4n_D}{3\hbar}|\chi_c \mu_{12}|^2 N_{\text{ph}}\left(\frac{\omega}{c}\right)^3.} \qquad (11.24)$$

This equation provides the rate of stimulated absorption. The rate of stimulated emission w_{em} is given by the same expression. Since both stimulated absorptions and emission are proportional to the density of the photons in the cavity, the Einstein coefficient $B(1 \to 2)$ is defined as the proportionality constant between the rate of stimulated transition $w(1 \to 2)$ and the field density inside the blackbody cavity

$$w(1 \to 2) = B(1 \to 2)\rho_\nu, \tag{11.25}$$

where

$$\rho_\nu = \frac{2(n_D\omega)^3\hbar}{\pi c^3}N_{\text{ph}} \tag{11.26}$$

is the radiation field density, $\rho_\nu d\nu = (2N_{\text{ph}}\hbar\omega/L^3)\rho(E)dE$. The Einstein coefficient becomes

$$B(1 \to 2) = \frac{2\pi}{3(n_D\hbar)^2}|\chi_c\mu_{12}|^2. \tag{11.27}$$

11.4   Extinction coefficient

Absorption of radiation is measured through the extinction coefficient $\epsilon(\nu)$ extracted from the Beer-Lambert law ($\omega = 2\pi\nu$). It establishes the drop of light's flux $I(d)$ relative to the flux from the source I_0 through the length of the absorbers d present in the molar concentration c_0 (M=mol/L)

$$\ln[I(d)/I_0] = -\ln 10\epsilon(\nu)c_0d. \tag{11.28}$$

If the cross-section for light absorption is $\sigma(\nu)$ (in cm^2), it provides an alternative form for the flux decay in terms of the number density ρ_0 of the absorbing molecules (given in cm^{-3})

$$\ln[I(d)/I_0] = -\sigma(\nu)\rho_0d. \tag{11.29}$$

Combining these two equations, one can obtain for $\epsilon(\bar{\nu})$ (in units M^{-1}cm^{-1})

$$\epsilon(\nu) = \sigma(\nu)10^{-3}N_A/(\ln 10 \text{ cm}^3), \tag{11.30}$$

where N_A is the Avogadro number.

The absorption cross section is obtained by dividing the energy absorbed at a given frequency per unit of time by the total incident flux of radiative energy. The energy absorbed is calculated from Eq. (11.4) by taking two steps. One first replaces E_0 with $E_0/2$ in the rotating wave approximation (see arguments after Eq. (11.17)) and then takes the average over the isotropic orientations of the transition dipole, $\langle(\hat{e} \cdot \mathbf{E}_0)^2\rangle = E_0^2/3$. This averaged transition rate is then multiplied by the energy of the photon $\hbar\omega$ to obtain the rate of energy absorption

$$\hbar\omega\langle w\rangle = \frac{\pi\omega}{6\hbar}|\chi_c\mu_{12}|^2E_0^2\text{FC}(\omega). \tag{11.31}$$

The function

$$\mathrm{FC}(\omega) = \langle \delta\,(\omega_{12} - \omega)\rangle, \quad \hbar\omega_{12} = E_2 - E_1. \tag{11.32}$$

is known as the Frank-Condon factor. It is given as the average of the delta-function, imposing the condition of energy conservation, over the thermal agitation of the medium and vibrational excitations in the light absorbing molecule.

The flux of radiation energy is

$$I_0 = (c/n_D)\rho_E, \quad \rho_E = n_D^2 E_0^2/(8\pi), \tag{11.33}$$

where c/n_D is the speed of light in the medium. Taking the ratio of Eqs. (11.31) and (11.33), one obtains for the absorption cross section

$$\sigma = \frac{4\pi^2\omega}{3\hbar c n_D}|\chi_c\mu_{12}|^2\mathrm{FC}(\omega). \tag{11.34}$$

Substituting this equation into the expression for the extinction coefficient in Eq. (11.30), one obtains

$$\frac{\epsilon(\nu)}{\nu} = \frac{8\pi^3 N_A}{3000\ln 10 n_D\hbar c}|\chi_c\mu_{12}|^2\mathrm{FC}(\omega). \tag{11.35}$$

In this equation, because of the transformation from the molar concentration to the number of particles in cm^3, the extinction coefficient $\epsilon(\nu)$ is given in cm^2mol^{-1}. Note that the Franck-Condon factor is defined in terms of the circular frequency ω according to Eq. (11.32). For the more common in spectroscopy definition in terms of frequency ν, $\omega = 2\pi\nu$, one needs to account for the normalization $\mathrm{FC}(\omega) = (2\pi)^{-1}\mathrm{FC}(\nu)$. Either definition can be integrated over frequencies with the account of frequency normalization of the Franck-Condon factor to determine the magnitude of the transition dipole

$$2\pi\int_{-\infty}^{\infty}\frac{d\nu}{\nu}\epsilon(\nu) = \frac{8\pi^3 N_A}{3000\ln 10 n_D\hbar c}|\chi_c\mu_{12}|^2. \tag{11.36}$$

This result can be written in a compact form for the transition dipole (in units of debye, D) in terms of the integrated absorption bands

$$\boxed{|\chi_c\mu_{12}| = 9.58\times 10^{-2}\ \mathrm{D}\ \left[n_D\int d\nu\epsilon(\nu)/\nu\right]^{1/2}.} \tag{11.37}$$

11.5   Solvatochromism by a classical bath

When a molecule capable of absorbing and/or emitting light is transferred from the gas phase to a liquid or solid, its absorption/emission line changes. One usually observes both the line shift and line broadening. Solvatochromism refers to the former [76]. On reports the change $\Delta\omega_{abs/em}$ of the maximum frequency compared to the corresponding transition in the gas phase (Fig. 11.3).

To understand spectral solvatochromism one has to appreciate the separation of time-scales in the solvent (either liquid or solid). Motions of electrons in the solvent molecules are fast, and their time-scale $\tau_e \simeq \omega_e^{-1}$ is often shorter than the time-scale of the electronic transition in the spectral probe, $\tau_0 \simeq \omega_0^{-1}$. This condition is the result of the experimental design requiring the solvent to be transparent in the range of frequencies in which the spectral probe absorbs/emits radiation. The usual situation is that one has $\omega_0 \ll \omega_e$.

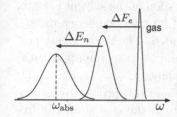

Fig. 11.3   Solvent-induced shift of the absorption line by electronic, ΔF_e, and nuclear, ΔE_n, polarization of the solvent.

Since electrons of the solvent are much faster than the electrons of the spectral probe, they respond instantaneously to the changes in the charge distribution of the molecule. This situation is similar to the instantaneous polarization of the electronic dipoles of the dielectric when a step voltage is applied to a plane capacitor (Fig. 8.10). In the dipolar approximation, the change in the dipole moment from the ground-state value $m_1 = \langle 1|\hat{\boldsymbol{\mu}}|1\rangle$ to the excited-state value $m_2 = \langle 2|\hat{\boldsymbol{\mu}}|2\rangle$ is of primary importance in considering the interactions with the medium. The dipole moment $m_{1,2}$ creates an electric field acting on the molecules of the medium, thus inducing dipoles and higher multipoles in response to that electric field. These induced multipoles in turn create their own electric field acting back on the spectral probe. This field is the reaction field discussed in Sec. 2.18.

It is reasonable to assume that the reaction field of the electrons in the solvent, R_e, is proportional to the magnitude of the dipole that caused it

$$R_e = \chi_e m. \tag{11.38}$$

Here, χ_e is the dipolar susceptibility of the electronic subsystem of the solvent. Since induced dipoles in the solvent are responsible for refraction of light and its refractive index, one represent R_e in terms of the refractive index of the solvent n_D linked to the high-frequency dielectric constant $\epsilon_\infty = n_D^2$. The subscript "∞" here is typically employed by dielectric relaxation measurements implying that optical frequencies lie outside the typical frequency window of dielectric spectroscopy, that is at $\omega \to \infty$. The result for R_e is (Eq. (2.200))

$$R_e = \frac{2m}{a^3} \frac{n_D^2 - 1}{2n_D^2 + 1}. \tag{11.39}$$

where a is the effective radius of the light absorbing molecule.

The free energy of the solute dipole interacting with the field R_e is (Eq. (2.201))

$$F_e = -\tfrac{1}{2}mR_e = -\tfrac{1}{2}\chi_e m^2. \tag{11.40}$$

One uses the free energy, instead of the energy, because of the separation of time-scales. When the electronic polarization of the solvent is faster than the transition time, both the solute-solvent interaction energy and the energy of polarizing the solvent need to be included. The latter, being positive, reduces the negative interaction energy and leads to the appearance of the factor of $1/2$ in the equation for the free energy.

The alteration of the dipole moment of the spectral probe, $m_1 \to m_2$, when the photon of radiation is absorbed leads to the change in the free energy of electronic polarization of the solvent

$$\Delta F_e = -\tfrac{1}{2}\chi_e(m_2^2 - m_1^2). \tag{11.41}$$

If $m_2 > m_1$, this is a negative change of the absorption energy corresponding to the "red shift" of the absorption line (Fig. 11.3).

The second component of the reaction field affecting the transition energy is caused by the nuclear polarization of the solvent. In most polar substances, this component is created by altered orientations of the permanent molecular dipoles in response to the electric field of the solute. In liquids, translations of molecules in response to the field of the solute also make a part of the response. The corresponding reaction field, R_n, is again linear in the solute dipole

$$R_n = \chi_n m \tag{11.42}$$

with the nuclear susceptibility χ_n. A specific form for χ_n depends on assumptions applied to separate the electronic and nuclear polarization of the solvent.

The nuclear polarization is dynamically frozen on the transition time scale τ_0 and the reaction field R_n stays constant. Therefore, the only energy change that occurs is the energy of interaction of the solute dipole with the frozen reaction field R_n

$$\Delta E_n = -\Delta m R_n = -\chi_n \Delta m m_1, \qquad (11.43)$$

where $\Delta m = m_2 - m_1$. One has to stress that in contrast to ΔF_e, which is a free energy difference, the nuclear component of the spectral shift ΔE_n is the energy difference. The nuclei cannot change their coordinates on the transition time and no entropy component should be included in ΔE_n.

Bringing the electronic and nuclear components of the spectral shift together, one obtains for the absorption frequency ω_{abs}

$$\hbar \omega_{abs} = \hbar \omega_{abs}^0 - \tfrac{1}{2} \chi_e (m_2^2 - m_1^2) - \chi_n \Delta m m_1, \qquad (11.44)$$

where ω_{abs}^0 denotes the absorption frequency in the gas phase. One can repeat the same steps for the emission transition $m_2 \to m_1$ starting with the spectral probe in the excited state with the dipole moment m_2. The result is

$$\hbar \omega_{em} = \hbar \omega_{em}^0 - \tfrac{1}{2} \chi_e (m_2^2 - m_1^2) - \chi_n \Delta m m_2, \qquad (11.45)$$

where ω_{em}^0 is the emission frequency in the gas phase.

The difference of the absorption and emission energies is known in spectroscopy as the Stokes shift, $\hbar \Delta \omega_{St}$. One obtains the Stokes shift in the gas phase $\hbar \Delta \omega_{St}^v$ (where "v" stands for the vibrational shift as explained below) and the solvent-induced Stokes shift

$$\hbar \Delta \omega_{St}^s = \chi_n \Delta m^2. \qquad (11.46)$$

The electronic (fast) component of the spectral shift cancels out in the difference.

The solvent-induced Stokes shift is fully determined by Δm and the nuclear susceptibility χ_n. Measuring the solvent-induced Stokes shift thus gives access to the reaction field by the nuclear polarization and the corresponding free energy of solvation (Sec. 2.18). According to fluctuation-dissipation arguments (Sec. 8.3), the susceptibility χ_n also characterizes the breadth of electrostatic fluctuations of the electric field $\mathbf{E}_s$ of the medium at the position of the molecule (Eqs. (8.36) and (12.31))

$$\chi_n = (\beta/3)\langle (\delta \mathbf{E}_s)^2 \rangle. \qquad (11.47)$$

These field fluctuations are responsible for the dielectric friction experienced by an ion diffusing in a polar liquid (Sec. 7.9). Overall, measuring the linear response of the medium to a change of some property of the molecular probe (dipole moment here) gives access to equilibrium thermal fluctuations of the conjugate variable (electric field for dipolar transitions in spectroscopy).

The total Stokes shift $\hbar\Delta\omega_{St}$ includes a gas-phase component, which is related to different molecular geometries of the spectral probe in the ground and excited states. It is given in terms of the reorganization energy of intramolecular vibrations, that is the energy required to stretch the vibrational oscillators from the initial to final positions. This vibrational component of the Stokes shift is considered next in the model of two shifted parabolas representing a single vibrational mode of the molecule.

11.6   *Two shifted parabolas

The classical model for spectroscopy of molecules is the picture of shifted parabolas. This model applies to many spectroscopies and also to radiationless transitions. The latter are transitions not involving absorption and emission of the radiation photon. Instead, they typically involve the transfer of a subatomic particle (electron or proton) or the transfer of excitation energy from one molecule to the other (exciton). Here, the basic results of the model of two shifted parabolas [52] are outlined without specifics of its physical application.

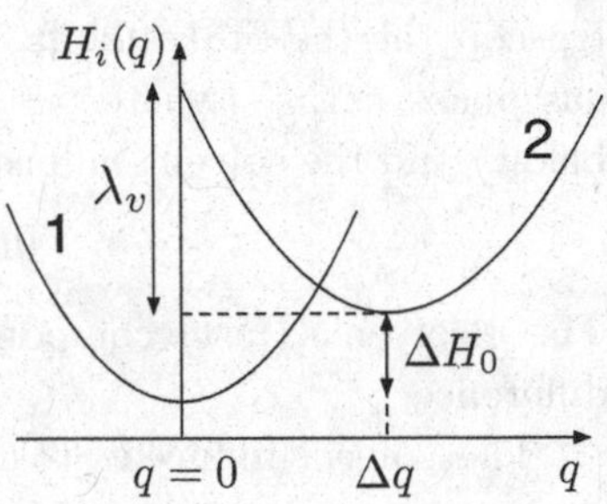

Fig. 11.4 Schematics of parabolas crossing along the nuclear coordinate q and shifted horizontally by Δq.

One considers two parabolas as functions of a nuclear coordinate q and shifted in their minima by Δq (Fig. 11.4)

$$\hat{H}_1(q) = H_{01} + \frac{\kappa}{2}\hat{q}^2,$$
$$\hat{H}_2(q) = H_{02} + \frac{\kappa}{2}(\hat{q} - \Delta q)^2. \tag{11.48}$$

With this model, the equation for the transition intensity $I(t)$ (Eq. (11.8)) can be written as follows

$$I(t) = \mu_{if}^2 \langle e^{i\hat{H}_1 t/\hbar} e^{-i\hat{H}_2 t/\hbar}\rangle_1, \tag{11.49}$$

where $\langle\ldots\rangle_1$ implies the statistical average over the thermally populated states in the initial state of the system with the Hamiltonian H_1. This equation assumes a Condon approximation in which the transition dipole is viewed as a number unaffected by coordinates of the molecule or the medium.

In order to move to the next step, one needs to deal with the product of two quantum operators (propagators)

$$\hat{U}_i(t) = \exp[-i\hat{H}_i t/\hbar] \qquad (11.50)$$

propagating quantum vibrational wave functions $\chi_i(q,t)$ in each harmonic well

$$\chi(q,t) = \hat{U}_i(t)\chi(q,0). \qquad (11.51)$$

More specifically, the product $U_1^\dagger(t)U_2(t)$ appears in the transition intensity. The goal one wants to achieve is to represent the quantum dynamics in the final harmonic well H_2 in terms of the wave function in the initial harmonic well H_1 from which the transition occurs. One therefore can write H_2 as

$$\hat{H}_2 = \hat{H}_1 + \Delta\hat{H}, \quad \Delta\hat{H} = \Delta H_0 + \lambda_v - \kappa\Delta q\hat{q}, \qquad (11.52)$$

where $\Delta H_0 = H_{20} - H_{10}$ (Fig. 11.4). Further, in this equation,

$$\lambda_v = \tfrac{1}{2}\kappa\Delta q^2 \qquad (11.53)$$

is known as the reorganization energy of the mode q (Fig. 11.4). The difference of two Hamiltonians ΔH can be identified as the "interaction". One can then use the interaction representation of quantum mechanics (Sec. 4.7), which leads to the following mathematical identity [77] (Eq. (4.81))

$$e^{i\hat{H}_1 t/\hbar}e^{-i(\hat{H}_1 + \Delta\hat{H})t/\hbar} = \hat{T}\exp\left[-\frac{i}{\hbar}\int_0^t \Delta\hat{H}(\tau)d\tau\right]. \qquad (11.54)$$

Here, $\hat{T}$ is the time-ordering operator and

$$\Delta\hat{H}(\tau) = e^{i\hat{H}_1 t/\hbar}\Delta\hat{H}e^{-i\hat{H}_1 t/\hbar} \qquad (11.55)$$

is the interaction operator in the interaction representation.

Applying this scheme to the shifted parabolas, one obtains

$$\left\langle e^{i\hat{H}_1 t/\hbar}e^{-i\hat{H}_2 t/\hbar}\right\rangle_1 = \left\langle \hat{T}\exp\left[-\frac{i}{\hbar}\int_0^t \Delta\hat{H}(\tau)d\tau\right]\right\rangle_1. \qquad (11.56)$$

One now can write ΔH in terms of the creation, $a^\dagger$, and annihilation, a, operators (Sec. 4.9) for the displacement of the harmonic oscillator

$$\Delta\hat{H} = \Delta H_0 + \lambda_v - \hbar\omega_0\sqrt{S}(\hat{a}^\dagger + \hat{a}). \qquad (11.57)$$

Here,

$$\omega_0 = \sqrt{\kappa/m} \tag{11.58}$$

is the frequency of vibrations in the harmonic well and

$$S = \lambda_v/(\hbar\omega_0) \tag{11.59}$$

is the Huang-Rhys factor. This parameter defines the number of vibrational quanta in the reorganization energy λ_v.

As is easy to appreciate from Fig. 11.4, the vertical transition energy from the bottom of the lower parabola has the energy of $\Delta H_0 + \lambda_v$. According to the Franck-Condon principle, the nuclei are much slower than light electrons and do not move during an electronic excitation of a molecule promoted by absorption of a photon. Therefore, in many cases, the vertical transition has the highest probability and establishes the maximum of the spectral line. The Huang-Rhys factor thus counts the number of vibrational quanta excited in a vertical transition from the bottom of the lowest surface given that $\hbar\omega = \Delta H_0 + \lambda_v$. In this vertical transition, S vibrational quanta are excited.

The vertical Franck-Condon transition is not the only transition allowed by quantum mechanics and non-vertical transitions always appear with a lower probability. Such transitions are often identified as "nuclear tunneling". They form the vibronic progression of transition lines contributing to the overall transition band-shape. The goal of this derivation is to establish such a band-shape for the simplest model of two shifted parabolas.

The transition band-shape is obtained by Fourier-transforming the transition intensity in Eq. (11.49)

$$I(\omega) = \mu_{12}^2 \int_{-\infty}^{\infty} \frac{dt}{2\pi\hbar} e^{\frac{it}{\hbar}(\hbar\omega - \Delta H_0 - \lambda_v) + g(t)} \tag{11.60}$$

with

$$e^{g(t)} = \left\langle \exp\left[i\omega_0\sqrt{S} \int_0^t (a^\dagger(\tau) + a(\tau))d\tau \right] \right\rangle_1. \tag{11.61}$$

One can next apply Bloch's identity [77] postulating that for any linear combination c of raising a lowering operators $\langle e^c \rangle = e^{\langle c^2 \rangle/2}$. Applying this relation to the term in angular brackets in Eq. (11.61) leads to the following result

$$g(t) = -\omega_0^2 S \int_0^t d\tau' \int_0^{\tau'} d\tau'' \phi(\tau' - \tau'') \tag{11.62}$$

with

$$\phi(\tau' - \tau'') = \bar{n}e^{i\omega_0(\tau'-\tau'')} + (\bar{n}+1)e^{-i\omega_0(\tau'-\tau'')}. \tag{11.63}$$

This equation applies the time dependence of the creation and annihilation operators in the interaction representation: $a(\tau) = e^{-i\omega_0\tau}a$ and $a^\dagger(\tau) = e^{i\omega_0\tau}a^\dagger$.

Integrating over times τ' and τ'' in Eq. (11.62) yields

$$g(t) = i\omega_0 St - S(2\bar{n}+1) + S\left[\bar{n}e^{i\omega_0 t} + (\bar{n}+1)e^{-i\omega_0 t}\right]. \tag{11.64}$$

Here, $\bar{n}$ is the equilibrium population for the harmonic oscillator

$$\bar{n} = Z_v^{-1}\sum_{n=0}^{\infty} ne^{-\beta\hbar\omega_0(n+1/2)} = \left[e^{\beta\hbar\omega_0} - 1\right]^{-1}, \tag{11.65}$$

where $Z_v = [2\sinh\chi_v]^{-1}$, $\chi_v = \beta\hbar\omega_0/2$ is the partition function for the harmonic oscillator.

The next step in the derivation involves calculating the time integral in Eq. (11.60). This is accomplished by transforming $\exp[g(t)]$ according to the mathematical identity [7]

$$e^{\frac{x}{2}(t+t^{-1})} = \sum_{k=-\infty}^{\infty} I_k(x)t^k, \tag{11.66}$$

where $I_k(x)$ is the modified Bessel function. In order to apply this expansion to $g(t)$, one notes that

$$S\left[\bar{n}e^{i\omega_0 t} + (\bar{n}+1)e^{-i\omega_0 t}\right] =$$

$$S\sqrt{\bar{n}(\bar{n}+1)}\left[\sqrt{\frac{\bar{n}}{\bar{n}+1}}e^{i\omega_0 t} + \sqrt{\frac{\bar{n}+1}{\bar{n}}}e^{-i\omega_0 t}\right], \tag{11.67}$$

which has the form suitable for the series expansion in Eq. (11.66). Applying this identity to $\exp[g(t)]$ leads to the following expression

$$e^{g(t)} = e^{i\omega_0 St - S(2\bar{n}+1)}\sum_{k=-\infty}^{\infty} I_k[2S\sqrt{\bar{n}(\bar{n}+1)}]\left(\frac{\bar{n}}{\bar{n}+1}\right)^{k/2}e^{ik\omega_0 t}. \tag{11.68}$$

When substituted to Eq. (11.60), one obtains for the transition rate

$$I(\omega) = \mu_{12}^2 e^{-S(2\bar{n}+1)}\sum_{k=-\infty}^{\infty} I_k[2S\sqrt{\bar{n}(\bar{n}+1)}]$$

$$\left(\frac{\bar{n}}{\bar{n}+1}\right)^{k/2}\delta\left(\hbar\omega + k\hbar\omega_0 - \Delta H_0\right). \tag{11.69}$$

One can further write $[\bar{n}/(\bar{n}+1)]^{(k/2)} = \exp[-k\chi_v]$ and $2S\sqrt{\bar{n}(\bar{n}+1)} = S/(\sinh\chi_v)$. For vibrations in the quantum domain of frequencies, $\chi_v = \beta\hbar\omega_0/2 \gg 1$, one can use the series expansion of the Bessel function [7]

$$I_k(2S\sqrt{\bar{n}(\bar{n}+1)}) \approx \frac{S^{|k|}}{|k|!}e^{-|k|\chi_v}. \tag{11.70}$$

Substituting this expansion to Eq. (11.69), one realizes that only terms with $k = -m$, $m > 0$ substantially contribute to the sum. This observation leads to a somewhat simplified expression for the transition intensity represented by a sum over separate vibronic transitions separated by the vibrational frequency ω_0. This sum yields the absorption band-shape

$$I_{abs}(\omega) = \mu_{12}^2 e^{-S} \sum_{m=0}^{\infty} \frac{S^m}{m!}\delta(\hbar\omega - m\hbar\omega_0 - \Delta H_0). \tag{11.71}$$

Each subsequent transition adds one more vibrational excitation $\hbar\omega_0$ and the peak heights follow the Poisson distribution with the Huang-Rhys factor S playing the role of the average number of vibrational excitations.

The average transition frequency comes out in agreement with the Franck-Condon interpretation

$$\hbar\langle\omega\rangle_{abs} = \hbar\int_{-\infty}^{\infty} d\omega\,\omega I(\omega)/\int_{-\infty}^{\infty} d\omega I(\omega) = \Delta H_0 + \hbar\omega_0 S. \tag{11.72}$$

This equation indicates that the average transition frequency is at the Franck-Condon vertical energy gap and S vibrational excitations $\hbar\omega_0$ are required to satisfy the Franck-Condon principle. The transition to $m = 0$ is known as the 0-0 transition or a zero-phonon line. This is a non-vertical transition that requires nuclear tunneling and produces no vibrational excitations (phonons for transition in the lattice).

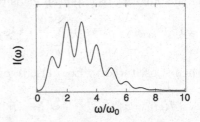

Fig. 11.5 Absorption intensity vs ω/ω_0 at $S = 2$. The maxima of individual quantum transitions are given by Eq. (11.75); $\langle\Delta H_0\rangle = 0$ is used in the plot.

The sum of delta-functions in Eq. (11.71) is not realistic since all quantum transitions involve line broadening. If the 0-0 energy gap is influenced by a Gaussian noise, the sum of infinitely narrow lines is replaced by a sum of Gaussian functions

$$I_{\text{abs}}(\omega) = \mu_{12}^2 e^{-S} \sum_{m=0}^{\infty} \frac{S^m}{m!} G(\hbar\omega - m\hbar\omega_0 - \langle\Delta H_0\rangle). \tag{11.73}$$

where the Gaussian function is given through the variance σ_G^2 as follows

$$G(x) = \left[2\pi\sigma_G^2\right]^{-1/2} \exp\left[-x^2/(2\sigma_G^2)\right]. \tag{11.74}$$

The resulting band-shape is illustrated in Fig. 11.5, where one can see the sequence of Gaussian-broadened vibronic transitions with maxima at

$$\hbar\omega_m = \langle\Delta H_0\rangle + m\hbar\omega_0. \tag{11.75}$$

11.7 Vibrational Stokes shift

The arguments presented above for light absorption can be repeated for the backward transition with light emission. The intensity is given by a similar expression in which enumeration of vibrational excitation quanta reverses sign $m \to -m$

$$I_{\text{em}}(\omega) = \mu_{12}^2 e^{-S} \sum_{m=0}^{\infty} \frac{S^m}{m!} G(\hbar\omega + m\hbar\omega_0 - \langle\Delta H_0\rangle). \tag{11.76}$$

The average emission energy becomes

$$\hbar\langle\omega\rangle_{\text{em}} = \Delta H_0 - \hbar\omega_0 S. \tag{11.77}$$

The vibrational Stokes shift is obtained by combining this equation with Eq. (11.72)

$$\hbar\Delta\omega_{\text{St}}^v = \hbar\left[\langle\omega\rangle_{\text{abs}} - \langle\omega\rangle_{\text{em}}\right] = 2\lambda_v. \tag{11.78}$$

One finds that the difference between the absorption and emission first spectral moments is equal to twice the vibrational reorganization energy λ_v.

The physical meaning of separation between two transition lines for absorption and emission is explained in Fig. 11.6. The most probable vertical Frank-Condon transition creates $S = \lambda_v/(\hbar\omega_0)$ vibrational excitations of the molecule. They relax to the vibrationally ground state through some mechanism of vibrational energy relaxation. If the time of vibrational relaxation (see Sec. 11.9 below) is shorter than the life-time of the electronically excited state, the molecule emits a photon from the vibrationally ground state with the energy $\hbar\omega_{\text{em}}$. As is clear from the diagram, it is lower than

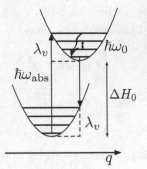

Fig. 11.6   Absorption and emission transitions in
the model of two shifted parabolas. The number
of vibrational quanta excited in a vertical transition is $S = \lambda_v/(\hbar\omega_0)$.

the absorption energy $\hbar\omega_{abs}$. The difference between absorption and emission energies is the Stokes shift given by Eq. (11.78). This picture shows
that photoexcitation of a molecule followed by emission to return to the
ground state has to proceed with the loss of energy to the surrounding
medium in the amount of $2\lambda_v$. This effect is responsible for the transfer of
radiation energy to the medium in the form of heat.

The position of the maximum of the spectral line is given by the first
spectral moment $\langle\omega\rangle_{abs,em}$ (Eqs. (11.72) and (11.77)). One can take a
next step and define the spectral width in terms of the second spectral
moment or the variance of the transition energies $\langle(\delta\omega)^2\rangle_{abs,em}$, where $\delta\omega =
\omega - \langle\omega\rangle_{abs,em}$. Similarly to Eq. (11.77), one calculates the variance on the
band-shape profiles $I_{abs,em}(\omega)$ for the absorption and emission transitions.
The result is the same in both cases and is given by the following relation

$$\hbar^2\langle(\delta\omega)^2\rangle = \sigma_G^2 + \hbar\omega_0\lambda_v, \tag{11.79}$$

where σ_G^2 is the Gaussian variance in Eq. (11.74). This variance represents
the breadth of medium fluctuations affecting the transition frequency. As
we have already seen in Eq. (11.47) the variance of the medium fluctuations defines the nuclear solvation susceptibility χ_n, which also specifies
the solvent-induced Stokes shift $\hbar\Delta\omega_{St}^s$. One can, therefore, rearrange Eq.
(11.79) to exclude the often unknown σ_G^2 and λ_v in favor of the solvent-
induced and vibrational Stokes shifts

$$\boxed{\hbar\langle(\delta\omega)^2\rangle = k_B T\Delta\omega_{St}^s + \tfrac{1}{2}\hbar\omega_0\Delta\omega_{St}^v.} \tag{11.80}$$

All parameters in this equation, except for ω_0, are directly accessible from
analyzing the absorption and emission band-shapes. One, therefore, gains
access to the effective vibrational frequency ω_0, which can be a weighted
average of a number of promoting frequencies affecting the shapes of absorption and emission lines.

11.8 Kubo's line shape

Equations (11.71)-(11.73) give the transition intensity for a quantum variable q. A similar result is derived for a classical variable $q(t)$ (Eq. (11.52))

$$I_{\text{abs}}(\omega) = \mu_{12}^2 \int_{-\infty}^{\infty} \frac{dt}{2\pi\hbar} e^{\frac{it}{\hbar}(\hbar\omega - \Delta H_0 - \lambda_v)} \left\langle e^{(i\kappa\Delta q/\hbar)\int_0^t \delta q(\tau)d\tau} \right\rangle_q. \quad (11.81)$$

If $\delta q(t)$ is a Gaussian variable with zero mean, the average over $\delta q(t)$ in the angular brackets is given in terms of the time correlation function

$$e^{g(t)} = \left\langle e^{(i\kappa\Delta q/\hbar)\int_0^t \delta q(\tau)d\tau} \right\rangle_q = \exp\left[-\frac{(\kappa\Delta q)^2}{\hbar^2} \int_0^t d\tau \int_0^\tau d\tau' C_q(\tau, \tau') \right]. \quad (11.82)$$

This result applies to an arbitrary time auto-correlation function

$$C_q(\tau, \tau') = \langle \delta q(\tau) \delta q(\tau') \rangle. \quad (11.83)$$

Exponential decay of correlations with the relaxation time τ_r is often a good approximation

$$C_q(\tau - \tau') = \langle \delta q^2 \rangle e^{-|\tau - \tau'|/\tau_r}. \quad (11.84)$$

In that case one obtains

$$\int_0^t d\tau \int_0^\tau d\tau' C_q(\tau, \tau') = \tau_r^2 \left[e^{-t/\tau_r} + t/\tau_r - 1 \right]. \quad (11.85)$$

For a classical harmonic oscillator, the equipartition theorem suggests

$$\kappa \langle \delta q^2 \rangle = k_B T. \quad (11.86)$$

Therefore, one can write for the broadening function [71, 72]

$$\boxed{g(t) = -(\Delta\tau_r)^2 \left[t/\tau_r - 1 + e^{-t/\tau_r} \right].} \quad (11.87)$$

where $\Delta^2 = 2k_B T \lambda_v / \hbar^2$ is the variance of the transition frequency produced by thermal agitation of the classical coordinate q. This type of functionality often appears in theories involving exponentially decaying correlation functions. We have previously encountered the same functional form in Eq. (8.113), where the mean-squared displacement of a Brownian particle was calculated from the exponentially decaying velocity autocorrelation function.

The time integral in Eq. (11.81) can be calculated only numerically, but it has two well-defined limits. When $t \gg \tau_r$ is taken in $g(t)$, $I(\omega)$ becomes

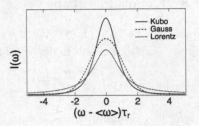

Fig. 11.7 Absorption intensity vs frequency at $\tau_r \Delta = 1$. "Kubo" marks numerical integration with the Kubo lineshape function (Eq. (11.87)).

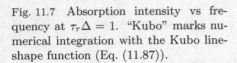

the Lorentzian band-shape. In order to achieve the solution in this limit, the time integral has to be understood as

$$\int_{-\infty}^{\infty} dt \cdots \to 2\mathrm{Re} \int_{0}^{\infty} dt \ldots, \tag{11.88}$$

which is the correct result of the time-dependent perturbation theory. One obtains

$$I_{\mathrm{abs}}(\omega) \propto \frac{2\Delta^2 \tau_r}{(\Delta^2 \tau_r)^2 + (\omega - \langle \omega \rangle_{\mathrm{abs}})^2}, \tag{11.89}$$

where the average Franck-Condon transition frequency is $\hbar \langle \omega \rangle_{\mathrm{abs}} = \Delta H_0 + \lambda_v$ (Fig. 11.6) and $2\Delta^2 \tau_r$ is the Lorentzian broadening. The Lorentzian shape, representing the phenomenon of motional narrowing, is the limit of fast modulation $\tau_r \Delta \ll 1$. The line shape depends in this limit on both the statistics, through Δ, and on the dynamics of the bath, through τ_r.

The opposite limit of slow modulation $\tau_r \Delta \gg 1$ leads to a static band-shape fully determined by statistical fluctuations of of the variable q. This limit is obtained when $t \ll \tau_r$ is taken in Eq. (11.87) and $g(t) \simeq -(t\Delta)^2/2$. The Gaussian band-shape follows

$$I_{\mathrm{abs}}(\omega) \propto \exp\left[-(\omega - \langle \omega \rangle_{\mathrm{abs}})^2/(2\Delta^2)\right]. \tag{11.90}$$

Figure 11.7 illustrates the result for $I(\omega)$ in the intermediate case $\Delta \tau_r = 1$ when neither the Gaussian nor Lorentzian limit applies.

11.9 Vibrational energy relaxation

We now consider the relaxation of vibrationally excited states shown by the wiggled line in Fig. 11.6. We start with the population relaxation from the vibrational quantum state n to the quantum state $n - 1$ (Fig. 11.8). An example would be a single bond with the reduced mass m_r and the vibrational frequency ω_0. The bond is initially in the state $|n\rangle$ with the occupation equal to unity. We want to calculate the rate of decay to the state $|n - 1\rangle$.

According to quantum-mechanical Fermi's golden rule, the rate of vibrational energy relaxation from the level n to the lower level $n-1$ is given by the following relation (Eq. (4.142))

$$k_{n,n-1} = \hbar^{-2} \int_{-\infty}^{\infty} dt e^{i\omega_0 t} \langle V(t)V \rangle. \tag{11.91}$$

This is a radiationless transition, i.e, there is no photon of radiation emitted in the transition ($\omega = 0$).

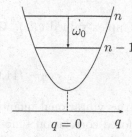

Fig. 11.8 Vibrational energy relaxation between states n and $n-1$ of a harmonic oscillator with the frequency ω_0.

The matrix element $V(t) = \langle n|\hat{H}'(t)|n-1 \rangle$ of the perturbation Hamiltonian $\hat{H}'(t)$ depends on time through classical motions of the thermal bath. The origin of these fluctuations is typically in electrostatic interactions, arising from the electric field of the medium acting on the dipole moment of the vibrating bond. However, it does not need to be specified here since the action of the medium field can be reduced to the friction coefficient.

One can expand the perturbation Hamiltonian $H'(q)$ in the quantum vibrational coordinate q around the equilibrium value $q = 0$ (Fig. 11.8)

$$\hat{H}'(q) = \hat{H}'(0) - f\hat{q}, \tag{11.92}$$

where $f = -(\partial H'/\partial q)_{q=0}$ is the force acting from the medium on the vibrating bond. Since $\langle n|\hat{H}'(0)|n-1 \rangle = \hat{H}'(0)\langle n|n-1 \rangle = 0$, one obtains

$$V(t) = -f(t)\langle n|\hat{q}|n-1 \rangle. \tag{11.93}$$

For a harmonic oscillator, one has (Sec. 4.9)

$$\langle n|q|n-1 \rangle = \sqrt{\frac{\hbar}{2m_r\omega_0}}\langle n|(\hat{a} + \hat{a}^\dagger)|n-1 \rangle = \sqrt{n}\sqrt{\frac{\hbar}{2m_r\omega_0}}. \tag{11.94}$$

The rate of decay is then given by Zwanzig's formula [78]

$$k_{n,n-1} = \frac{n}{2m_r\hbar\omega_0} \int_{-\infty}^{\infty} dt e^{i\omega_0 t} \langle f(t)f \rangle. \tag{11.95}$$

The section on mobility of ions (Sec. 7.9) already showed that the force-force correlation function can be associated with friction. The extension of Eq. (7.92) is to determine the frequency-dependent friction

$$\tilde{\zeta}(\omega) = \beta \int_0^\infty dt e^{i\omega t} \langle f(t)f \rangle. \tag{11.96}$$

The frequency integral in Eq. (11.95) becomes

$$\int_{-\infty}^\infty dt e^{i\omega_0 t} \langle f(t)f \rangle = 2k_B T \zeta'(\omega_0), \tag{11.97}$$

where $\zeta'(\omega) = \text{Re}[\tilde{\zeta}(\omega)]$ is the real part of the complex-valued $\tilde{\zeta}(\omega)$. One finally obtains for the rate of decay

$$k_{n,n-1} = \frac{n\zeta'(\omega_0)}{m_r \beta \hbar \omega_0} \propto nT. \tag{11.98}$$

The resulting decay rate is proportional to the vibrational quantum number n and to temperature T. If the force-force correlation function decays exponentially with the relaxation time τ_F, one obtains from Eqs. (11.97) and (11.98)

$$k_{n,n-1} = \frac{\zeta}{m_r \beta \hbar \omega_0} \frac{n}{1 + (\omega_0 \tau_F)^2}, \tag{11.99}$$

where $\zeta = \zeta'(0)$ is the hydrodynamic friction which can be connected to the diffusion constant (Eq. (7.8)) or to the equilibrium variance of the force

$$\zeta = \beta \tau_F \langle f^2 \rangle. \tag{11.100}$$

One can use the rates for specific transitions to derive the relaxation time for the time-dependent total energy deposited to the oscillator

$$E(t) = \hbar \omega_0 \sum_n \rho_n(t)(n + \tfrac{1}{2}), \tag{11.101}$$

where $\rho_n(t)$ is the population of the vibrational quantum state n. The derivation presented here assumes $\beta \hbar \omega \gg 1$ for simplicity. This is a fairly typical case of a quantum oscillator in a classical thermal bath.

The time derivative of the energy $E(t)$ is given by the master equation specifying the all channels of energy decay from downward transitions and the energy increase from upward transitions. Since each step up or down implies the change of energy by $\hbar \omega_0$, one obtains

$$\frac{dE(t)}{dt} = \hbar \omega_0 \left[\sum_{n=0}^\infty \rho_n(t)k_{n,n+1} - \sum_{n=1}^\infty \rho_n k_{n,n-1} \right]. \tag{11.102}$$

Detailed balance requires $k_{n,n+1} = e^{-\beta\hbar\omega_0}k_{n+1,n}$, and the upward transitions can be neglected at $\beta\hbar\omega_0 \gg 1$. One then applies scaling of the transition probability with the quantum number n to define the population relaxation time T_1

$$k_{n,n-1} = nT_1^{-1}. \tag{11.103}$$

Equation (11.102) becomes

$$\frac{dE(t)}{dt} = -T_1^{-1}\sum_n \left[\hbar\omega_0(n+\tfrac{1}{2}) - \tfrac{1}{2}\hbar\omega_0\right] = -T_1^{-1}\sum_n (E(t) - E_{eq}),$$

$$\tag{11.104}$$

where $E_{eq} = \tfrac{1}{2}\hbar\omega_0$ is the equilibrium ground-state energy at $\beta\hbar\omega_0 \gg 1$. This is the kinetic equation for the energy relaxation with the characteristic relaxation time

$$\boxed{\frac{1}{T_1} = \frac{\zeta'(\omega_0)}{m_r\beta\hbar\omega_0}.} \tag{11.105}$$

The relaxation time of a hot vibrational state of a molecule turns out to reflect the friction imposed by the medium on the molecule at the frequency of vibrational excitation.

11.10 Scattering

This section discusses the basics of scattering of thermal neutrons by disordered materials [73, 74]. Neutron scattering is chosen to have a more focused discussion and to connect to the concepts of time correlation functions and structural properties in reciprocal space (intermediate scattering functions, etc) discussed in previous Chapters. The principles discussed here are largely applicable to other scattering techniques.

Consider the flux of neutrons arriving from the source to a sample with the surface area S and the length L (Fig. 11.9). Its volume is $V = S \times L$. The incoming flux, i.e., the number of neutrons passing through the unit area per unit time is I_0. The cross-section for neutron scattering σ is defined in terms of the number of scattering events per second I_s: $I_s = \sigma I_0$. The cross-section obviously carries the units of length squared. The integrated cross-section σ loses lots of information about specific scattering events, which depends on the scattering angle and the energy of scattered neutrons. In order to provide the angular information one considers the number of scattering events dI_s into the element of the solid angle $d\Omega$. The corresponding differential cross-section becomes

$$\frac{d\sigma}{d\Omega} = \frac{dI_s}{I_0 d\Omega}. \tag{11.106}$$

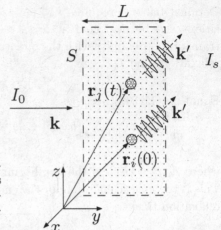

Fig. 11.9 Geometry of neutron scattering. Interference of individual waves scattered at time $t = 0$ and time t contributes to the overall observed scattering counts I_s.

The next step in discriminating scattering events is to distinguish them based on the energy $\hbar\omega$ transferred from the neutrons to the sample. A thermal non-relativistic neutron travels with the energy $E = (\hbar k)^2/(2m_n)$, where k is the wave vector and m_n is the mass of the neutron. The energy transferred to the sample is due to the change in the wave vector $\mathbf{k} \to \mathbf{k}'$ in the scattering event

$$\hbar\omega = \frac{(\hbar k')^2}{2m_n} - \frac{(\hbar k)^2}{2m_n}. \tag{11.107}$$

The wave vector changes not only its magnitude, but also direction. The angle between $\mathbf{k}'$ and $\mathbf{k}$ is the scattering angle.

If one wants to determine the number of scattering events within the range of energies $dE = \hbar d\omega$, the second derivative of the cross-section follows

$$\frac{d^2\sigma(\Omega)}{d\Omega d\omega} = \frac{dI_s}{I_0 d\Omega d\omega}. \tag{11.108}$$

One now needs to count the number of scattering events as the neutrons travel through the sample on the time $t = L/v = Lm_n/(\hbar k)$ with the velocity v. If the probability of scattering per unit time (scattering rate) is W, the number of scattering events must be $I_s = I_0 SWt = I_0 VW m_n/(\hbar k)$. One wants to be more specific and define the fraction of scattering events from the incoming wave vector $\mathbf{k}$ to a specific scattering vector $\mathbf{k}'$. If the corresponding probability is $W_{\mathbf{k},\mathbf{k}'}$, the number of scattering events is

$$\frac{dI_s}{I_0} = V\frac{m_n}{\hbar k}W_{\mathbf{k},\mathbf{k}'}\rho_{\mathbf{k}}d\mathbf{k}', \tag{11.109}$$

where $\rho_{\mathbf{k}} = V/(2\pi)^3$ is the density of wave vectors. Given that $d\mathbf{k}' = (k')^2 dk' d\Omega'$, one obtains for the differential cross-section

$$\frac{d^2\sigma}{d\Omega d\omega} = \frac{V^2 m_n (k')^2}{(2\pi)^3 \hbar k} \frac{dk'}{d\omega} W_{\mathbf{k},\mathbf{k}'}. \qquad (11.110)$$

From Eq. (11.107), one obtains $dk'/d\omega = m_n/(\hbar k')$ and

$$\frac{d^2\sigma}{d\Omega d\omega} = \frac{1}{(2\pi)^3} \left(\frac{V m_n}{\hbar}\right)^2 \frac{k'}{k} W_{\mathbf{k},\mathbf{k}'}. \qquad (11.111)$$

Neutrons are scattered from the atomic nuclei. The scattering length is very short, of the order of 10^{-12} cm. Therefore, the cross-section is of the order of 10^{-24} cm^2, which is the scattering unit of 1 barn. Given that the scattering length is much shorter than any length of molecular dimension, scattering is described by the scattering length b_j entering Fermi's pseudopotential placed at the position $\mathbf{r}_j$ of each scattering nucleus

$$V(r) = \frac{2\pi\hbar^2}{m_n} \sum_j b_j \delta(\mathbf{r} - \mathbf{r}_j). \qquad (11.112)$$

The scattering length b_j depends on the quantum spin state of the scattering nucleus and on the spin polarization of the incoming neutron. The matrix element in the Fermi's golden rule (Sec. 4.8) is now constructed on the incoming and scattered plane waves

$$|\psi_{\mathbf{k}}\rangle = |\mathbf{k}\rangle = \frac{1}{\sqrt{V}} e^{i\mathbf{k}\cdot\mathbf{r}}, \quad |\psi_{\mathbf{k}'}\rangle = |\mathbf{k}'\rangle = \frac{1}{\sqrt{V}} e^{i\mathbf{k}'\cdot\mathbf{r}}. \qquad (11.113)$$

One obtains for the matrix element of Fermi's pseudopotential

$$\langle \mathbf{k}|\hat{V}|\mathbf{k}'\rangle = \frac{2\pi\hbar^2}{m_n V} \sum_j b_j e^{i\mathbf{q}\cdot\mathbf{r}_j}, \qquad (11.114)$$

where $\mathbf{q} = \mathbf{k}' - \mathbf{k}$ is the scattering vector. Substituting this equation in the golden-rule transition probability leads to the following expression for the differential cross-section

$$\frac{d^2\sigma}{d\Omega d\omega} = \frac{k'}{k} \sum_{i,j} \int_{-\infty}^{\infty} \frac{dt}{(2\pi)} \langle b_i b_j e^{i\mathbf{q}\cdot\mathbf{r}_j(t)} e^{-i\mathbf{q}\cdot\mathbf{r}_i} \rangle e^{-i\omega t}. \qquad (11.115)$$

The angular brackets in this expression include both the average over the quantum states of the nuclei and the average over their positions. Because of the different length scale, these averages can be separated

$$\langle b_i b_j e^{i\mathbf{q}\cdot\mathbf{r}_j(t)} e^{i\mathbf{q}\cdot\mathbf{r}_i} \rangle = \langle b_i b_j \rangle_s \langle e^{i\mathbf{q}\cdot\mathbf{r}_j(t)} e^{-i\mathbf{q}\cdot\mathbf{r}_i} \rangle, \qquad (11.116)$$

where $\langle b_i b_j \rangle_s$ involves only the average over the spin states of the nuclei for which we can write

$$\langle b_i b_j \rangle_s = \delta_{ij} \langle (\delta b)^2 \rangle_s + \langle b \rangle_s^2. \tag{11.117}$$

The first term comes from the assumption that quantum nuclear states of separate nuclei are independent. This term requires $i = j$ in the double sum over i and j in Eq. (11.115) and is called incoherent scattering. The second term in Eq. (11.117), which allows correlated (coherent) scattering from different nuclei, is responsible for coherent scattering. One defines an intermediate scattering function for each case

$$F_c(q, t) = N^{-1} \sum_{i,j} \left\langle e^{i\mathbf{q} \cdot \mathbf{r}_j(t)} e^{-i\mathbf{q} \cdot \mathbf{r}_i} \right\rangle,$$

$$F_{\text{inc}}(q, t) = N^{-1} \sum_j \left\langle e^{i\mathbf{q} \cdot \mathbf{r}_j(t)} e^{-i\mathbf{q} \cdot \mathbf{r}_j} \right\rangle. \tag{11.118}$$

The differential cross-section becomes

$$\boxed{\frac{d^2\sigma}{d\Omega d\omega} = N \frac{k'}{8\pi^2 k} \left[\sigma_{\text{inc}} S_{\text{inc}}(q, \omega) + \sigma_c S_c(q, \omega) \right].} \tag{11.119}$$

Here, $\sigma_{\text{inc}} = 4\pi \langle (\delta b)^2 \rangle_s$ is the incoherent cross-section and $\sigma_c = 4\pi \langle b \rangle_s^2$ is the coherent cross-section. Further, N is the total number of scattering nuclei and $S_a(q, \omega)$, $a = (\text{c, inc})$ are the time Fourier transforms

$$S_a(q, \omega) = \int_{-\infty}^{\infty} dt F_a(q, t) e^{-i\omega t}. \tag{11.120}$$

Coherent scattering $a = \text{c}$ makes $S_c(q, \omega) = S(q, \omega)$ equal to the dynamic structure factor (Sec. 8.8). Correspondingly, incoherent scattering $a = \text{inc}$ leads to the self-structure factor $S_{\text{inc}}(q, \omega) = S_s(q, \omega)$.

Hydrogen is a highly important example of a strong incoherent scatterer with $\sigma_{\text{inc}} = 80.3$ barn and $\sigma_c = 1.76$ barn. Next section considers incoherent neutron scattering, with the focus on extracting information about the dynamics of condensed materials. For an immobile particle, $F_{\text{inc}}(q, t) = 1$ and incoherent scattering leads to a sharp elastic line $\propto \delta(\omega)$. If the particle moves on the observation time of the instrument τ_{obs}, this motion contributes to broadening of the elastic line of zero energy transfer recorded by quasielastic neutron scattering. The observation time is connected to the energy resolution of the spectrometer ΔE by the energy uncertainty relation, $\tau_{\text{obs}} \Delta E \simeq \hbar$. Recording the width of quasielastic scattering provides information about relaxation events occurring within the time scale τ_{obs}.

11.11   Incoherent scattering

Here a specific case of a single scatterer is considered within a molecule undergoing both translational and rotational motions. Incoherent scattering is considered as a model of scattering from a single hydrogen atom (Fig. 11.10).

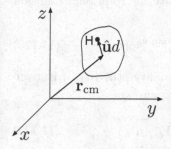

Fig. 11.10   Hydrogen atom H in a molecule with the center of mass at $\mathbf{r}_{cm}$. The hydrogen atom is located at the distance d from the center of mass; the unit vector $\hat{\mathbf{u}}$ specifies orientation.

Since all hydrogen atoms can be viewed as equivalent, the expression for the incoherent intermediate scattering function can be written as

$$F_{inc}(q,t) = \left\langle e^{i\mathbf{q}\cdot\Delta\mathbf{r}(t)} \right\rangle, \tag{11.121}$$

where $\Delta\mathbf{r}(t) = \mathbf{r}(t) - \mathbf{r}$ is the displacement of the proton over time t as the result of both molecular translation and rotation. The angular brackets in this equation include both the average over the statistical configurations at the initial time $\mathbf{r} = \mathbf{r}(0)$ and over all scatterers in the liquid.

One can next represent the displacement as the sum of the center-of-mass translation $\mathbf{r}_{cm}$ and rotation of the unit vector $\hat{\mathbf{u}}$ moving the proton at the distance d from the center of mass within the molecule (Fig. 11.10)

$$\Delta\mathbf{r}(t) = \Delta\mathbf{r}_{cm}(t) + \Delta\hat{\mathbf{u}}(t)d. \tag{11.122}$$

One can next assume that translations and rotations are dynamically decoupled and write the intermediate scattering function as a product of translational and rotational components

$$F_{inc}(q,t) = F_T(q,t) \times F_R(q,t), \tag{11.123}$$

with

$$F_T(q,t) = \left\langle e^{i\mathbf{q}\cdot\Delta\mathbf{r}_{cm}(t)} \right\rangle,$$
$$F_R(q,t) = \left\langle e^{id\mathbf{q}\cdot\Delta\hat{\mathbf{u}}(t)} \right\rangle. \tag{11.124}$$

The translational part is the van Hove self-correlation function (Eq. (8.119)) the solution for which is particularly simple in the case of isotropic translational diffusion with the diffusion constant D

$$F_T(q,t) = e^{-Dq^2t}. \tag{11.125}$$

Calculation of the rotational component of the intermediate scattering function requires using the transition probability for rotational diffusion derived in Sec. 7.7. One writes

$$F_R(q,t) = \int d\hat{\mathbf{u}} \, d\hat{\mathbf{u}}_0 \, e^{idq\cdot\hat{\mathbf{u}}} p(\hat{\mathbf{u}},t|\hat{\mathbf{u}}_0,0)e^{-idq\cdot\hat{\mathbf{u}}_0} p(\hat{\mathbf{u}}_0), \tag{11.126}$$

where $p(\hat{\mathbf{u}}_0) = (4\pi)^{-1}$ and the transition probability $p(\hat{\mathbf{u}},t|\hat{\mathbf{u}}_0,0)$ is given by Eq. (7.58). One then applies the Rayleigh expansion

$$e^{idq\cdot\hat{\mathbf{u}}} = 4\pi \sum_{\ell=0}^{\infty} \sum_{m=-\ell}^{\ell} i^\ell j_\ell(qd) Y_{\ell m}(\hat{\mathbf{u}}) Y_{\ell m}^*(\hat{\mathbf{q}}), \tag{11.127}$$

where $j_\ell(x)$ is the spherical Bessel function (Sec. 1.9). By using orthogonality of spherical harmonics and Eq. (7.58) for the transition probability, one arrives at

$$F_R(q,t) = \sum_{\ell=0}^{\infty} (2\ell+1) j_\ell^2(qd) e^{-\ell(\ell+1)D_r t}, \tag{11.128}$$

where D_r is the rotational diffusion constant (Eq. (7.53)).

Combining the results for translations and rotations, one calculates the incoherent scattering function by taking the time Fourier transform of $F_{\mathrm{inc}}(q,t)$ in Eq. (11.123). This is done by taking the real part of the one-sided Fourier-Laplace transform: $S_{\mathrm{inc}}(q,\omega) = 2\mathrm{Re}[\tilde{F}_{\mathrm{inc}}(q,\omega)]$. The result is

$$\boxed{S_{\mathrm{inc}}(q,\omega) = \sum_{\ell=0}^{\infty} (2\ell+1) j_\ell^2(qd) \frac{2\tau_\ell(q)}{1+(\omega\tau_\ell(q))^2},} \tag{11.129}$$

where the relaxation time at each ℓ depends on the wave vector through molecular diffusivity

$$\tau_\ell(q) = \left(q^2 D + \ell(\ell+1)D_r\right)^{-1}. \tag{11.130}$$

The consistency of this result is easily tested. Integration over frequencies yields

$$\int_{-\infty}^{\infty} \frac{d\omega}{(2\pi)} S_{\mathrm{inc}}(q,\omega) = F_{\mathrm{inc}}(q,0) = \sum_{\ell=0}^{\infty} (2\ell+1) j_\ell^2(qd) = 1. \tag{11.131}$$

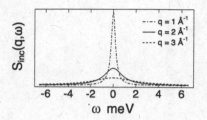

Fig. 11.11 Quasielastic incoherent scattering function for water (Eq. (11.129)).

The last identity is a sum rule valid for spherical Bessel functions [7] (Eq. (1.84)).

The resulting scattering law is a sum of Lorentzian lines centered at $\omega = 0$. Given that $\omega = 0$ implies no energy transfer from the neutrons to the sample, this type of scattering is called quasielastic scattering. The physical origin of line broadening is in molecular motions altering positions of scatterers on the resolution time τ_{obs} of the instrument. Quasielastic scattering provides access to molecular dynamics, but modeling is required to extract the dynamical parameters. An important result delivered by Eq. (11.130) is that multiple relaxation times are captured in the scattering spectrum when both translations and rotations affect the motion of the scatterer.

Figure 11.11 illustrates Eqs. (11.129) and (11.130) with the diffusion constants of water at 298 K: $D = 0.2$ Å^2/ps and $D_r = 0.08$ ps^{-1} (the rotational diffusion constant is related to the rotational relaxation time by Eq. (7.64)). The scattering profiles become broader with increasing q, with the half-intensity width $\Gamma(q) \propto \tau_\ell(q)^{-1}$ scaling linearly with q^2 (Eq. (11.130)). The experimental width for quasielastic scattering from water is in fact nonlinear at high q because the assumption of continuous diffusion becomes inadequate at the molecular length scale [74].

Chapter 12

Solvation

Much of chemistry and all biology happen in solutions. Free energy of *solvation* is an important concept underlying much of solution chemistry and kinetics of chemical reactions in condense materials. Formally, the solvation free energy is defined as the reversible work required to bring a particle from the gas phase to solution and it might appear that this is not a very practical question to be addressed by chemical dynamics. The importance of this concept lies, however, in the fact that the chemical potential of a molecule in solution changes in the course of a chemical transformation and much of this change is the alteration of the free energy of solvation. These changes usually occur for two reasons: (i) the size or shape of the molecule changes and (ii) the distribution of molecular charge is altered by chemistry. These two contributions are viewed separately as (i) the free energy of expulsion of the solvent from the repulsive core of the solute and (ii) the free energy of long-range interactions. Both are considered here in very general terms to show how the length scale of the interaction potential necessitates in each case a separate theoretical formulation to tackle the problem.

12.1   Solvation free energy

Consider solute molecules (e.g., molecules of methane gas) in the gas phase in equilibrium with the solution phase where they are dissolved in a liquid (Fig. 12.1). The number density of the solute molecules is ρ_0^g in the gas phase, and it becomes ρ_0^l in the liquid phase. The ratio of these densities can be measured in the laboratory and is called the Ostwald partition coefficient

$$K_{\text{eq}} = \rho_0^l / \rho_0^g. \tag{12.1}$$

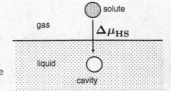

Fig. 12.1 Transition of the solute molecule from the
gas phase to solution.

The equilibrium condition is established by equating the chemical potentials of solutes in the gas phase and in the liquid

$$\mu_0^g = \mu_0^l. \tag{12.2}$$

The solute molecules in the gas phase can be viewed as an ideal gas at a sufficiently low concentration. This is not true in the liquid phase because of many interactions of the solute with the liquid. However, the ideal gas component always exists in the chemical potential since it comes from the kinetic energy in the Hamiltonian producing the ideal-gas factor in the partition function. One can write the equilibrium condition as (see Eq. (5.73))

$$\ln[\rho_0^g \lambda^3] = \ln[\rho_0^l \lambda^3] + \beta \Delta \mu_0. \tag{12.3}$$

The excess chemical potential in the liquid phase

$$\beta \Delta \mu_0 = - \ln K_{\text{eq}} \tag{12.4}$$

is the main target of solvation theories.

The excess chemical potential represents the reversible work, performed at a constant pressure, to move a solute with zero kinetic energy from the gas phase to the liquid. Since this reversible work is equal to the change in free energy, $\Delta \mu_0$ is often called the solvation free energy. One has to realize that this is strictly speaking a chemical potential, i.e., the Gibbs energy per particle in the limit of infinite dilution or the corresponding derivative of the Helmholtz free energy over the number of solutes N_0 with the limit $N_0 \to 0$ taken after calculating the derivative (infinite dilution).

Following the ideas of van der Waals leading to his famous equation of state (Sec. 6.4), the intermolecular interactions in bulk materials are often separated into the short-range repulsion and the long-range attraction. They are treated by different theoretical approximations exactly because of the different length scales on which they change substantially. The same approach turned out to be fruitful in application to solvation. One views the solute-solvent interaction as composed of a short-range repulsion and a long-range attraction. The short-range component is replaced by the idealized potential of hard-sphere repulsion, which is infinite within the solute radius a and zero at $r > a$. In contrast, long-range attractions are accounted for explicitly with a separately developed techniques some of which are considered below.

Two components in the interaction potential require two steps in performing the reversible work. One first creates a cavity in the liquid to accommodate the solute (Fig. 12.1). The corresponding change in the chemical potential, $\Delta\mu_{HS}$, is the free energy of cavity formation. The subscript "HS" is referring to the hard-sphere repulsive potential assigned to the repulsive core of the solute. One next allows the long-range electrostatic and dispersion forces to polarize liquid multipoles and shift their positions to minimize the free energy of the entire solute-solvent system. Since the interactions are long-ranged, the force acting from the solute on each molecule of the solvent is relatively weak and can be viewed as a small perturbation. This assumption leads to linear theories of solvation considered below. They are statistical (time-independent) versions of the linear response theory covered in Sec. 8.2. Overall, the excess chemical potential is a sum of the short-range, $\Delta\mu_{HS}$, and long-range, $\Delta\mu$, contributions

$$\Delta\mu_0 = \Delta\mu_{HS} + \Delta\mu. \tag{12.5}$$

A subscript specifying the long-range part of the excess chemical potential is dropped for brevity.

Based on the splitting of the solvation chemical potential into two parts carrying different physical meaning, we will consider them separately. Motivated by the history of the subject, we start with theories of long-range solvation. They were developed first driven by interest in solvation of small ions and addressed initially by the Born model of ion solvation.

12.2 Born formula for ion solvation

Dielectric susceptibility:
$4\pi\chi_s = \epsilon_s - 1.$

Born equation provides the free energy of electrostatic solvation for a spherical ion placed in a polar liquid (solvent). The only solvent parameter entering the equation is the static dielectric constant ϵ_s (Eq. (2.73)). Born theory is a linear solvation theory, which means that the electrostatic potential of the solvent ϕ_b at the position of the ion is proportional to, or linear in, the ion charge q. The electrostatic potential ϕ_b is caused by the bound charges of the solvent (Fig. 12.2). It is constant inside the repulsive core of the ion, which also means that the field of bound charges is zero within the ion (see similar arguments in Sec. 2.5).

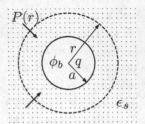

Fig. 12.2   Solvation of an ion in the dielectric.

The property calculated by the theory is the average energy of the solute-solvent electrostatic interaction $\langle u \rangle$ or the chemical potential of electrostatic solvation $\Delta\mu$. The linearity of the theory, discussed in more general terms in the next section, allows one to represent both of them in terms of the average electrostatic potential ϕ_b

$$\langle u \rangle = q\phi_b, \quad \Delta\mu = \tfrac{1}{2}q\phi_b. \tag{12.6}$$

As we discuss in more detail in Sec. 12.4, the average solute-solvent interaction energy $\langle u \rangle$ is not equal to the thermodynamic solvation energy following from the rules of thermodynamics

$$\Delta e = (\partial(\beta\Delta\mu)/\partial\beta)_V . \tag{12.7}$$

On the other hand, $\Delta\mu$ is the true thermodynamic potential of the solute (its electrostatic component) in the limit of infinite dilution.

The problem of all linear solvation theories is how to relate ϕ_b to the size and shape of the ion and to the properties of the solvent. In other words, one wants to calculate the solvation susceptibility χ_{solv} connecting ϕ_b and q

$$\phi_b = q\chi_{\text{solv}}. \tag{12.8}$$

The chemical potential of solvation becomes

$$\Delta\mu = \tfrac{1}{2}q^2\chi_{\text{solv}}. \tag{12.9}$$

The standard solution of the problem is to solve the Poisson equation for ϕ_b given the boundary conditions specified by Maxwell's electrostatics. A somewhat more intuitive approach is offered here by directly considering the polarization density field of the solvent $\mathbf{P}$ created in the solvent dielectric in response to the field of the ion

$$\mathbf{E}_0 = q\frac{\mathbf{r}}{r^3}. \tag{12.10}$$

The ion plays the role of the source of external field $\mathbf{E}_0 = -\nabla\phi_0$ with the electrostatic potential ϕ_0. The logic of the solution offered here follows the steps of a similar problem of a void in the dielectric polarized by an external field considered in Sec. 2.7.

The formulation in terms of the medium polarization density is based on applying the empirical approximation of locality of dielectric polarization. It is this approximation that physically determines this solution as "dielectric continuum". We used this approximation in Chap. 2 to derive a number of well-established results of dielectric theories effectively bypassing the traditional boundary-value problem. Sec. 12.6 shows how this assumption can be lifted in terms of nonlocal susceptibility functions provided by microscopic electrostatics. Here, the locality assumption is applied again to derive the Born formula for ion solvation.

The assumption of locality of the inhomogeneous Maxwell field implies that the polarization density at each point of space is proportional to the Maxwell field at the same point (the meaning of "locality")

$$\mathbf{P} = \chi_s \mathbf{E}, \tag{12.11}$$

where χ_s is the dielectric susceptibility (Eq. (2.74)). The average interaction energy and the solvation free energy are calculated by integrating $\mathbf{P}$ with $\mathbf{E}_0$ (Eq. (2.63)) over the volume Ω occupied by the dielectric

$$\langle u \rangle = -\int_\Omega d\mathbf{r}\, \mathbf{E}_0 \cdot \mathbf{P}. \tag{12.12}$$

It is worth reminding here that locality of the medium polarization in the uniform Maxwell field is supported by empirical evidence. In contrast, locality of the polarization field induced by an inhomogeneous Maxwell field is an approximation that is bound to fail at the length scale comparable to the size of the molecule of the medium.

The main difficulty of solvation theories, which is why solving the Poisson equation is typically required, is that the field of the ion $\mathbf{E}_0$ is what we know, but the locality condition is established for the Maxwell field

$$\mathbf{E} = \mathbf{E}_0 + \mathbf{E}_b, \tag{12.13}$$

which additionally contains the field $\mathbf{E}_b$ of bound charges. The latter is created by all partial atomic charges distributed with the charge density ρ_b

$$\mathbf{E}_b = -\nabla \int d\mathbf{r}'\, \frac{\rho_b(\mathbf{r}')}{|\mathbf{r} - \mathbf{r}'|}. \tag{12.14}$$

As we discussed previously in connection with the electric field produced by a polarized dielectric (Eqs. (1.70) and (2.69)), there is a direct proportionality between $\mathbf{E}_b$ and the longitudinal projection $\mathbf{P}_L$ of the polarization density

$$\mathbf{E}_b = -4\pi\mathbf{P}_L. \tag{12.15}$$

This result modifies the relation between the polarization density and the external field

$$\mathbf{P} = \chi_s(\mathbf{E}_0 - 4\pi\mathbf{P}_L). \tag{12.16}$$

Generally, $\mathbf{P}$ involves both longitudinal and transverse projections and the above equation does not solve the problem. However, for a spherical ion, only longitudinal polarization contributes by symmetry, $\mathbf{P} = \mathbf{P}_L$. One then solves for $\mathbf{P}_L$

$$\mathbf{P}_L = \frac{\chi_s}{1 + 4\pi\chi_s}\mathbf{E}_0. \tag{12.17}$$

This result solves the solvation problem since $\mathbf{P}$ is now given in terms of $\mathbf{E}_0$. By applying the relation between the dielectric susceptibility and dielectric constant, the solvation free energy is found by substituting Eq. (12.17) to Eq. (12.12) and then to Eq. (12.6)

$$\Delta\mu = -\frac{1}{8\pi}\left(1 - \frac{1}{\epsilon_s}\right)\int_\Omega E_0^2 d\mathbf{r}. \tag{12.18}$$

The integral $(8\pi)^{-1}\int E_0^2 d\mathbf{r}$ is the free energy of electrostatic field of the ion in vacuum. When the field outside of the ion's repulsive core is considered, one calculates the integral for $r > a$, where a is the ionic radius. The result is

$$\frac{1}{8\pi}\int_\Omega E_0^2 d\mathbf{r} = \frac{q^2}{2}\int_a^\infty \frac{dr}{r^2} = \frac{q^2}{2a}. \tag{12.19}$$

Combining these two equations, one gets the Born equation for the free energy of solvation

$$\Delta\mu = \Delta\mu_B = -\frac{q^2}{2a}\left(1 - \frac{1}{\epsilon_s}\right). \tag{12.20}$$

The solvation susceptibility for this problem is obtained by comparing to Eq. (12.9)

$$\boxed{\chi_{\text{solv}} = -\frac{1}{a}\left(1 - \frac{1}{\epsilon_s}\right).} \tag{12.21}$$

12.3 Linear electrostatic solvation

Born formula is a specific case of a general result for the static limit of the linear response approximation applied to the electrostatic interaction of a charged ion with a polar liquid. One can formulate the linear-response approximation for solvation in general terms by applying the protocol of thermodynamic integration.

We want to establish general rules relating the solvation chemical potential $\Delta\mu$ to the interaction potential energy u. This interaction energy does not have to include all interactions between the solute and the liquid. For small ions, the focus is on electrostatic interactions since they dominate in the solvation thermodynamics.

The chemical potential $\Delta\mu$ caused by the interaction energy u is given in terms of the ratio of two partition functions: the one after u has been added to the system Hamiltonian H and the partition function before the addition of u

$$e^{-\beta\Delta\mu} = \frac{\int e^{-\beta u - \beta H} d\Gamma}{\int e^{-\beta H} d\Gamma}, \tag{12.22}$$

where Γ denotes the entire phase space of the system. We next scale u with the variable $0 \leq \lambda \leq 1$: $u \to \lambda u$ and then write the derivative

$$\frac{\partial \beta \Delta\mu(\lambda)}{\partial \lambda} = -\frac{\partial}{\partial \lambda} \ln \left[\int e^{-\beta \lambda u - \beta H} d\Gamma \right]. \tag{12.23}$$

One gets

$$\frac{\partial \Delta\mu(\lambda)}{\partial \lambda} = \langle u \rangle_\lambda, \tag{12.24}$$

where on the right-hand side one gets the average in the presence of the scaled potential

$$\langle \ldots \rangle_\lambda = \frac{\int \ldots e^{-\beta \lambda u - \beta H} d\Gamma}{\int e^{-\beta \lambda u - \beta H} d\Gamma}. \tag{12.25}$$

The chemical potential $\Delta\mu = \Delta\mu(1)$ corresponding to $\lambda = 1$ is obtained by integrating both sides of Eq. (12.24)

$$\boxed{\Delta\mu = \int_0^1 d\lambda \langle u \rangle_\lambda.} \tag{12.26}$$

Equation (12.26) is known as thermodynamic integration. It is exact, allowing one to develop approximations by expanding $\langle u \rangle_\lambda$ around some reference value λ_0. Truncation of this expansion after the second term is what constitutes the linear-response approximation. One starts with a Taylor expansion in $\lambda - \lambda_0$ in Eq. (12.25) by expanding both the nominator and the denominator. The result is

$$\langle u \rangle_\lambda = \langle u \rangle_{\lambda_0} - (\lambda - \lambda_0)\beta \langle (\delta u)^2 \rangle_{\lambda_0}, \qquad (12.27)$$

where $\delta u = u - \langle u \rangle_{\lambda_0}$. The Taylor expansion in Eq. (12.25) requires evaluating the first derivative of the average energy over λ

$$\frac{d}{d\lambda} \langle u \rangle_\lambda = \frac{d}{d\lambda} \frac{\int u e^{-\beta \lambda u - \beta H} d\Gamma}{\int e^{-\beta \lambda u - \beta H} d\Gamma}. \qquad (12.28)$$

The parameter λ enters both the nominator and denominator. Taking the derivative of both parts leads to the second term in Eq. (12.27)

$$\frac{d}{d\lambda} \langle u \rangle_\lambda = -\beta \langle u^2 \rangle_\lambda + \beta \langle u \rangle_\lambda^2 = -\beta \langle (\delta u)^2 \rangle_\lambda. \qquad (12.29)$$

The expansion in Eq. (12.27) can now be used in Eq. (12.26) to obtain for the chemical potential by integrating over λ

$$\Delta \mu = \langle u \rangle_{\lambda_0} + (\lambda_0 - \tfrac{1}{2})\beta \langle (\delta u)^2 \rangle_{\lambda_0}. \qquad (12.30)$$

Two choices for λ_0 are usually implemented: (i) $\lambda_0 = 0$ and (ii) $\lambda_0 = 1$. The first choice implies that one considers the fictitious solute with the interaction energy u turned off (uncharged solute) as the reference system. In many cases, the fact that u depends on the angular distribution of molecules in the liquid allows one to simplify the problem by considering $\lambda_0 = 0$. If the uncharged core of the solute does not create preferential orientations of the molecules of the solvent, $\langle u \rangle_0 = 0$ and one gets a simple result

$$\boxed{\Delta \mu = -\tfrac{1}{2}\beta \langle (\delta u)^2 \rangle_0.} \qquad (12.31)$$

Here the variance of the solute-solvent interaction energy yields the chemical potential of solvation. For an ion carrying charge q, $u \propto q$ and $\Delta \mu \propto q^2$. This is the scaling with the solute charge anticipated by the Born equation (Eq. (12.20)). This result also makes the electrostatic potential of the medium bound charges proportional to the ionic charge, $\phi_b \propto q$, whence the term "linear solvation".

For the choice $\lambda_0 = 1$, one performs averages with the full solute-solvent interaction energy and both the average interaction energy and the energy variance are required

$$\Delta \mu = \langle u \rangle_1 + \tfrac{1}{2}\beta \langle (\delta u)^2 \rangle_1. \qquad (12.32)$$

One, however, can achieve an additional simplification by realizing that neglecting the higher-order expansion terms is equivalent to assuming the variance to be independent of the state chosen for the reference

$$\langle (\delta u)^2 \rangle_0 = \langle (\delta u)^2 \rangle_1. \tag{12.33}$$

This condition requires

$$\boxed{\Delta \mu = \tfrac{1}{2} \langle u \rangle_1.} \tag{12.34}$$

This result provides a significant simplification of solvation calculations, in particular in numerical simulations. The average $\langle u \rangle_1$ converges much faster than the variance and this route leads to a significant speedup of calculations for the linear free energy of solvation.

12.4   *Energy and entropy of solvation

Equations for the chemical potential based on thermodynamic integration lead to general results for the thermodynamic energy, Δe, and entropy, Δs, of solvation by long-range interactions. Combined, they form the chemical potential .

$$\Delta \mu = \Delta e - T \Delta s. \tag{12.35}$$

The first step in this derivation is to arrive at Widom's formula [45,79]. This is achieved by taking the logarithm of both sides in Eq. (12.22)

$$\boxed{\beta \Delta \mu = -\ln \left\langle e^{-\beta u} \right\rangle_0,} \tag{12.36}$$

where $\langle \dots \rangle_0$ corresponds to the statistical average at $\lambda = 0$ in Eq. (12.25). Solvation energy is calculated by applying the thermodynamic relation (12.7) to Widom's formula. After some algebra, one arrives at the result

$$\Delta e = (\partial \beta \Delta \mu / \partial \beta)_V = \langle (u + H) \rangle_1 - \langle H \rangle_0. \tag{12.37}$$

One can identify the average $e_{0s} = \langle u \rangle_1$ as the average energy of the solute-solvent interaction. The solvation energy becomes

$$\Delta e = e_{0s} + e_{ss}, \tag{12.38}$$

where the second term on the right-hand side is the change in the energy of the solvent produced by turning the solute-solvent interaction on

$$e_{ss} = \langle H \rangle_1 - \langle H \rangle_0. \tag{12.39}$$

This energy is often called the reorganization energy of the solvent in solvation.

One can arrive at an alternative equation for e_{ss} by applying the identity

$$\langle H \rangle_1 - \langle H \rangle_0 = \int_0^1 d\lambda \frac{\partial \langle H \rangle_\lambda}{\partial \lambda}. \qquad (12.40)$$

One next takes the derivative over λ

$$\frac{\partial \langle H \rangle_\lambda}{\partial \lambda} = -\beta \langle \delta u \delta H \rangle_\lambda. \qquad (12.41)$$

The final result for the component e_{ss} is

$$e_{ss} = -\beta \int_0^1 d\lambda \langle \delta u \delta H \rangle_\lambda. \qquad (12.42)$$

This fluctuation relation is more convenient than Eq. (12.39) in numerical simulations where the correlation between fluctuations of the solute-solvent and solvent-solvent energies can be directly calculated from configurations produced by computer simulations.

One can write Widom's formula in an alternative form

$$e^{-\beta \Delta \mu} = \langle e^{-\beta u} \rangle_0 = \frac{\int d\Gamma e^{-\beta u - \beta H}}{\int d\Gamma e^{\beta u - \beta u - \beta H}} = \frac{1}{\langle e^{\beta u} \rangle_1}. \qquad (12.43)$$

The chemical potential thus becomes

$$\beta \Delta \mu = \ln \langle e^{\beta u} \rangle_1 = \beta \langle u \rangle_1 + \ln \langle e^{\beta \delta u} \rangle_1, \qquad (12.44)$$

where $\delta u = u - \langle u \rangle_1$. Since $e_{0s} = \langle u \rangle_1$ is the average energy of the solute-solvent interaction, the second term is often interpreted as the non-thermodynamic solvation entropy in the equation

$$\Delta \mu = e_{0s} - T s_{0s}. \qquad (12.45)$$

The meaning of the entropic term

$$T s_{0s} = -\beta^{-1} \ln \langle e^{\beta \delta u} \rangle_1 \qquad (12.46)$$

is that it involves all possible cumulants involving the deviation δu of the interaction energy from its average value. An important result of this derivation is that the chemical potential of solvation is affected only by the statistics of solute-solvent interactions. This conclusion does not apply to the thermodynamic energy and entropy of solvation. The latter is found by taking the temperature derivative of $\Delta \mu$

$$\Delta s = -(\partial \Delta \mu / \partial T)_V. \qquad (12.47)$$

Repeating the same steps as are done for Δe, one obtains

$$T \Delta s = T s_{0s} + e_{ss}. \qquad (12.48)$$

It is clear that both Δe and $T \Delta s$ contain the same term e_{ss} reflecting the modification of the solvent-solvent interactions induced by the solute. However, this term cancels identically from $\Delta \mu$ [80].

12.5   Work at finite rate

Equations (12.26) and (12.36) are two extreme cases of switching the interaction energy on. In Eq. (12.26), the interaction energy is increased infinitely slowly to its full value while maintaining the equilibrium between the solute and the surrounding medium. In contrast, the interaction energy is switched on instantaneously in Eq. (12.36). Both protocols result in the same chemical potential $\Delta\mu$. One might wonder if the same chemical potential can be reached by switching the potential with a finite rate. This is indeed possible and the corresponding result is known as Jarzynski equality.

Consider the interaction potential $u(\mathbf{x}, \mathbf{p})$ switched on over a finite time τ with the rate $\dot{\lambda} = \tau^{-1}$. The work done in this process is

$$W = \int_0^\tau dt \dot{\lambda} u(\mathbf{x}(t), \mathbf{p}(t)). \tag{12.49}$$

This equation is a special case of the general connection between the work and the system Hamiltonian $H(\mathbf{x}(t), \mathbf{p}(t), t)$ which is often used to find nonequilibrium work

$$W = \int_0^\tau dt \partial_t H(\mathbf{x}(t), \mathbf{p}(t), t). \tag{12.50}$$

Equation (12.49) follows from this equation by taking $H(t) = H_0 + \lambda(t)u$. This representation assumes that the Hamiltonian does not contain a function of time only, such as $H(t) = H_0 + \lambda(t)u + f(t)$, since in that case the work becomes ill-defined.

The interaction potential in Eq. (12.49) depends on all phase-space variables of the system which evolve according to Liouville's equation when the interaction energy is introduced. The evolution of the system transfers the initial phase-space positions and momenta $\mathbf{x}_0, \mathbf{p}_0$ to $\mathbf{x}(\tau), \mathbf{p}(\tau)$. According to Liouville's theorem (Sec. 9.1), the probability density in the phase space satisfies the equation $f(\mathbf{x}^0, \mathbf{p}^0) d\Gamma_0 = f(\mathbf{x}(t), \mathbf{p}(t)) d\Gamma_t$ and, additionally, the conservation of the phase-space volume, $d\Gamma_0 = d\Gamma_t$. Further, the work w done on each single trajectory during the switching time τ is simply the change of the Hamiltonian, which is $u(\mathbf{x}(\tau), \mathbf{p}(\tau))$. One can next write the following average

$$\left\langle e^{-\beta W} \right\rangle = \int d\Gamma_\tau f_\tau e^{-\beta w} = \int d\Gamma_0 f_0 e^{-\beta u} = \left\langle e^{-\beta u} \right\rangle_0. \tag{12.51}$$

According to Eq. (12.36), the last equality provides the chemical potential $\Delta\mu$ and one can write

$$e^{-\beta \Delta\mu} = \left\langle e^{-\beta W} \right\rangle. \tag{12.52}$$

This is the Jarzynski equality, which establishes the connection between the change in the free energy of the system and the work done over the finite switching time. The example considered here is based on switching the interaction energy between the solute and the medium, but the derivation can be extended to any type of finite-rate work done on a system. A substantial advantage of this result is that only the distribution f_0 needs to be the equilibrium function. The distribution f_τ characterizing the system after the switching time τ can be arbitrary. It is usually a nonequilibrium distribution when this equation is used in experiment to establish the equilibrium free energy change from repeated nonequilibrium trajectories, each characterized by the work W [81].

12.6 *Microscopic electrostatics

The Born equation, as well as all solutions based on Maxwell's electrostatics of dielectrics, relies on locality of the polarization response to the Maxwell electric field. This approximation is an extrapolation of the locality of the homogeneous Maxwell field, established in homogeneous dielectric experiments (Sec. 2.12), to inhomogeneous fields found in liquid interfaces. An inhomogeneous Maxwell field cannot be directly probed by experiment and assuming its locality must have some limitations. In fact, molecular hydrodynamics operating in terms of linear response functions in the wave-vector and frequency domains (Chap. 8) makes it clear that the electrostatic response to an inhomogeneous external field is characterized by some intrinsic length scale and thus cannot be local. The definition of the k-dependent dielectric constant in Eq. (8.155) is a clear evidence of nonlocality of the solvent polarization for a general spatially varying external field polarizing the liquid.

Given these general observations, theories of polar solvation need to be extended to include nonlocal response of the dielectric. The path to such a microscopic formulation of electrostatics is allowed by considering the medium polarization through a convolution relation already shown in Eq. (2.152)

$$\mathbf{P}_L(\mathbf{r}) = \int_V d\mathbf{r}' \chi_{\text{solv}}(\mathbf{r}, \mathbf{r}') \cdot \mathbf{E}_0(\mathbf{r}'). \qquad (12.53)$$

The external field is propagated to produce the polarization density by a nonlocal solvation susceptibility $\chi_{\text{solv}}(\mathbf{r}, \mathbf{r}')$. In contrast to the susceptibility function in the bulk material, which depends on $\mathbf{r} - \mathbf{r}'$, the solvation problem is inhomogeneous and the solvation response in general depends on two coordinates separately. This complication is neglected here to arrive at a simple solution allowing physical insight. We therefore assume $\chi_{solv}(\mathbf{r}, \mathbf{r}') = \chi_{solv}(\mathbf{r} - \mathbf{r}')$. This assumption allows us to use the convolution theorem and arrive at the integral involving the electric field of the solute and the Fourier transform of the susceptibility function $\tilde{\chi}_{solv}(\mathbf{k})$ as shown in Eq. (2.155)

$$\Delta\mu = -\tfrac{1}{2} \int \frac{d\mathbf{k}}{(2\pi)^3} \tilde{\mathbf{E}}_0(\mathbf{k}) \cdot \tilde{\chi}_{solv}(\mathbf{k}) \cdot \tilde{\mathbf{E}}_0^*(\mathbf{k}). \tag{12.54}$$

Here, $\tilde{\mathbf{E}}_0^*(\mathbf{k})$ denotes the complex conjugate of the Fourier transformed electric field of the ion. Complex conjugate is required for complex-valued fields to arrive at the real value of the solvation chemical potential $\Delta\mu$.

For a spherical ion with the radius a, the Fourier transform of its electric field has to be calculated in the volume Ω ($r > a$) occupied by the polar liquid

$$\tilde{\mathbf{E}}_0(\mathbf{k}) = \int_\Omega d\mathbf{r}\, \mathbf{E}_0 e^{i\mathbf{k}\cdot\mathbf{r}}. \tag{12.55}$$

This integral can be written as

$$\tilde{\mathbf{E}}_0(\mathbf{k}) = \frac{q}{i}\nabla_\mathbf{k} \int_\Omega \frac{d\mathbf{r}}{r^3} e^{i\mathbf{k}\cdot\mathbf{r}} = \frac{4\pi q}{i}\nabla_\mathbf{k} \int_{ka}^\infty dx \frac{\sin x}{x^2}, \tag{12.56}$$

where $\nabla_\mathbf{k}$ is the del operator in reciprocal space. By taking the derivative, one obtains

$$\tilde{\mathbf{E}}_0(\mathbf{k}) = \frac{4\pi i q}{k^2}\mathbf{k} j_0(ka), \tag{12.57}$$

where $j_0(x) = x^{-1}\sin x$ is the zeroth-order spherical Bessel function (Sec. 1.9). It is clear that $\tilde{\mathbf{E}}_0 \propto \mathbf{k}$ carries longitudinal symmetry. This field couples to the solvent polarization $\mathbf{P}_L$ of the same longitudinal symmetry.

The main challenge of microscopic electrostatic calculations is the evaluation of the solvation susceptibility. The properties of the interface have to affect this function in realistic calculations. Here, for the sake of simplicity, we replace the interfacial susceptibility with its analog taken from the calculations in the bulk. Since the reciprocal-space electric field carries longitudinal symmetry, only the longitudinal component of the susceptibility needs to be accounted for. For the bulk dielectric, it is given by the relation

$$\chi_{solv}^L(\mathbf{k}) = \frac{\beta}{V}\langle |P_\mathbf{k}^L|^2 \rangle. \tag{12.58}$$

We view here the polarization density as the density of molecular dipoles (first term in Eq. (2.51))

$$\mathbf{P} = m \sum_j \hat{\mathbf{u}}_j \delta(\mathbf{r} - \mathbf{r}_j), \tag{12.59}$$

where we have separated the unit vector $\hat{\mathbf{u}}_j$ for the molecular dipole j from its magnitude m (see also Eq. (2.210)). One obtains for the longitudinal projection of the polarization

$$P_{\mathbf{k}}^L = m \sum_j \hat{\mathbf{k}} \cdot \hat{\mathbf{u}}_j e^{i\mathbf{k}\cdot\mathbf{r}_j} \tag{12.60}$$

and for the longitudinal susceptibility function in Eq. (12.58)

$$\chi_{\text{solv}}^L(\mathbf{k}) = \frac{3y}{4\pi} S_L(k). \tag{12.61}$$

Here, $y = (4\pi/9)\beta\rho m^2$ is the scaled density of dipoles appearing in standard formulations of dielectric theories (Eq. (2.193)). Further, the longitudinal structure factor $S_L(k)$ is given by the equation [39]

$$S_L(k) = \frac{3}{N} \sum_{i,j=1}^N \left\langle (\hat{\mathbf{k}} \cdot \hat{\mathbf{u}}_j)(\hat{\mathbf{u}}_i \cdot \hat{\mathbf{k}}) e^{i\mathbf{k}\cdot(\mathbf{r}_i - \mathbf{r}_j)} \right\rangle. \tag{12.62}$$

The convenience of separating the structure factor is that it depends only on orientations of dipoles and is not affected by their magnitudes (Fig. 12.3). Similar considerations lead to the separation of the Kirkwood factor in the Kirkwood-Onsager equation (Eq. (2.212)). It is also worth noting that the longitudinal structure factor describes fluctuations of dipolar orientations at a given length scale, in contrast to the more traditional density structure factor (Eq. (6.42)) describing fluctuations of the microscopic density.

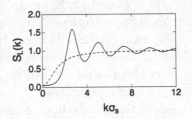

Fig. 12.3 Structure factor $S_L(k)$ of a polar liquid and its approximation by Eq. (12.65). The wave vector k is scaled with the molecular solvent diameter σ_s.

Substituting all these results into Eq. (12.54) leads to a single integral representing the chemical potential of ion solvation

$$\Delta\mu = -\frac{3q^2 y}{2a} I, \quad I = \frac{2a}{\pi} \int_0^\infty dk j_0^2(ka) S_L(k). \tag{12.63}$$

It is clear that this integral should lead to the Born solvation energy $\Delta\mu_B$ (Eq. (12.20)) when nonlocality of the response is neglected. This is achieved when the structure factor is viewed as a constant $S_L(k) = S_L(0)$. The k-integral in Eq. (12.63) becomes $I = S_L(0)$. To convert Eq. (12.63) to the Born equation (Eqs. (12.20) and (12.21)), one needs to require

$$S_L(0) = \frac{\epsilon_s - 1}{3y\epsilon_s}. \tag{12.64}$$

The microscopic structure factor is a more complex function passing through a number of oscillations before reaching the limit $S_L(k) \to 1$ at $k \to \infty$ (Fig. 12.3). The reason for oscillations in $S_L(k)$ is the microscopic structure of the polar liquid and specific length scales of dipolar correlations in the liquid. One can introduce a single correlation length Λ by adopting an analytical approximation to $S_L(k)$ interpolating between the $k = 0$ and $k \to \infty$ limits (Fig. 12.3)

$$S_L(k) = \frac{S_L(0) + \Lambda^2 k^2}{1 + \Lambda^2 k^2}. \tag{12.65}$$

This simple approximation allows one to calculate the k-integral in Eq. (12.63), with the result

$$\Delta\mu = \Delta\mu_B \left[1 + \frac{\Lambda}{2a} \left(\frac{1}{S_L(0)} - 1\right) \left(1 - e^{-2a/\Lambda}\right)\right]. \tag{12.66}$$

Here, the Born solvation term $\Delta\mu_B$ is multiplied by the term in the brackets, which involves the ratio of the correlation length Λ to the ionic radius a. As expected, the brackets term tends to unity in the limit of solvation by a local (continuum) dielectric requiring $a \gg \Lambda$. Equation (12.66) is derived here as an illustration of the general result that microscopic theories convert to dielectric calculations when the size of the solute significantly exceeds the length of dipolar correlations in the medium.

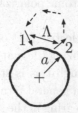

Fig. 12.4 Mutual correlations between medium dipoles interacting with the ion.

The physical picture of mutual correlations between the medium dipoles interacting with an ion is illustrated in Fig. 12.4. Consider an ion polarizing (orienting) dipole 1 in the liquid. This alteration in the orientation of a given dipole cannot be viewed without accounting for its effect on orientations of other dipoles in the medium. Correlations between dipoles propagate through chains of dipolar orientations such that the orientation of dipole 1 is correlated to the orientation of dipole 2 through the characteristic correlation length Λ. The longitudinal structure factor in fact accounts for the collective effects of all such dipole-dipole correlations, mostly propagating through dipolar chains. They become irrelevant when the solute size a far exceeds Λ and all dipolar chains can be collapsed to a point characterized by the dielectric susceptibility. The picture of a local polarization field specific to dielectric theories is recovered in this limit.

Chains of correlated dipoles store more electrostatic energy in the medium than anticipated by the dielectric theories. Given that $S_L(k) > S_L(0)$ at $k > 0$ (Fig. 12.3), microscopic electrostatics allows more solvation stabilization than dielectric solvation leading to the Born equation. That is why the second term in the brackets in Eq. (12.66) is positive, thus making $\Delta\mu$ more negative than $\Delta\mu_B$. The practice of applying the Born equation to solvation problems balances this deficiency out by adjusting the effective ionic radius.

12.7   Nonpolar solvation

Nonpolar solvation broadly describes thermodynamics of interactions between solutes and materials made of molecules carrying no permanent dipoles. In contrast, polar solvation is typically assigned to dipolar liquids or solids, but also includes solvation by higher multipole moments, predominantly by molecular quadrupoles. The case of solvation by dipolar liquids was considered in the previous section. This section describes analytical results for solvation of solutes by induced electronic dipoles in the liquid. Those are dipoles associated with the electronic polarizability of the molecules (Sec. 4.14).

Induced dipoles are caused by the polarizing electric field of the solute. The stabilization free energy appearing from polarizing the material is often called the induction interaction energy, even though this is strictly a free energy (see Sec. 11.5 for a detailed discussion). In general terms, nonpolar solvation also includes dispersion interactions appearing in the second-order quantum-mechanical perturbation theory (Sec. 4.15). These interactions are not explicitly considered here. Nevertheless, the formalism outlined here follows the same steps for both types of interactions. The goal here is to show how to construct a perturbation theory for nonpolar solvation.

The focus is on conceptual steps and physical consequences of the short-range interaction potentials involved, instead of a detailed description of the computational formalism.

If a molecule of the liquid carries the isotropic polarizability α, this molecule will be polarized by an external electric field $\mathbf{E}_0$ to acquire the induced molecular dipole

$$\mathbf{p} = \alpha \mathbf{E}_0. \tag{12.67}$$

The free energy of this dipole in the external field follows from Eq. (2.151) and reads

$$f = -\tfrac{1}{2}\mathbf{p} \cdot \mathbf{E}_0 = -\tfrac{1}{2}\alpha E_0^2. \tag{12.68}$$

This interaction energy is known as induction interaction [15]. If the field $\mathbf{E}_0 = -\nabla(q/r)$ is produced by a single charge q, the induced dipole scales as $p(r) \propto r^{-2}$ and the interaction energy decays with the distance r as r^{-4} (Fig. 12.5). In contrast, dispersion interactions universally scale as r^{-6} at sufficiently short distances (Sec. 4.15). Therefore, induction interactions always dominate over dispersion interactions when charged solutes are involved. The case of a single ion in a nonpolar liquid is considered here.

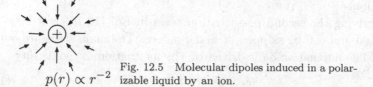

$$p(r) \propto r^{-2}$$

Fig. 12.5   Molecular dipoles induced in a polarizable liquid by an ion.

From the induction interaction energy of the field source with a single induced dipole one can construct the solute-solvent interaction energy with the entire liquid. If the density profile of the liquid around the solute is $\rho(\mathbf{r})$, one obtains

$$u = -(\alpha/2)\int_{\Omega} d\mathbf{r}\rho(\mathbf{r})E_0^2(\mathbf{r}). \tag{12.69}$$

With this interaction energy, thermodynamic integration considered in Eqs. (12.26) and (12.27) leads to the following equation for the chemical potential of solvation when the expansion in u is truncated after the second term

$$\Delta\mu = \langle u \rangle_0 + (\beta/2)\langle(\delta u)^2\rangle_0. \tag{12.70}$$

The expansion here is performed with the reference system corresponding to $\lambda = 0$ in the thermodynamic integration, i.e., the system with the interaction potential u turned off.

We further simplify the mathematical description by replacing the microscopic density profile $\rho(r)$ with a step function $\rho(r) = \rho h(r - a)$ ($h(x)$ is the Heaviside function) allowing the bulk density ρ right at the solute radius a. More accurate calculations need to include the radial distribution function $g(r)$ to determine the interfacial density profile $\rho(r) = \rho g(r)$. With this simplified problem setup, one obtains

$$\langle u \rangle_0 = -2\pi\alpha\rho q^2 \int_a^\infty \frac{dr}{r^2} = -\frac{n_D^2 - 1}{n_D^2 + 2} \frac{3q^2}{2a}. \tag{12.71}$$

In the second part of this equation, Clausius-Mossotti equation (Eq. (2.173)) was used to replace $(4\pi/3)\rho\alpha$ with the Clausius-Mossotti function of the refractive index n_D. The refractive index squared gives the dielectric constant of a nonpolar liquid (see Table 2.3 for examples). The important result of this first step of the derivation is that the average energy $\langle u \rangle_0$ scales with the properties of the ion as q^2/a, which has the same order of magnitude as the Born solvation energy (Eq. (12.12)). Nonpolar solvation due to $\langle u \rangle_0$ is comparable in magnitude to polar solvation (Table 12.1). Given that polarizability is a universal molecular property for all solvating media, this term will always be a significant part of solvation stabilization of an ion.

Deriving the second perturbation term in Eq. (12.70) is a little more involved and can be skipped at first reading. The main steps are outlined here. The fluctuation δu is driven by the fluctuation of the density

$$\delta u = -\tfrac{1}{2}\alpha \int_\Omega d\mathbf{r}\, \delta\rho E_0^2. \tag{12.72}$$

The variance hence becomes

$$\tfrac{1}{2}\beta\langle(\delta u)^2\rangle_0 = \frac{\beta\alpha^2}{8} \int_\Omega d\mathbf{r} d\mathbf{r}' \langle \delta\rho(\mathbf{r})\delta\rho(\mathbf{r}')\rangle_0 E_0^2(\mathbf{r}) E_0^2(\mathbf{r}')$$
$$= \frac{\beta\rho\alpha^2}{8} \int \frac{d\mathbf{k}}{(2\pi)^3} S(k) \left|E_0^2(\mathbf{k})\right|^2. \tag{12.73}$$

The transformation from the first to the second line in this equation assumes that the correlation of the density fluctuations $\langle \delta\rho(\mathbf{r})\delta\rho(\mathbf{r}')\rangle_0 = K(\mathbf{r} - \mathbf{r}')$ should be a function of $\mathbf{r} - \mathbf{r}'$ only. By using the convolution theorem (Sec. 1.7), one converts the direct-space convolution to the reciprocal-space integral. For an arbitrary space field $\phi(\mathbf{r})$, one obtains

$$\int_\Omega d\mathbf{r} d\mathbf{r}' \phi(\mathbf{r})\phi(\mathbf{r}') K(\mathbf{r} - \mathbf{r}') = \int \frac{d\mathbf{k}}{(2\pi)^3} |\tilde{\phi}_\mathbf{k}|^2 \tilde{K}(\mathbf{k}), \tag{12.74}$$

Table 12.1 Components of the solvation free energy (eV) of an ion with the radius $a = 3$ Å and $q = e$ in polar solvents at 298 K.

Solvent	ϵ_s	β_T, (GPa)$^{-1}$	$-\Delta\mu_B$[a]	$-\Delta\mu_{np}$[b]	$\Delta\mu_{HS}$[c]
Water	78.5	0.457	2.37	1.42	1.30
Methanol	35.9	1.256	2.32	1.39	0.69
Acetonitrile	35.9	1.112	2.33	1.44	0.59
Chloroform	4.9	1.033	1.89	1.80	0.65

[a]Born solvation free energies according to Eq. (12.20). [b]Nonpolar solvation free energies from Eq. (12.80). [c]Free energy of cavity formation from Eq. (12.96).

where integration over the volume Ω occupied by the liquid is shifted to the Fourier transform

$$\tilde{\phi}_{\mathbf{k}} = \int_\Omega d\mathbf{r}\phi(\mathbf{r})e^{i\mathbf{k}\cdot\mathbf{r}}. \tag{12.75}$$

To calculate the Fourier transform $\tilde{K}(\mathbf{k})$ in Eq. (12.73), one can consider the correlation of Fourier-transformed density fields $\rho_{\mathbf{k}}$ and notice that that it should be proportional to $\delta(\mathbf{k} + \mathbf{k}')$ for the corresponding correlation in direct space $\langle\delta\rho(\mathbf{r})\delta\rho(\mathbf{r}')\rangle_0$ to depend on $\mathbf{r} - \mathbf{r}'$

$$\langle\rho_{\mathbf{k}}\rho_{\mathbf{k}'}\rangle_0 = \tilde{K}(\mathbf{k})(2\pi)^3\delta(\mathbf{k} + \mathbf{k}'), \quad \tilde{K}(\mathbf{k}) = V^{-1}\langle|\rho_{\mathbf{k}}|^2\rangle_0. \tag{12.76}$$

The need for the factor V^{-1} can be proved by noting that

$$\int d\mathbf{r}\langle\delta\rho(\mathbf{r})\delta\rho(\mathbf{r}')\rangle_0 = N^2/V = \tilde{K}(0). \tag{12.77}$$

One can next observe that $\tilde{K}(k) = \rho S(k)$, and this connection is responsible for the appearance of the density-density structure factor $S(k)$ (Eq. (6.42)) in Eq. (12.73). The situation here is very similar to microscopic electrostatics (Eq. (12.63)), where the structure factor of dipolar orientations enters the variance of the interaction energy. For nonpolar media, correlated density fluctuations play a physically identical role.

If the solute is sufficiently large, one can replace $S(k)$ with its $k = 0$ value and convert back to direct space for the integral of the electric field (see Sec. 5.14 for more details)

$$\tfrac{1}{2}\beta\langle(\delta u)^2\rangle_0 = \frac{\beta\rho\alpha^2 S(0)}{8}\int_\Omega d\mathbf{r}E_0^4(\mathbf{r}). \tag{12.78}$$

One can next substitute $\beta S(0) = \rho\beta_T$ with β_T standing for the medium isothermal compressibility (Eq. (6.44)). For a spherical solute with $E_0^2 = q^2/r^4$ the result of integration is

$$\tfrac{1}{2}\beta\langle(\delta u)^2\rangle_0 = \frac{\pi(\alpha\rho)^2\beta_T q^4}{10a^5}. \tag{12.79}$$

One can further substitute the Clausius-Mossotti equation for $\alpha\rho$ (Eq. (2.173)) and combine Eqs. (12.71) and (12.79) in Eq. (12.70) to obtain the chemical potential of nonpolar solvation (subscript "np")

$$\Delta\mu_{\rm np} = -\frac{n_D^2 - 1}{n_D^2 + 2}\frac{3q^2}{2a}\left[1 - \frac{3q^2\beta_T}{80\pi a^4}\frac{n_D^2 - 1}{n_D^2 + 2}\right]. \tag{12.80}$$

The second term in the brackets, arising from the variance of the interaction potential, scales as a^{-4} and vanishes for sufficiently large solutes. Since fluctuations of the potential are caused by density fluctuations, the result is proportional to the medium compressibility β_T. An incompressible solvent will not allow density fluctuations and only the average interaction energy given by $\langle u \rangle_0$ will be present in $\Delta\mu_{\rm np}$.

Transition to reciprocal space was a critical step in the derivation presented here since it allowed us to express the result in terms of a collective property of density fluctuations (isothermal compressibility). Collective properties are accessed in the $k \to 0$ limit of the corresponding response functions. Likewise, transition to $k = 0$ in the orientational (longitudinal) structure factor of the polar liquid (previous section) allowed us to express the result in terms of the collective longitudinal response scaling as $(1 - \epsilon_s^{-1})$ (Sec. 8.14) and recovering the Born formula. The ability of the theory to formulate the result in terms of collective properties and corrections to them arising from the microscopic structure of the liquid is a great advantage of working with reciprocal-space response functions.

The distinction in the relative contributions of the first and second statistical moments of the interaction potential to the solvation free energy is worth mentioning. The first statistical moment, $\langle u \rangle_0$, is zero for electrostatic interactions with the solvent dipoles (but $\langle u \rangle_1$ is non-zero) and the solvation free energy is fully determined by the second moment, the variance $\langle (\delta u)^2 \rangle_0$ (Eq. (12.31)). In contrast, fluctuations are typically small for nonpolar interactions and the solvation free energy is dominated by the first moment of the interaction potential. Since the magnitude of a statistical moment decreases with increasing order (e.g., going from the first to the second statistical moment), ion-dipole (second order) and induction (first order) interactions are of similar order of magnitude (Table 12.1). This result comes despite the fact that the interaction of an ion with a permanent dipole is stronger and decays slower with the distance than the induction interaction with an induced dipole.

From these differences in statistics of electrostatic interactions of permanent charges, $u_{\rm el}$, and the induction interaction with induced dipoles, $u_{\rm ind}$, one can write an approximate solution for the total solvation free energy. By combining the results presented here with Eq. (12.34), one obtains

$$\Delta\mu \simeq \tfrac{1}{2}\langle u_{\mathrm{el}}\rangle_1 + \langle u_{\mathrm{ind}}\rangle_1. \tag{12.81}$$

The second statistical average can be estimated at $\lambda = 0$ as well since $\langle u_{\mathrm{ind}}\rangle_1 \simeq \langle u_{\mathrm{ind}}\rangle_0$ for induction interactions.

12.8 Free energy of cavity formation

The Widom's formula can be applied to derive the free energy required to create an empty space in a liquid to place the molecule in. The question asked here is what is the amount of reversible work required to carve a cavity in the liquid for the molecule to be solvated. This is the free energy of cavity formation.

Consider a solute represented by a hard billiard ball of the radius a. One assumes that the repulsion energy $u(r)$ is infinite at $r < a$ and there is no interaction at $r > a$: $u(r > a) = 0$. Widom's formula predicts the change of the chemical potential of the solute achieved by turning the repulsion potential on

$$\beta\Delta\mu_{\mathrm{HS}}(a) = -\langle e^{-\beta u}\rangle_0 = -\langle h(r-a)\rangle_0, \tag{12.82}$$

where $h(r - a)$ is the Heaviside step function. One can next take the derivative of $\Delta\mu_{\mathrm{HS}}(a)$ to obtain

$$\frac{\partial\beta\Delta\mu_{\mathrm{HS}}(a)}{\partial a} = \langle\delta(r-a)\rangle_0 = 4\pi\rho a^2 G(a). \tag{12.83}$$

Here, $G(a)$ is the solute-solvent radial distribution function calculated at the contact between the repulsive core of the solute with the radius a and the surrounding liquid. This function has a meaning different from the standard radial distribution function $g(r)$ (Sec. 6.2), which defines the probability of finding a molecule at the distance r from the center of a solute of fixed size. Here, changing the variable a means having a new solute for each radius a, which is why a different notation is used.

To determine $\Delta\mu_{\mathrm{HS}}(a)$, one needs to integrate over a in Eq. (12.83). It is, however, clear that a cannot reach the value of zero since even for a solute of zero size the liquid molecules cannot approach such a geometric point closer than the radius $\sigma_s/2$ of their repulsive cores (Fig. 12.6). The cavity radius is therefore given as

$$a = \sigma_s\frac{1+d}{2}, \tag{12.84}$$

where $d = \sigma_0/\sigma_s$ is the ratio of diameters of the solute and solvent molecules. Since the lowest value of d is zero, it is more convenient to formulate the theory in terms of this dimensionless parameter. Equation (12.83) can therefore be written as

$$\frac{\partial \beta \Delta \mu_{HS}(d)}{\partial d} = 3\eta(1+d)^2 G(d), \tag{12.85}$$

where

$$\eta = \rho v_0 = (\pi/6)\rho \sigma_s^3 \tag{12.86}$$

is the packing fraction of the liquid equal to the fraction of the volume of the molecules to the volume of the container (liquid volume V) containing these molecules. Integrating Eq. (12.85) in terms of the variable d leads to the reversible work required to grow the solute from zero size to d

$$\boxed{\beta \Delta \mu_{HS}(d) = \beta \Delta \mu_{HS}(0) + 3\eta \int_0^d dx (1+x)^2 G(x).} \tag{12.87}$$

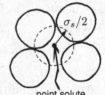

Fig. 12.6   Point solute making the sphere with the radius $\sigma_s/2$ inaccessible to the solvent molecules.

One might think that $\Delta\mu_{HS}(0)$ is zero, but this is incorrect. Inserting a solute of zero size adds information content and thus must be accompanied by a loss of entropy. This entropy comes from the knowledge that centers of the molecules with the diameter σ_s cannot occupy the sphere with the radius of $\sigma_s/2$ surrounding the zero-size solute (Fig. 12.6). The volume excluded from the access by the liquid molecules is $v_0 = (\pi/6)\sigma_s^3$. This volume restriction must lead to a negative entropy change Δs compared to the ideal gas. A positive chemical potential $\Delta\mu_{HS}(0) = -T\Delta s$ given in terms of the probability of finding a zero-size solute in the liquid (see Eq. (5.37))

$$\beta \Delta \mu_{HS}(0) = -\ln(1 - \rho v_0) = -\ln(1 - \eta). \tag{12.88}$$

Equations (12.87) and (12.88) provide the complete solution of the problem provided one has access to the contact value of the solute-solvent radial distribution function $G(d)$. Since this is not the case in most practical cases, approximations are required. A successful solution of this problem can be achieved by interpolating between known geometric and thermodynamic restrictions on $G(d)$. This approach is realized in the scaled-particle theory [19, 39].

In addition to the chemical potential at $d = 0$, one can add another geometric consideration to establish that $G(0)$ is also well defined

$$G(0) = \frac{1}{1 - \eta}. \tag{12.89}$$

This result follows from the mathematical observation that in the unphysical region $-1 < d \leq 0$ only one molecule of the solvent can appear within the sphere with the radius a given by Eq. (12.84) and excluding the volume $(4\pi/3)a^3$ from access by the solvent molecules (Fig. 12.6). The chemical potential in this range is

$$\beta \Delta \mu_{\text{HS}}(d) = - \ln \left(1 - \eta(1 + d)^3\right). \tag{12.90}$$

Substituting this equation to Eq. (12.85), one arrives at Eq. (12.89) at $d = 0$. This relation puts an exact constraint on the derivative of the chemical potential

$$\left. \frac{\partial \beta \Delta \mu_{\text{HS}}(d)}{\partial d} \right|_{d=0} = \frac{3\eta}{1 - \eta}. \tag{12.91}$$

In the opposite limit, $d \to \infty$, the reversible work of creating a macroscopic void in the liquid is equal to the volume work $\beta P V_0 = (\eta Z)d^3$, where the compression factor $Z = \beta P/\rho$ is available from the equation of state of a particular liquid ($Z = 1$ for the ideal gas). With these exact constraints, one constructs a polynomial interpolation

$$\beta \Delta \mu_{\text{HS}}(d) = \beta \Delta \mu_{\text{HS}}(0) + a_1 d + \tfrac{1}{2} a_2 d^2 + \eta Z d^3. \tag{12.92}$$

In addition to Eq. (12.89), one can assume that the contact value $G(1) = g_s$ for the radial distribution function in the bulk solvent can be separately established. This creates an additional constraint from Eq. (12.85)

$$a_1 + a_2 + 3\eta Z = 12\eta g_s. \tag{12.93}$$

Combining this relation with the result for the derivative at $d = 0$ (Eq. (12.91)), one arrives at the free energy of cavity formation

$$\beta \Delta \mu_{\text{HS}}(d) = \beta \Delta \mu_{\text{HS}}(0) + \frac{3\eta}{1 - \eta} d + \left(6\eta g_s - \tfrac{3}{2}\eta Z - \frac{3\eta}{2(1 - \eta)}\right) d^2 + \eta Z d^3. \tag{12.94}$$

The problem is reduced to an expression without unknown parameters for a special case when only entropic effects of repacking the hard cores of

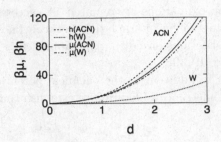

Fig. 12.7 Chemical potential (μ) and enthalpy (h) of cavity formation from Eqs. (12.96) and (12.97) for water (W) and acetonitrile (ACN).

the solvent molecules are considered when growing the cavity. In this case, Carnahan-Starling equation of state for the fluid of hard spheres can be applied to get Z and g_s in Eq. (12.94) (Eqs. (6.67) and (6.68))

$$Z^{\text{CS}} = \frac{1 + \eta + \eta^2 - \eta^3}{(1 - \eta)^3}, \quad g_s^{\text{CS}} = \frac{2 - \eta}{2(1 - \eta)^3}. \tag{12.95}$$

Substituting these values to Eq. (12.94), one obtains

$$\beta \Delta \mu_{\text{HS}}(d) = -\ln(1 - \eta)$$
$$+ \frac{3\eta}{1 - \eta} d + \frac{3\eta(2 - \eta)(1 + \eta)}{2(1 - \eta)^2} d^2 + \frac{\eta(1 + \eta + \eta^2 - \eta^3)}{(1 - \eta)^3} d^3. \tag{12.96}$$

The choice of g_s and Z from Eq. (12.95) is a convenient approximation since it leads to the result fully determined in terms of the solvent density and the relative cavity size. However, both parameters might be available from numerical simulations of liquids, scattering experiments, and empirical equations of state. More realistic choices can be made to improve the accuracy of calculations.

As mentioned above, the free energy of cavity formation mostly goes to rearrange the repulsive cores of the solvent around the solute. The use of the results for hard-sphere fluids in deriving Eq. (12.96) leads to a purely entropic chemical potential $\Delta \mu_{\text{HS}} \propto T$ at constant density (cf. to the chemical potential of the ideal gas, Eq. (5.73)). This result does not mean that the free energy of cavity formation does not have an enthalpy contribution. Liquid expansivity requires a nonzero enthalpy

$$\Delta h_{\text{HS}} = T \alpha_p \eta \left(\partial \Delta \mu_{\text{HS}} / \partial \eta \right)_T, \tag{12.97}$$

where $\alpha_p = V^{-1}(\partial V / \partial T)_p$ is the isobaric expansivity of the liquid.

Figure 12.7 illustrates the free energy and enthalpy of cavity formation in water and acetonitrile calculated from Eqs. (12.96) and (12.97). Very low expansivity α_p of water (0.26×10^{-3} K^{-1}) is responsible for predominantly entropic cavity formation. Larger expansivities typical for most molecular liquids (1.38×10^{-3} K^{-1} for acetonitrile) make enthalpy the main component of $\Delta\mu_{HS}$.

It is also worth noting that the free energy of cavity formation is positive, while solvation free energies related to long-range electrostatic interactions with permanent and induced dipoles are negative (Table 12.1). The total solvation free energy comes as a result of compensation between these components. It is negative and dominated by electrostatic interactions for ions in polar liquids, but gains an overall positive value, dominated by the cavity formation, for solvation of nonpolar solutes.

Bibliography

[1] H. Jeffreys, *Cartesian Tensors*. Cambridge University Press, Cambridge, UK (1931).

[2] R. Aris, *Vectors, Tensors, and the Basic Equations of Fluid Mechanics*. Dover Publications, Inc., Mineola, N. Y. (1989).

[3] R. W. Soutas-Little, *Elasticity*. Dover, Mineola, N. Y. (1999).

[4] A. J. M. Spencer, *Continuum Mechanics*. Dover Publications, Mineola, N. Y. (1980).

[5] L. Eyges, *The Classical Electromagnetic Field*. Dover Publications, New York (1972).

[6] J. D. Jackson, *Classical Electrodynamics*. Wiley, New York (1999).

[7] M. Abramowitz and I. A. Stegun (eds.), *Handbook of Mathematical Functions*. Dover, New York (1972).

[8] L. D. Landau and E. M. Lifshitz, *Electrodynamics of Continuous Media*. Pergamon, Oxford (1984).

[9] J. A. Stratton, *Electromagnetic Theory*. McGraw-Hill Inc., New York (1941).

[10] C. J. F. Böttcher, *Theory of Electric Polarization, Vol. 1: Dielectrics in Static Fields*. Elsevier, Amsterdam (1973).

[11] H. Fröhlich, *Theory of Dielectrics*. Oxford University Press, Oxford (1958).

[12] W. K. H. Panofsky and M. Phillips, *Classical Electricity and Magnetism*. Dover Publications, Mineola, N.Y. (2005).

[13] R. P. Feynman, R. B. Leighton and M. Sands, *The Feynman lectures on physics, Vol. II: Mainly electromagnetism and matter*. Addison-Wesley, Reading, MA (1964).

[14] R. Schmid and D. V. Matyushov, Entropy of attractive forces and molecular nonsphericity in real liquids: A measure of structural ordering, *J. Phys. Chem.* **99**, p. 2393 (1995).

[15] C. G. Gray and K. E. Gubbins, *Theory of Molecular Liquids*, Vol. 1: Fundamentals. Clarendon Press, Oxford (1984).

[16] Y. Marcus, *Ions in Solution and their Solvation*. Wiley, New Jersey (2015).

[17] D. V. Matyushov, Nonlinear dielectric response of polar liquids, in R. Richert (ed.), *Nonlinear Dielectric Spectroscopy*. Springer, Cham, pp. 1–34 (2018).

[18] B. K. P. Scaife, *Principles of Dielectrics*. Clarendon Press, Oxford (1998).

[19] C. G. Gray, K. E. Gubbins and C. G. Joslin, *Theory of Molecular Liquids*, Vol. 2: Applications. Oxford University Press, Oxford (2011).

[20] L. D. Landau and E. M. Lifshitz, *Mechanics, Course of theoretical Physics*, Vol. 1, 3rd edn. Elsevier, Butterworth Heinemann, Amsterdam (2007).

[21] R. P. Feynman, R. B. Leighton and M. Sands, *The Feynman lectures on physics, Vol. I: Mainly mechanics, radiation, and heat.* Addison-Wesley, Reading, MA (1963).

[22] H. Goldstein, *Classical Mechanics.* Addison-Wesley, Reading, MA (1964).

[23] W. Yourgrau and S. Mandelstam, *Variational Principles in Dynamics and Quantum Theory.* Dover Publications, Inc., Mineola, N. Y. (1979).

[24] L. Susskind and G. Hrabovsky, *The Theoretical Minimum.* Basic Books, New York (2013).

[25] L. D. Landau and E. M. Lifshits, *Quantum Mechanics: Non-Relativistic Theory.* Pergamon Press, Oxford (1977).

[26] S. Gasiorowicz, *Quantum Physics*, 3rd edn. Wiley (2003).

[27] D. A. McQuarrie, *Quantum Chemistry.* University Science Books, Mill Valey, CA (2008).

[28] A. I. M. Rae and J. Napolitano, *Quantum Mechanics*, 6th edn. CRC Press (2016).

[29] M. E. Rose, *Elementary Theory of Angular Momentum.* Dover Publications, Inc., New York (1995).

[30] G. Fischer, *Vibronic Coupling.* Academic Press, London (1984).

[31] H. Kleinert, *Path Integrals in Quantum Mechanics, Statistics, Polymer Physics, and Financial Markets*, 3rd edn. World Scientific, New Jersey (2004).

[32] R. P. Feynman and A. R. Hibbs, *Quantum Mechanics and Path Integrals.* Dover Publications, Inc., Mineola, N. Y. (2005).

[33] D. Chandler, *Introduction to Modern Statistical Mechanics.* Oxford University Press, Oxford (1987).

[34] L. D. Landau and E. M. Lifshits, *Statistical Physics.* Pergamon Press, New York (1980).

[35] M. Toda, R. Kubo and N. Saito, *Statistical Physics I.* Springer-Verlag, New York (1982).

[36] D. A. McQuarrie, *Statistical Mechanics.* University Science Books, Sausalito, CA (2000).

[37] A. Katz, *Principles of Statistical Mechanics.* W. H. Frieman and Co., San Francisco (1967).

[38] M. Rubinstein and R. H. Colby, *Polymer Physics.* Oxford University Press, Oxford (2003).

[39] J.-P. Hansen and I. R. McDonald, *Theory of Simple Liquids*, 4th edn. Academic Press, Amsterdam (2013).

[40] N. H. March and M. P. Tosi, *Coulomb Liquids.* Academic Press, London (1984).

[41] U. Balucani and M. Zoppi, *Dynamics of the Liquid Phase.* Oxford Science Publications, Oxford (1994).

[42] J. P. Boon and S. Yip, *Molecular Hydrodynamics.* McGraw-Hill Inc. (1980).

[43] H. L. Friedman, *A Course in Statistical Mechanics*. Prentice-Hall, New Jersey (1985).

[44] D. V. Matyushov and R. Schmid, Calculation of Lennard-Jones energies of molecular fluids, *J. Chem. Phys.* **104**, pp. 8627–8638 (1996).

[45] B. Widom, *Statistical Mechanics: A Concise Introduction for Chemists* . Cambridge University Press, Cambridge, UK (2002).

[46] D. V. Matyushov and R. Schmid, Properties of liquids at the boiling point: Equation of state, internal pressure and vaporization entropy, *Ber. Bunsenges. Phys. Chem.* **98**, 12, pp. 1590–1595 (1994).

[47] C. W. Gardiner, *Handbook of Stochastic Methods*. Springer, Berlin (1997).

[48] N. G. V. Kampen, *Stochastic Processes in Physics and Chemistry*, 3rd edn. Elsevier (2007).

[49] P. M. Chaikin and T. C. Lubensky, *Principles of Condensed Matter Physics*. Cambridge University Press, Cambridge (1995).

[50] J. K. G. Dhont, *An Introduction to Dynamics of Colloids, Studies in Interface Science*, Vol. 2. Elsevier Science (1996).

[51] B. J. Berne and R. Pecora, *Dynamic Light Scattering*. Dover Publications, Inc., Mineola, N.Y. (2000).

[52] A. Nitzan, *Chemical Dynamics in Condensed Phases: Relaxation, Transfer and Reactions in Condensed Molecular Systems*. Oxford University Press, Oxford (2006).

[53] A. Einstein, *Investigations on the Theory of the Brownian Movement*. BN Publishing (2011).

[54] R. Kubo, The fluctuation-dissipation theorem, *Rep. Prog. Phys.* **29**, pp. 255–284 (1966).

[55] S. Das, *Statistical Physics of Liquids at Freezing and Beyond*. Cambridge University Press, Cambridge, UK (2011).

[56] D. J. Evans and G. Morriss, *Statistical Mechanics of Nonequilibrium Liquids*, 2nd edn. Cambridge University Press, Cambridge, UK (2008).

[57] R. Kubo, Some aspects of the statistical-mechanical theory of irreversible processes, in *Lectures in Theoretical Physics*, Vol. 1. Interscience Publishers, Inc., New York, p. 120 (1959).

[58] C. J. F. Böttcher, *Theory of Electric Polarization. Deielctrics in Time-Dependent Fields*, Vol. 2. Elsevier (1973).

[59] L. D. Landau and E. M. Lifshitz, *Theory of Elasticity*, Vol. 7 of Course of Theoretical Physics. Elsevier, Oxford (1986).

[60] D. Forster, *Hydrodynamic Fluctuations, Broken Symmetry, and Correlation Functions*. W. A. Benjamin, Reading, MA (1975).

[61] P. Nelson, *Biological Physics. Energy, Information, Life*. W. H. Frieman and Co., New York (2008).

[62] S. Kim and S. J. Karrila, *Microhydrodynamics: Principles and Selected Applications*. Butterworth, Boston (1991).

[63] C. Kittel, *Quantum Theory of Solids*. Wiley, New York (1963).

[64] R. M. Christensen, *Theory of Viscoelasticity*. Dover Publications, Inc., Mineola, N. Y. (2003).

[65] D. Boal, *Mecanics of the Cell*, 2nd edn. Cambridge University Press, Cam-

bridge, UK (2012).

[66] W. B. Russel, D. A. Saville and W. R. Schowalter, *Colloidal Dispersions.* Cambridge University Press, Cambridge, UK (1989).

[67] H. Falkenhagen, *Electrolytes.* Clarendon Press, Oxford (1934).

[68] G. C. Schatz and M. A. Ratner, *Quantum Mechanics in Chemistry*, 1st edn. Dover, New York (2002).

[69] D. P. Craig and T. Thirunamachandran, *Molecular Quantum Electrodynamics.* Academic Press, London (1984).

[70] C. H. Wang, *Spectroscopy of Condensed Media.* Academic Press, Orlando (1985).

[71] S. Mukamel, *Principles of Nonlinear Optical Spectroscopy.* Oxford University Press, New York (1995).

[72] P. Hamm and M. Zanni, *Concepts and Methods of 2D Infrared Spectroscopy.* Cambridge University Press, Cambridge, UK (2011).

[73] S. W. Lovesey, *Theory of Neutron Scattering from Condensed Matter*, Vol. 1. Clarendon Press, Oxford (1984).

[74] A. Furrer and J. M. anf T. Strässel, *Neutron Scattering in Condensed Matter Physics, Series on Neutron Scattering and Applications*, Vol. 4. World Scientific (2009).

[75] M. Bee, *Quasielastic Neutron Scattering, Principles and Applications in Solid State Chemistry, Biology and Materials Science .* Adam Hilger, Bristol, UK (1988).

[76] N. Mataga and T. Kubota, *Molecular Interactions and Electronic Spectra.* Marcel Dekker, New York (1970).

[77] G. D. Mahan, *Many-Particle Physics.* Plenum Press, New York (1990).

[78] R. Zwanzig, *Nonequilibrium Statistical Mechanics.* Oxford University Press, Oxford (2001).

[79] T. L. Beck, M. E. Paulaitis and L. R. Pratt, *The Potential Distribution Theorem and Models of Molecular Solutions.* Cambridge University Press, Cambridge, UK (2006).

[80] A. Ben-Naim, *Molecular Theory of Water and Aqueous Solutions: Understanding Water.* World Scientific Publishing Co Pte Ltd (2014).

[81] R. Spinney and I. Ford, Fluctuation relations: A pedogical overview, in K. Klages, W. Just and C. Jarzynski (eds.), *Nonequilibrium Statistical Physics of Small Systems.* Wiley-VCH, Weinheim, Germany (2013).

Index

Printed in the United States
by Baker & Taylor Publisher Services